Mathematical Sciences Research Institute
Publications

25

Mathematical Sciences Research Institute
Publications

Selman Akbulut   Henry King

Topology of
Real Algebraic Sets

Springer-Verlag

New York Berlin Heidelberg London Paris
Tokyo Hong Kong Barcelona Budapest

Selman Akbulut
Department of Mathematics
Michigan State University
East Lansing, MI 48824
USA

Henry King
Department of Mathematics
University of Maryland
College Park, MD 20742
USA

Sci
QA
247
A42
1992

The Mathematical Sciences Research Institute wishes to acknowledge the support
by the National Sciences Foundation.

Mathematical Subject Classifications: 14P25, 57N80, 32C05, 58A35

Library of Congress Cataloging-in-Publication Data
Akbulut, Selman, 1948-
    Topology of real algebraic sets / Selman Akbulut, Henry King.
    p. cm -- (Mathematical Sciences Research Institute
publications : 25)
    Includes bibliographical references and index.
    ISBN 0-387-97744-9 (Springer-Verlag New York Berlin Heidelberg :
acid-free paper). -- ISBN 3-540-97744-9 (Springer-Verlag Berlin
Heidelberg New York : acid-free paper)
    1. Ordered fields. 2. Geometry, Algebraic. I. King, Henry,
1948- . II. Title. III. Series.
QA247.A42 1992
516.3'5--dc20   91-37834

Printed on acid-free paper.

Production managed by Karen Phillips, Manufacturing supervised by Robert Paella.
Camera-ready copy prepared by the Mathematical Sciences Research Institute using
$\mathcal{A}\mathcal{M}\mathcal{S}$-TEX.
Printed and bound by Braun-Brumfield, Ann Arbor, MI.
Printed in the United States of America.

9 8 7 6 5 4 3 2 1

ISBN 0-387-97744-9 Springer-Verlag New York Berlin Heidelberg
ISBN 3-540-97744-9 Springer-Verlag Berlin Heidelberg New York

Preface

In the Fall of 1975 we started a joint project with the ultimate goal of topologically classifying real algebraic sets. This has been a long happy collaboration
(c.f., [K2]). In 1985 while visiting M.S.R.I. we organized and presented our
classification results up to that point in the M.S.R.I. preprint series [AK14]
-[AK17]. Since these results are interdependent and require some prerequisites
as well as familiarity with real algebraic geometry, we decided to make them self
contained by presenting them as a part of a book in real algebraic geometry.
Even though we have not arrived to our final goal yet we feel that it is time to
introduce them in a self contained coherent version and demonstrate their use
by giving some applications.

Chapter I gives the overview of the classification program. Chapter II has all
the necessary background for the rest of the book, which therefore can be used
as a course in real algebraic geometry. It starts with the elementary properties
of real algebraic sets and ends with the recent solution of the Nash Conjecture.
Chapter III and Chapter IV develop the theory of resolution towers. Resolution
towers are basic topologically defined objects generalizing the notion of manifold.
They enable us to study singular spaces in an organized way. Chapter V shows
how to obtain algebraic sets from resolution towers. Chapter VI shows how to
put resolution tower structures on real or complex algebraic sets. Chapter VII
applies this theory to real algebraic sets of dimension ≤ 3 by giving their topological characterization. An impatient reader can go directly to Chapter VII
from Chapter I in order to get motivated for the results of Chapter III through
Chapter VI .

We would like to thank National Science Foundation, the Institute for Ad-
vanced Study, the Max-Planck Institute, the Mathematical Sciences Research
Institute, the General Research Board of the University of Maryland as well as
our respective universities: Michigan State University and University of Maryland for generous support while this work has been in progress. Also we would

like to thank Lowell Jones and Elmer Rees for timely advice. The first named author would like thank his advisor R.Kirby for introducing him to the subject, and his teachers: Fahrettin Akbulut, İrfan Barış, S.S.Chern, Tom Farrell, Moe Hirsch, Dennis Sullivan, Larry Taylor for inspiration, and TUBITAK (Turkish scientific research institute) for the initial support. The second named author would also like to thank Dick Palais for teaching him much about real algebraic geometry and Dennis Sullivan for general mathematical stimulation. We would like to thank J. Bochnak and W. Kucharz for their helpful comments on preliminary versions of this book. We thank D. Glaubman for helping us with some of the computer generated figures. We would like to thank Tammy Hatfield, Cindy Smith, and Cathy Friess for doing a great job of typesetting this book in LaTeX. Finally we would like to thank Margaret Pattison for her help in preparing this book for publication.

We now fix some notation, some of it nonstandard, which we will use throughout the book. We let $\mathbf{R}$ and $\mathbf{C}$ denote the real and complex numbers. We let I denote the closed interval $[0, 1]$ in $\mathbf{R}$. If A is a subset of a topological space then we let $\mathrm{Cl}(A)$ denote the closure of A and let $\mathrm{Int}(A)$ denote its interior. If A and B are sets, then $A - B$ denotes their difference. If $f\colon M \to N$ is a smooth map between smooth manifolds, we let $df\colon TM \to TN$ denote the induced mapping on tangent spaces. The expression $A \sqcup B$ means the disjoint union. A closed manifold means a compact manifold without boundary.

We now introduce some nonstandard notation. If $f\colon M \to N$ is a function and $S \subset M$ we will let $f|$ denote the restriction $f|_S$ if S is clear from context. This is useful if S is some complicated expression which would only clutter up a formula and make it more unreadable. If X is a topological space, we let $\mathfrak{c}X$ denote the cone on X, so $\mathfrak{c}X = X \times [0,1]/X \times 0$, the quotient space of $X \times [0,1]$ with $X \times 0$ crushed to a point. If x (or y or z etc.) is a point in $\mathbf{R}^n$ then x_i (or y_i or z_i) will denote the i-th coordinate of x. We let $\mathbf{R}^n_i$ denote the coordinate hyperplane $\{\, x \in \mathbf{R}^n \mid x_i = 0 \,\}$.

The end of a proof is marked by a sign thusly: $\square$

We have tried to organize long proofs in a hierarchical manner. In the midst of a proof we may make an assertion, which we then proceed to prove. The reader might prefer to skip this proof on first reading or do it as an exercise if she is energetic. The hope is that this will make the overall argument clearer by hiding some of the details. To set off the proof of an assertion from the rest of the proof we mark its end thusly: $\square$

Contents

List of Figures

CHAPTER I

INTRODUCTION

1. Overview

A real algebraic set is the set of solutions of a collection of polynomial equations in real variables. A fundamental problem of algebraic geometry is to topologically classify real algebraic sets, to give a topological characterization of all topological spaces which are homeomorphic to real algebraic sets.

In the case of nonsingular real algebraic sets this problem has been solved. A nonsingular real algebraic set is by definition a smooth manifold. In 1952 J. Nash showed that every closed smooth manifold can occur as a component of a nonsingular real algebraic set [**N**]. Later, A. Tognoli generalized this result by proving that every closed smooth manifold can in fact occur as a whole nonsingular real algebraic set [**To1**]. Then finally in [**AK1**] we topologically classified nonsingular real algebraic sets by showing that up to diffeomorphism,

$$\left\{ \begin{array}{c} \text{Nonsingular real} \\ \text{algebraic sets} \end{array} \right\} = \left\{ \begin{array}{c} \text{Interiors of smooth} \\ \text{compact manifolds} \end{array} \right\}$$

As far as singular real algebraic sets are concerned, those with isolated singularities were also topologically classified in [**AK1**]. The result is that a space is homeomorphic to a real algebraic set with isolated singularities if and only if it is obtained in the following way. Take a compact smooth manifold W with boundary, divide up the boundary components in some way $\partial W = \bigcup_{i=1}^{r} M_i$ so that each M_i bounds, then crush some M_i's to points and delete the remaining M_i's. An example is given in Figure I.1.1.

Furthermore, according to [**AK1**] a topological space is homeomorphic to a real algebraic set if and only if its one point compactification is homeomorphic to a real algebraic set. Therefore it suffices to topologically classify compact real algebraic sets. The extension of the classification to noncompact spaces will be automatic.

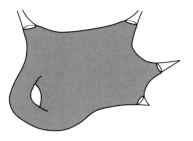

FIGURE I.1.1. A noncompact real algebraic set

There is a nice generalization of the above result classifying isolated singularities. We inductively define a class of stratified spaces: Let A_0-*spaces* be compact smooth manifolds. Inductively let A_k-*spaces* be spaces of the form $X = X_0 \cup \bigcup_\partial N_i \times \text{cone}(\Sigma_i)$ where X_0 is a compact A_{k-1}-space, Σ_i are boundaries of compact A_{k-1}-spaces and N_i are compact smooth manifolds. The union is taken along codimension zero subsets of ∂X_0 and $N_i \times \Sigma_i \subset N_i \times \text{cone}(\Sigma_i)$. We let the boundary of X be $\partial X = (\partial X_0 - \bigcup N_i \times \Sigma_i) \cup \bigcup \partial N_i \times \text{cone}(\Sigma_i)$. We say X is a closed A_k-space if $\partial X = \emptyset$.

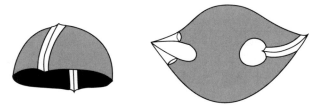

FIGURE I.1.2. A_1 spaces with and without boundary

In [**AK2**] it was shown that every closed A_k-space is homeomorphic to a real algebraic set such that the natural stratification of this algebraic set coincides with the stratification of the A_k-space. In [**AT**] it was shown that the set of A_k-spaces is large enough to contain all closed P.L. manifolds. Hence we have inclusions:

$$\left\{ \begin{array}{c} \text{closed P.L.} \\ \text{manifolds} \end{array} \right\} \subset \left\{ \begin{array}{c} \text{closed } A_k\text{-spaces} \\ k = 0, 1, 2, \cdots \end{array} \right\} \subset \left\{ \begin{array}{c} \text{compact real} \\ \text{algebraic sets} \end{array} \right\}$$

We refer the reader to [**AK2**] for a more detailed summary of these results. Despite the great generality of A_k-spaces we cannot hope that they capture the topology of all real algebraic sets. For example the space in Figure I.1.3 is not an A_k-space but it is an algebraic set. It can be described by the equation $(x^2 + y^2 - 2x)^2(x^2 + y^2 - 1) + z^2 = 0$ in $\mathbf{R}^3$. Hence we must generalize A_k-spaces if we hope to topologically classify algebraic sets. Notice A_k-spaces are defined in such a way that they admit 'topological resolutions' to smooth manifolds. If

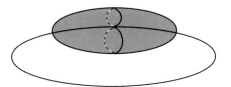

FIGURE I.1.3. An algebraic set which is not an A_k-space

$Z_k = X_0 \cup \bigcup_\partial N_i \times \text{cone}(\Sigma_i)$ is an A_k-space, then there are A_{k-1}-spaces W_i with $\partial W_i = \Sigma_i$. Then we define an A_{k-1} space by $Z_{k-1} = X_0 \cup \bigcup_\partial N_i \times W_i$, and a map $\pi_k : Z_{k-1} \to Z_k$ induced by $W_i \to \text{cone}(\Sigma_i)$ (crushing a spine of W_i to the cone point). Iteration of this process gives a topological resolution to a smooth manifold Z_0

$$Z_0 \to Z_1 \to \cdots \to Z_{k-1} \to Z_k.$$

We can consider Z_k to be a quotient of the smooth manifold Z_0.

So, rather than trying to generalize the definition of A_k-spaces to more complicated stratified spaces we prefer to generalize this notion of 'topological resolution'. In this way we would be defining Z_k via the smooth manifold Z_0 and 'collapsing data' $\pi : Z_0 \to Z_k$. An A_k-space resolution $\pi : Z_0 \to Z_k$ resolves not only the space Z_k but also gives a recipe of how to resolve the various strata of Z_k. This information has to be extracted from π in order to get a deeper understanding of resolutions. To see this let us study an example which we illustrate in Figure I.1.4.

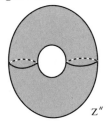 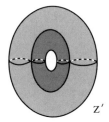 

FIGURE I.1.4. A topological resolution

Let Y be the figure 8 and let $Z = (Y \times I) \cup_\partial \text{cone}(Y \times \dot{I})$ where $I = [0,1]$ and $\dot{I} = \partial I = \{0,1\}$. Z is an A_2-space and it can be resolved to the A_1-space $Z' = Y \times I \cup_\partial Y \times I = Y \times S^1$. We can see Z' is an A_1-space by writing $Z' = S^1 \times J \cup_\partial S^1 \times \text{cone}\{a,b,c,d\}$ where $J = \widehat{ab} \cup \widehat{cd}$ and a,b,c,d are distinct ordered points on S^1, and $\widehat{ab}$ denotes the arc segment between a and b. Finally Z' can be resolved to the smooth manifold

$$Z'' = S^1 \times J \cup S^1 \times J' \text{ where } J' = \widehat{bc} \cup \widehat{da}$$
$$Z'' = S^1 \times (J \cup J') = S^1 \times S^1$$

Notice that we can determine the topology of Z from the following data of smooth manifolds and maps where $x_i \in S^1, x_1 \neq x_2, p_{0i}$ are the constant maps and p_{12} projects $S^1 \times \{x_1, x_2\}$ to the first factor.

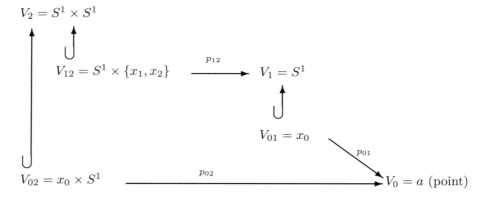

$V_2 = S^1 \times S^1$

$V_{12} = S^1 \times \{x_1, x_2\}$ $\xrightarrow{\ p_{12}\ }$ $V_1 = S^1$

$V_{01} = x_0$

$V_{02} = x_0 \times S^1$ $\xrightarrow{\ p_{02}\ }$ $V_0 = a$ (point)

p_{01}

Then Z is obtained by gluing V_0, V_1 and V_2 together by the maps p_{01}, p_{02}, p_{12}. Hence $Z = \bigcup_{i=0}^2 V_i /_{x \sim p_{ji}(x)}$ is the quotient space. So the smooth data $\mathfrak{T} = \{V_i, p_{ji}\}$ determines the topology of Z. This is the view that gives us the best generalizations of A_k-spaces, it allows us to avoid defining structures on singular spaces.

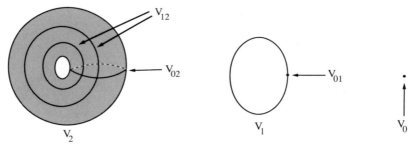

V_{12}

V_{02}

V_2

V_{01}

V_1

V_0

FIGURE I.1.5. A resolution tower for Z

We would like to view the set of real algebraic sets as an 'extension' of the set of smooth manifolds by monomial functions, just as algebraic numbers are extensions of rational numbers. We will first define the notion of topological monomial maps between smooth manifolds. Then we will define objects $\mathfrak{T} = \{V_i, p_{ji}\}$ where V_i are smooth manifolds and $p_{ji} \colon V_i \to V_j$ are topological monomial maps. We will show that the realization $|\mathfrak{T}| = \bigcup V_i /_{x \sim p_{ji}(x)}$ of $\mathfrak{T}$ (i.e., the set obtained by gluing the smooth manifolds V_i by the maps p_{ji}) describes any algebraic set. Then we will prove a certain converse to this characterization. But before defining topological monomial maps we define the notions of a stratified set and of a tico (short for transversally intersecting codimension one submanifolds).

2. Stratified Sets

In studying the topology of rather nice singular objects such as algebraic sets, stratified sets are useful. A *stratified set* is a topological space X together with a decomposition of X into a locally finite number of disjoint locally closed subspaces which happen to be manifolds. These subspaces are called *strata*. In this book, all of our stratified spaces will be smooth, i.e., the strata are all given a smooth structure. We also allow the strata to be manifolds with boundary. The boundary of a stratified set will be the union of the boundaries of the strata.

The *i-skeleton* of a stratified set is the union of all strata of dimension less than or equal to i. We will require our stratified sets to satisfy a frontier condition: the frontier of an i-dimensional stratum is contained in the $i-1$ skeleton. We also want the frontier of the boundary of an i-dimensional stratum to be contained in the boundary of the $i-1$ skeleton, i.e., it is contained in the union of boundaries of strata of lower dimension. In particular, the boundary of a stratified set will also satisfy the frontier condition so it will be a stratified set.

One example to keep in mind is a polyhedron, the strata are the interiors of simplices and there is no boundary. Another example to keep in mind is the singular stratification of an algebraic set X, we let $X_n = X$, $X_{n-1} = \operatorname{Sing} X$, and in general $X_i = \operatorname{Sing} X_{i+1}$. Eventually we get some X_k empty. Then we let the strata be the smooth manifolds $X_i - X_{i-1}$ (thus the X_i's are the skeleta although of course i need not be the dimension). See Chapter II for further explanations.

The *dimension* of a stratified set is the maximum dimension of its strata (or ∞ if no maximum exists). An *isomorphism* of stratified sets is a homeomorphism which maps each stratum onto a stratum and in fact restricts to a diffeomorphism on each stratum.

Note that an open subset of a stratified set has a natural structure of a stratified set. Likewise the cone on a compact stratified set has a natural structure of a stratified set.

Sometimes we want nicer stratified sets. For example a *locally conelike stratified set* is one in which for every point x in a k-stratum S there is a open neighborhood U of x in X which is isomorphic to $V \times Z$ where Z is the open cone on a compact stratified set L and V is a neighborhood of x in S. We call L the *link* of S in X. The link is not actually well-defined (although an elementary invertible cobordism argument shows that for a connected k-dimensional stratum any two links are isomorphic after crossing with $\mathbf{R}^{k+1}$). One can show using Whitney stratifications that any real algebraic set X has a locally conelike stratification, in general not the same as the singular stratification. In this case there is a canonical link, the intersection of X with a small sphere in the plane normal to the stratum. We will look at stratified sets more closely in section 3 of Chapter IV.

If we have two stratifications $\mathcal{A}$ and $\mathcal{B}$ on a set X, we say that $\mathcal{A}$ is *coarser* than $\mathcal{B}$ if all of $\mathcal{B}$'s strata are smooth submanifolds of $\mathcal{A}$'s strata. We also say that $\mathcal{B}$ is a *refinement* of $\mathcal{A}$. Notice that there may not be a coarsest smooth stratifications of a smooth stratified set. For topological stratifications there is a coarsest topological stratification (see [**K**], [**Si**]). We will only use this fact in Chapter VII where the dimensions are so low that it becomes an elementary exercise.

Sometimes we need a weaker notion of equivalence than isomorphism. Recall two polyhedra are PL isomorphic if they have isomorphic subdivisions. In analogy with that, we say that two stratified sets are *PL isomorphic* if they have isomorphic refinements.

3. Ticos

A *tico* $\mathcal{A}$ in a smooth manifold M is a finite collection of properly immersed closed smooth codimension one submanifolds of M in general position. We study them in Chapter III. A basic example is the tico $\{\mathbf{R}_i^n\}_{i=1}^n$ in $\mathbf{R}^n$ where $\mathbf{R}_i^n = \{x \in \mathbf{R}^n \mid x_i = 0\}$. We consider a tico as a global coordinatization of the manifold. We call the elements of $\mathcal{A}$ *sheets*. We say $\mathcal{A}$ is a *regular tico* if each sheet is a properly embedded submanifold. The *realization* of $\mathcal{A}$ is $|\mathcal{A}| = \bigcup_{s \in \mathcal{A}} S$. A tico gives a natural stratification of the manifold. If M is a nonsingular algebraic set we call a tico $\mathcal{A}$ in M an *algebraic tico* if each sheet is a nonsingular algebraic subset of M. Clearly algebraic ticos are regular. We will sometimes refer to a manifold with a tico $(M, \mathcal{A})$ simply as a tico. Now we can define the notion of topological monomials between manifolds with ticos.

Let $(M, \mathcal{A})$, $(N, \mathcal{B})$ be smooth manifolds with ticos. A *tico map* $f : (M, \mathcal{A}) \rightarrow (N, \mathcal{B})$ is a smooth map from M to N with the following local property. For any $p \in M$ and charts $\psi : (\mathbf{R}^m, 0) \rightarrow (M, p)$ and $\theta : (\mathbf{R}^n, 0) \rightarrow (N, f(p))$ with $\psi^{-1}(|\mathcal{A}|) = \bigcup_{j=1}^a \mathbf{R}_j^m$ and $\theta^{-1}(|\mathcal{B}|) = \bigcup_{i=1}^b \mathbf{R}_i^n$, then the i-th coordinate $f_i(x)$ of $\theta^{-1} \circ f \circ \psi(x)$ is in the form:

$$f_i(x) = \left(\prod_{j=1}^a x_j^{\alpha_{ij}} \right) \varphi_i(x) \quad \text{for} \quad i = 1, 2, \ldots, b$$

where α_{ij} are nonnegative integers and $\varphi_i : \mathbf{R}^m \rightarrow \mathbf{R}$ are smooth functions with $\varphi_i(0) \neq 0$. It follows that up to permutations of i and j the exponents α_{ij} above depend only on f and $p \in M$, not on the local charts ψ and θ.

If $\mathcal{A}$ and $\mathcal{B}$ are regular then for $S \in \mathcal{A}$ and $T \in \mathcal{B}$ we can define $\alpha_{ST} : S \rightarrow \mathbf{Z}$ by $\alpha_{ST}(p) = 0$ if $f(p) \notin T$ and $\alpha_{ST}(p) = \alpha_{ij}$ if $f(p) \in T$ and $\psi^{-1}(S) = \mathbf{R}_j^m$ and $\theta^{-1}(T) = \mathbf{R}_i^n$. This exponent α_{ST} is locally constant, hence constant on each component of S. Then $f^{-1}(T) = \bigcup_{s \in \mathcal{A}} \alpha_{ST}^{-1}(\mathbf{Z} - 0)$, so a tico map pulls back each sheet to a union of components of sheets.

A tico map $f : (M, \mathcal{A}) \to (N, \mathcal{B})$, is *type N* if for every $p \in M$ there are charts $\psi : U \to M$ and $\theta : (\mathbf{R}^n, 0) \to (N, f(p))$ where $U \subset \mathbf{R}^m$ is an open subset, such that $\psi(U)$ is a neighborhood of p, $\psi^{-1}(|\mathcal{A}|) = (\bigcup_{j=1}^a \mathbf{R}_j^m) \cap U$, $\theta^{-1}(|\mathcal{B}|) = \bigcup_{i=1}^b \mathbf{R}_i^n$, for some $c \geq a$ we have

$$f_i(x) = \prod_{j=1}^c x_j^{\alpha_{ij}} \quad \text{for} \quad i = 1, 2, \ldots, b$$

for all $x \in U$, and the $(b \times c)$-matrix (α_{ij}) is onto, i.e., it has rank b. We say f is *submersive* if in addition $m - c \geq n - b$, and $f_i(x) = x_{i-n+m}$ for all $i = b + 1, \ldots, n$ and $x \in U$. Basically a type N tico map is a tico map which becomes a pure monomial under some choice of coordinates, and the exponent map is onto. A submersive tico map is essentially a type N tico map which submerses each stratum of $(M, \mathcal{A})$ to some stratum of $(N, \mathcal{B})$, although it is actually slightly stronger than this in a subtle way.

We can also define partial tico maps as follows. If $(M, \mathcal{A})$ and $(N, \mathcal{B})$ are smooth manifolds with ticos and $\mathcal{C} \subset \mathcal{A}$, then a *mico* is a map $f : |\mathcal{C}| \to N$ which locally extends to a tico map. In other words there is an open neighborhood U of $|\mathcal{C}|$ in M and a tico map $g : (U, U \cap (\mathcal{A} - \mathcal{C})) \to (N, \mathcal{B})$ such that $f = g|_{|\mathcal{C}|}$.

4. Resolution Towers

Now we are ready to define the promised structures which we use in our attempt to classify real algebraic sets. In Chapter IV we study resolution towers. A compact *(topological) resolution tower* $\mathfrak{T} = \{V_i, \mathcal{A}_i, p_i\}_{i=0}^n$ is a collection of compact smooth manifolds with ticos $(V_i, \mathcal{A}_i)$, $i = 0, \ldots, n$ and collections of maps $p_i = \{p_{ji}\}_{j<i}$ with $p_{ji} : V_{ji} \to V_j$ where $V_{ji} = |\mathcal{A}_{ji}|$ for some $\mathcal{A}_{ji} \subset \mathcal{A}_i$ and

   I. $p_{ji}(V_{ji} \cap V_{ki}) \subset V_{kj}$ for $k < j < i$.

   II. $p_{kj} \circ p_{ji}| = p_{ki}|V_{ji} \cap V_{ki}$ for $k < j < i$.

   III. $p_{ji}^{-1}(\bigcup_{k<m} V_{kj}) = (\bigcup_{k<m} V_{ki}) \cap V_{ji}$ for $m \leq j < i$.

   IV. $\mathcal{A}_i$ is the disjoint union of the $\mathcal{A}_{ji}$'s.

   V. $p_{ji}^{-1}(\partial V_j) = \partial V_i \cap V_{ji}$ for $j < i$.

(For convenience, our definition in Chapter IV will allow more general indexing sets, but we can always reindex to the above form.)

We define $\partial \mathfrak{T} = \{\partial V_i, \partial \mathcal{A}_i, p_i|\}$ where $\partial \mathcal{A}_i = \{\partial S \mid S \in \mathcal{A}_i\}$ and $p_i|$ is the restriction. Condition V insures that $\partial \mathfrak{T}$ is also a resolution tower. An *algebraic resolution tower* is a resolution tower $\mathfrak{T}$ such that each $(V_i, \mathcal{A}_i)$ is an algebraic tico and each p_{ji} is an entire rational function (a rational function with nowhere zero denominator). There are some extra properties which resolution towers may satisfy. The important ones are

R	-	Each $\mathcal{A}_i$ is regular
M	-	Each p_{ji} is a mico
N	-	Each p_{ji} is a type N mico
E	-	all exponent maps α_{ST} for all p_{ji} are constant
S	-	Each p_{ji} is a submersive mico
U	-	$p_{ji}|_S$ is a submersion on each stratum S of $V_{ji} - \bigcup_{k<j} V_{ki}$
F	-	Each $(V_i, \mathcal{A}_i)$ is full.

The term full means that if S is V_i or any intersection of sheets of $\mathcal{A}_i$ then the homology $H_*(S; \mathbf{Z}_2)$ is generated by embedded smooth submanifolds of S.

We define the *realization* of $\mathfrak{T}$ to be the set

$$|\mathfrak{T}| = \bigcup_{i=0}^{n} V_i /_{x \sim p_{ji}(x)}.$$

That is $|\mathfrak{T}|$ is the quotient space of the disjoint union $\bigcup_{i=0}^{n} V_i$. $|\mathfrak{T}|$ is a stratified space with strata $\{V_i - \bigcup_{j<i} V_{ji}\}$. We define $\dim \mathfrak{T} = \dim |\mathfrak{T}|$.

For example, consider the following resolution tower. Let $\mathfrak{T} = \{V_i, \mathcal{A}_i, p_i\}_{i=0}^{2}$, $V_2 = $ a surface of genus 2, $V_1 = $ a circle, $V_0 = $ two points $\{\alpha, \beta\}$, $\mathcal{A}_{02} = \{A, A', B, B'\}$, $\mathcal{A}_{12} = \{C\}$, $\mathcal{A}_{01} = \{a, b\}$, p_{02} collapses $A \cup A'$ to α, $B \cup B'$ to β, p_{12} folds C onto the arc $\widehat{ab}$, p_{01} maps a to α; b to β. See Figure I.4.1.

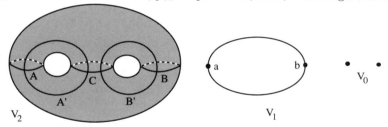

FIGURE I.4.1. A resolution tower

The space X in Figure I.4.2 is the realization $|\mathfrak{T}|$ of the resolution tower $\mathfrak{T}$.

The advantage of considering $\mathfrak{T}$ as opposed to X is that $\mathfrak{T}$ is given by smooth data and it has more structure.

Let $\mathfrak{TR}$ and $\mathcal{AR}$ denote the set of all compact topological and algebraic resolution towers, respectively. When we put subscripts R, M, N, E, S, U or F on $\mathfrak{TR}$ or $\mathcal{AR}$ we mean the set of resolution towers of that type. For example $\mathfrak{TR}_{RS}$ is a set of topological resolution towers of type R and S. Clearly $\mathcal{AR} = \mathcal{AR}_R$, $\mathfrak{TR}_S \subset \mathfrak{TR}_{NU} \subset \mathfrak{TR}_{MU}$, $\mathcal{AR}_S \subset \mathcal{AR}_{NU} \subset \mathcal{AR}_{MU}$, and there is a forgetful inclusion $\mathcal{AR} \to \mathfrak{TR}$. If $\mathcal{C}$ is any subset of $\mathfrak{TR}$ or $\mathcal{AR}$ then denote $|\mathcal{C}| = \{|\mathfrak{T}| \mid \mathfrak{TR} \in \mathcal{C}\}$.

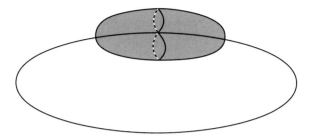

FIGURE I.4.2. Realization of a resolution tower

We can modify resolution towers to nicer ones without changing the realizations. For example Proposition 4.2.7 and Theorem 4.2.8 imply the following Theorem.

Theorem (4.2.8) *If $\mathfrak{T} \in \mathcal{TR}$ then there is a $\mathfrak{T}' \in \mathcal{TR}_R$ with $|\mathfrak{T}| = |\mathfrak{T}'|$. Moreover, if $\mathfrak{T} \in \mathcal{TR}_*$ for $* = M, N, S, U, F, E$, then $\mathfrak{T}' \in \mathcal{TR}_*$ also. In addition if $\mathfrak{T} \in \mathcal{TR}_M$ we may conclude $\mathfrak{T}' \in \mathcal{TR}_{RNE}$. In the algebraic case, if $\mathfrak{T} \in \mathcal{AR}_M$ then $\mathfrak{T}' \in \mathcal{AR}_{NE}$.*

Let $\mathcal{S}_{PL}$ be the set of PL isomorphism classes of smooth stratified sets and let $\mathcal{ALG}$ be the PL isomorphism classes of singular stratifications of compact real algebraic sets. We can now state our main results:

Theorem (5.3.3) *If $\mathfrak{T} \in \mathcal{AR}_U$ then $|\mathfrak{T}|$ is isomorphic to an algebraic set as a stratified space. In particular $|\mathcal{AR}_U| \subset \mathcal{ALG}$.*

Theorem (6.4.2) $|\mathcal{AR}_{FUN}| = \mathcal{ALG}$ *(In fact by Theorem 6.0 the complex version of this is also true).*

Theorem (5.0) *If $\mathfrak{T} \in \mathcal{TR}_{FS}$ then there is a $\mathfrak{T}' \in \mathcal{AR}_{FS}$ such that $|\mathfrak{T}| = |\mathfrak{T}'|$. In particular $|\mathfrak{T}|$ is isomorphic, as a stratified space, to an algebraic set.*

We can summarize these results in the following diagram

$$
\begin{array}{ccccc}
\mathcal{ALG} & = & |\mathcal{AR}_{FUN}| & = & |\mathcal{AR}_U| \\
& & & & \uparrow i \\
& & |\mathcal{AR}_S| & \xrightarrow{\ k\ } & |\mathcal{TR}_S| \\
& & \uparrow j & & \uparrow \\
& & |\mathcal{AR}_{FS}| & \xrightarrow{onto} & |\mathcal{TR}_{FS}|
\end{array}
$$

where vertical maps are inclusions and horizontal maps are forgetful maps. Hence proving i and j onto, (or alternatively i and k onto) would topologically characterize real algebraic sets by proving that the image of the forgetful map

$$\mathcal{ALG} \to \mathcal{S}_{PL}$$

is $|\mathcal{TR}_{FS}|$ (or $|\mathcal{TR}_S|$). This turns out to be the case for algebraic sets of dimension ≤ 3. In Chapter VIIthe topological classification of 3-dimensional real algebraic sets is given.

One of the nice properties of resolution towers is that they behave like manifolds, for example this allows us to define cobordism groups of real algebraic sets. In Chapter VII we calculate cobordism groups in dimension ≤ 2. These cobordism groups are used in Chapter VII as in [**AK9**] where we outline an obstruction theory to decide when a stratified set X is a realization $|\mathcal{T}|$ of some resolution tower $\mathcal{T}$.

Exercise: Show that any A_k-space is the realization of a resolution tower of type S. (Hint: see the proof of Proposition 2.8.13 or 7.3.2).   ◇

Because of some special properties of the resolution tower structures on A_k-spaces, they can be made algebraic sets without the type F condition. This is why all compact P.L. manifolds are homeomorphic to real algebraic sets [**AK2**].

We end this summary by giving an example to demonstrate how to find a realization of resolution tower structure of an algebraic set. In calculating this example we will use some blowup terminology which we introduce in Chapter II. For our example, we take the real algebraic set

$$
\begin{aligned}
X &= V((x^2 - y^3)^2(y^2 + x^2 - 2) + z^2) \\
&= \{(x, y, z) \in \mathbf{R}^3 \mid (x^2 - y^3)^2(y^2 + x^2 - 2) + z^2 = 0\}.
\end{aligned}
$$

The algebraic set X is obtained by a blowing down operation which we will formalize in section 6 of Chapter II. It is obtained by collapsing the sphere $x^2 + y^2 + z^2 = 2$ in $\mathbf{R}^3$ to the curve $x^2 = y^3$ in $\mathbf{R}^2$ via the projection $\pi : \mathbf{R}^3 \to \mathbf{R}^2$.

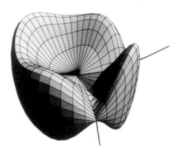

FIGURE I.4.3. X

We have the singular stratification on X given by $X_2 = X$, $X_1 = \mathrm{Sing}\,X = \{(x, y, 0) \mid x^2 = y^3\}$ and $X_0 = \mathrm{Sing}(\mathrm{Sing}X) = \{(0, 0, 0)\}$.

Our first step is to resolve the singularities of X. There are many ways to do this, we will describe one of them. The knowledgable reader will notice that in the interest of reducing the number of required blowups we do not follow Hironaka's recipe in [**H**], for example centers might not have only normal crossings with inverse images of centers. Our first step is to blow up the point $(0,0,0)$. The algebraic set we obtain is still singular, its singular set is the union of two smooth curves, one is the strict transform of X_1 and the other is a circle which is the inverse image of $(0,0,0)$. The reader interested in the actual equations of this algebraic set can find them at the end of this chapter. Our next step is to blow up the singular circle. After doing this we get an algebraic set whose singularity set is a smooth curve: the strict transform of X_1. We now blow up this curve and obtain a nonsingular algebraic set V_2 diffeomorphic to the Klein bottle. The composition of the blowup maps is an entire rational function $\pi : V_2 \to X$.

The next step is to make sure $\pi^{-1}(X_1)$ is a tico. It turns out that $\pi^{-1}(X_1)$ is not a tico, it is the union of three nonsingular curves, two of them are disjoint circles C_1 and C_2 with nontrivial normal bundle and the third D intersects each of the other two in a single point. However the curves are tangent at their points of intersection which keeps $\pi^{-1}(X_1)$ from being a tico. We draw half of V_2 below in Figure I.4.4, the other half is just the same so V_2 is obtained by doubling

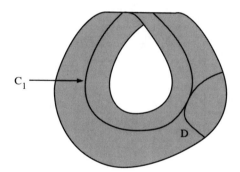

FIGURE I.4.4. Tangential intersection

the figure. The blowup map π maps C_1 and C_2 to $(0,0,0)$ and maps D onto $X_1 \cap \{x^2 + y^2 \le 2\}$.

To make $\pi^{-1}(X_1)$ a tico we do some more blowing up. We first blow up the two points $C_1 \cap D$ and $C_2 \cap D$. So as not to clutter up the notation, we let V_2 now be the blowup of the old V_2, let C_i and D be the strict transforms of the old curves, etc. The blowup adds two circles C_3 and C_4 to $\pi^{-1}(X_1)$. Although the circles in $\pi^{-1}(X_1)$ are now pairwise transverse, they still do not make a tico since there are two points where three of them intersect. Figure I.4.5 is a picture of half of the new V_2. Again V_2 is obtained by doubling Figure I.4.5.

We take care of the threefold intersections by blowing them up again. Finally $\pi^{-1}(X_1)$ is a tico. Figure I.4.6 shows half of V_2. There are two new circles in $\pi^{-1}(X_1)$ which we denote C_5 and C_6.

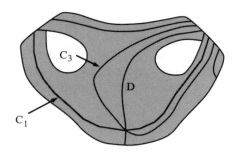

FIGURE I.4.5. $\pi^{-1}(X_1)$ pairwise transverse, but not a tico

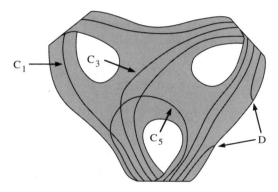

FIGURE I.4.6. $\pi^{-1}(X_1)$ now a tico

The next step is to refine the stratification of X to make π restricted to $\pi^{-1}(X_1 - X_0)$ be a submersion. In our example, this will be true if we change X_0 to be the three points $(0, 0, 0)$, $(1, 1, 0)$ and $(-1, 1, 0)$.

We now must make sure $\pi^{-1}(X_0)$ is a tico. Now $\pi^{-1}((0, 0, 0))$ is the union of the circles C_i which is a tico, but $\pi^{-1}((\pm 1, 1, 0))$ is two points which is not a tico since it has codimension two. We easily make it a tico by blowing up these two points, obtaining Figure I.4.7.

The next step is to resolve the singularities of X_1. This is easily done by blowing up the point $(0, 0, 0)$ to obtain an algebraic set V_1 which is in fact $\mathbf{R}$. The blowup map $p_1 : V_1 \to X_1$ is just $t \mapsto (t^3, t^2, 0)$. We want to make sure that $p_1^{-1}(X_0)$ is a tico in V_1, in our case it is, but in general one would need to blow up some more to assure this just as we did for V_2 (of course in this dimension that would never be necessary). Let V_{12} be the closure of $\pi^{-1}(X_1 - X_0)$, i.e., the sheets of the tico $\pi^{-1}(X_1)$ which are not in the tico $\pi^{-1}(X_0)$. Thus in our example, $V_{12} = D$. We now wish to lift the map $\pi| : V_{12} \to X_1$ to an entire rational function $p_{12} : V_{12} \to V_1$. As it turns out, in our example there is a lifting but in general one would need to blow up V_2 some more to get a lifting. We also want p_{12} to be a mico, but again in our case this is already true so we do not

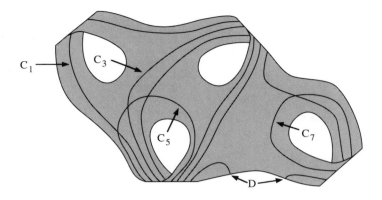

FIGURE I.4.7. $\pi^{-1}(X_0)$ now a tico

need to blow up more.

At this point we have our resolution tower. We just set $V_{12} = D$, $V_{02} = \pi^{-1}(X_0) = \cup C_i$, $V_0 = X_0$, $p_{02} = \pi|$, $V_{01} = p_1^{-1}(X_0)$ and $p_{01} = p_1|$.

For those interested in the details of the blowup, we will now provide them. For the purpose of computing examples of blowups, it is worthwhile using another point of view of algebraic sets. This is because our standard model of the blowup of an affine algebraic set is another affine algebraic set, but one in a much higher dimensional Euclidean space. The example above would probably end up in dimension several hundred.

To get around this, we look at the blowup as being what we call an algebraic space: the union of affine algebraic sets (called charts) glued together via birational isomorphisms of Zariski open subsets. Thus it is the algebraic analogue of a manifold. We will not bother to develop the theory of algebraic spaces here, the elementary notions are in [**AK4**] or can be done as an exercise. The advantage for calculating blowups in this way is that the ambient Euclidean space dimension grows slowly or sometimes not at all, so the number of variables is kept in check. We use a shorthand notation to describe the charts of the blowup which we will describe as we go along.

Our first blowup of the point $(0, 0, 0)$ has three charts at first glance although we will presently see that two of them will cover the blowup. These charts are represented as follows:

A	$x^2(1 - xy^3)^2(x^2 + x^2y^2 - 2) + z^2 = 0$	(x, xy, xz)
B	$y^2(x^2 - y)^2(y^2 + x^2y^2 - 2) + z^2 = 0$	(xy, y, yz)
$C*$	$z^2(x^2 - zy^3)^2(z^2(x^2 + y^2) - 2) + 1 = 0$	(xz, yz, z)

The first column just names the charts, the second column gives the equation of the algebraic set and the third column gives the blowup map down to $\mathbf{R}^3$. The star after C means that this chart is not necessary, any point in it is contained

in one of the other two charts. Hence we forget the chart C. The gluing maps
between charts can be obtained from the third column, just map down to $\mathbf{R}^3$
and follow by the inverse of the map from another chart. For example the gluing
map from A to B is $(1/y, xy, z/y)$.

Our next blowup was of the inverse image of $(0,0,0)$. In A this is $\{x = 0\} \cap A = \{x = z = 0\}$. In B this is $\{y = 0\} \cap B = \{y = z = 0\}$. We represent
these blowups as follows:

$$D > A \qquad (1 - xy^3)^2(x^2 + x^2y^2 - 2) + z^2 = 0 \qquad (x, xy, x^2z)$$

$$E* > A \qquad x^2(1 - xzy^3)^2(x^2z^2(1 + y^2) - 2) + 1 = 0 \quad (xz, xyz, xz^2)$$

$$F > B \qquad (x^2 - y)^2(y^2 + x^2y^2 - 2) + z^2 = 0 \qquad (xy, y, y^2z)$$

$$G* > B \qquad y^2(x^2 - yz)^2(y^2z^2(1 + x^2) - 2) + 1 = 0 \quad (xyz, yz, yz^2)$$

We have a new symbol, $> A$ (or $> B$) means that the chart was obtained by
blowing up the chart A (or B). The map from D to A for example is (x, y, xz).
Note we still have only two charts D and F since we can forget E and G. From
now on we will not bother to compute the third column of ignored charts, since
we will never need it.

The next step is to blow up the strict transform of X_1. In chart D this is
the Zariski closure of the inverse image of $X_1 - X_0$. The inverse image of X_0 is
$\{x = 0\} \cap D = \{x = 0, z^2 = 2\}$ and the inverse image of X_1 is $\{x^2 = (xy)^3, x^2z = 0\} \cap D = \{x = 0, z^2 = 2\} \cup \{z = 0, xy^3 = 1\}$. Hence the strict transform of X_1
in D is the curve $\{z = 0, xy^3 = 1\}$. Likewise, the part of the strict transform of
X_1 which lies in F is $\{z = 0, y = x^2\}$. We now blow these up, obtaining four
charts.

$$H > D \qquad x^2 + x^2y^2 - 2 + z^2 = 0 \qquad\qquad\qquad (x, xy, x^2z(1 - xy^3))$$

$$I* > D \qquad u^2(x^2 + x^2y^2 - 2) + 1 = 0, \ uz = 1 - xy^3$$

$$J > F \qquad y^2 + x^2y^2 - 2 + z^2 = 0 \qquad\qquad\qquad (xy, y, y^2z(x^2 - y))$$

$$K* > F \qquad u^2(y^2 + x^2y^2 - 2) + 1 = 0, \ uz = y - x^2$$

Notice that because we are blowing up something more complicated than just
coordinate planes, in charts I and K it was necessary to add a new variable u
(although in K we could also eliminate y). Recall that at this point our algebraic
set was nonsingular and in fact diffeomorphic to a Klein bottle. We can see this
from the equations for H and J. Each of these is a cylinder (for example in H,
for each fixed y we get an ellipse, likewise in J for fixed x). The gluing map from
J to H is just $(xy, 1/x, z)$ and this is easily seen to glue the two cylinders into a
Klein bottle.

In H, $\pi^{-1}(X_1)$ is $\{x = 0, \ z^2 = 2\} \cup \{1 = xy^3, \ z^2 = 2 - y^{-4} - y^{-6}\}$ which
is four nonintersecting nonsingular noncompact curves. Notice that H contains
all of $\pi^{-1}(X_1 - X_0)$, it is $\{1 = xy^3, \ z^2 = 2 - y^{-4} - y^{-6}\}$. In J, $\pi^{-1}(X_1)$ is

$\{y = 0,\ z^2 = 2\} \cup \{y = x^2,\ z^2 = 2 - x^4 - x^6\}$ which is two noncompact nonsingular curves union a nonsingular circle. There are two points of intersection, both not transverse, at $\{x = y = 0,\ z^2 = 2\}$. Our next step is to blow up these points which could be done one at a time, but to reduce the number of charts we blow up both points together by blowing up the ideal $\langle x, y \rangle$ in J. We obtain the following charts:

$$L > J \qquad x^2 y^2 + x^4 y^2 - 2 + z^2 = 0 \qquad\qquad (x^2 y, xy, x^3 y^2 z (x - y))$$

$$M > J \qquad y^2 + x^2 y^4 - 2 + z^2 = 0 \qquad\qquad (xy^2, y, y^3 z (x^2 y - 1))$$

Recall that we also still have the chart H, so we actually have three charts now, In M, $\pi^{-1}(X_1)$ is $\{y = 0,\ z^2 = 2\} \cup \{x^2 y = 1,\ z^2 = 2 - y^2 - y^3\}$ which is a tico. However in L, $\pi^{-1}(X_1)$ is $\{x = 0,\ z^2 = 2\} \cup \{y = 0,\ z^2 = 2\} \cup \{x = y,\ z^2 = 2 - x^4 - x^6\}$ which is not a tico because of the three lines coming together at $(0, 0, \pm\sqrt{2})$. Hence we blow up the ideal $\langle x, y \rangle$ in L and obtain the following new charts N and O as well as our old charts M and H.

$$N > L \qquad x^4 y^2 + x^6 y^2 - 2 + z^2 = 0 \qquad\qquad (x^3 y, x^2 y, x^6 y^2 z (1 - y))$$

$$O > L \qquad x^2 y^4 + x^4 y^6 - 2 + z^2 = 0 \qquad\qquad (x^2 y^3, xy^2, x^3 y^6 z (x - 1))$$

$$H > D \qquad x^2 + x^2 y^2 - 2 + z^2 = 0 \qquad\qquad (x, xy, x^2 z (1 - xy^3))$$

$$M > J \qquad y^2 + x^2 y^4 - 2 + z^2 = 0 \qquad\qquad (xy^2, y, y^3 z (x^2 y - 1))$$

We now must see where π restricted to $\pi^{-1}(X_1 - X_0)$ is a submersion. Recall that all of $\pi^{-1}(X_1 - X_0)$ is contained in H, it is $\{1 = xy^3,\ z^2 = 2 - y^{-4} - y^{-6}\}$. We see then that π will restrict to a submersion everywhere except when $x = y = \pm 1$, $z = 0$ in H. We then enlarge X_0 by adding the image of these points which is $(\pm 1, 1, 0)$. We now need to make $\pi^{-1}((\pm 1, 1, 0))$ a tico. So we blow up these two points $\pi^{-1}((\pm 1, 1, 0))$. This means we blow up the ideal $\langle y - 1, z \rangle$ in N, $\langle x - 1, z \rangle$ in O, $\langle 1 - xy^3, z \rangle$ in H and $\langle x^2 y - 1, z \rangle$ in M. These generators work since, for example,

$$1 - x^2 = \left((y + 1)x^4 (1 + x^2)/(2 + 2x^2 + x^4) \right) (y - 1) + \left(z/(2 + 2x^2 + x^4) \right) z$$

in chart N.

After doing this blowup we obtain the following charts:

$$P > N \qquad x^4 y^2 + x^6 y^2 - 2 + (y - 1)^2 z^2 = 0 \qquad (x^3 y, x^2 y, -x^6 y^2 z (1 - y)^2)$$

$$Q > N \qquad x^4 y^2 + x^6 y^2 - 2 + z^2 = 0,\ y = 1 + uz \qquad (x^3 y, x^2 y, -x^6 y^2 u z^2)$$

$$R > O \qquad x^2 y^4 + x^4 y^6 - 2 + (x - 1)^2 z^2 = 0 \qquad (x^2 y^3, xy^2, x^3 y^6 z (x - 1)^2)$$

$$S > O \qquad x^2 y^4 + x^4 y^6 - 2 + z^2 = 0,\ x = 1 + uz \qquad (x^2 y^3, xy^2, x^3 y^6 u z^2)$$

$$T > H \qquad x^2 + x^2 y^2 - 2 + (1 - xy^3)^2 z^2 = 0 \qquad (x, xy, x^2 z (1 - xy^3)^2)$$

$$U > H \qquad x^2 + x^2 y^2 - 2 + z^2 = 0,\ 1 - xy^3 = uz \qquad (x, xy, x^2 u z^2)$$

$V > M$ $y^2 + x^2 y^4 - 2 + (x^2 y - 1)^2 z^2 = 0$ $(xy^2, y, y^3 z(x^2 y - 1)^2)$

$W > M$ $y^2 + x^2 y^4 - 2 + z^2 = 0,\ x^2 y = 1 + uz$ $(xy^2, y, y^3 u z^2)$

Finally we must check that $\pi|_{V_{12}}$ lifts to a tico map to V_1. It does. The following table gives the equations of V_{12} and the lifted map for each chart in which V_{12} lies.

Q $y = 1,\ u = 0,\ z^2 = 2 - x^4 - x^6$ (x)

S $x = 1,\ u = 0,\ z^2 = 2 - y^4 - y^6$ (y)

U $x = y^{-3},\ u = 0,\ z^2 = 2 - y^{-4} - y^{-6}$ (y^{-1})

W $y = x^{-2},\ u = 0,\ z^2 = 2 - x^{-4} - x^{-6}$ (x^{-1})

In fact, we see that V_{12} is wholly contained in either Q or S. So to check that p_{12} is a tico it suffices to check in Q. In Q for example, $V_{12} \cap V_{02}$ is the points $u = x = 0$, $y = 1$, $z^2 = 2$ where p_{12} submerses and also the points $q_{\pm}$ given by $y = 1$, $u = z = 0$, $x = \pm 1$. Near $q_{\pm}$, V_{12} has z for local coordinates, $dx/dz = -2z/(6x^5 + 4x^3)$ which is 0 at $q_{\pm}$. Also $d^2 x/dz^2 = \mp 0.2$ at $q_{\pm}$ so $p_{12}(z) = z^2 \varphi(z)$ near $q_{\pm}$ with $\varphi(q_{\pm}) = \mp 0.1 \neq 0$ so p_{12} is a tico map.

In drawing pictures of our algebraic set we claimed that we could always divide it into two pieces which were symmetric and thus get away with showing only half the set. To be more precise we have an involution $(x, y, z) \mapsto (-x, y, -z)$ on X and we are always showing a fundamental domain of liftings of this involution. For example we could take Figure I.4.4 to be the points in H where $z \geq x$ and the points on J where $z \geq xy$. The other figures are just the topological strict transforms of this. (The more obvious choice of $z \geq 0$ in both H and J would not do because when we blow up to make charts P through W we would blow up two points on the boundary, and thus end up with a strict transform which is not a manifold with boundary).

In the next chapter we will establish all the background material before seriously studying the resolution towers.

CHAPTER II

ALGEBRAIC SETS

In this chapter we start out with elementary properties of real algebraic sets and gradually go on to discuss more recent results in real algebraic geometry. We begin with a review of the basic properties of algebraic sets, we will avoid proving some of the more well-known elementary results by referring the reader to [**W**] and [**M**]. Also looking at [**AK6**] at times would be useful. Other useful books for the introductory material are [**Sh**], [**P**], [**B**], [**BCR**].

1. Basic Properties of Algebraic Sets

Definitions: Let k be a field. We let $k[x_1, x_2, \ldots, x_n]$ denote the ring of polynomials in n variables with coefficients in k. If $J \subset k[x_1, \ldots, x_n]$ is a set of polynomials then we define

$$\mathcal{V}(J) = \{ (x_1, \ldots, x_n) \in k^n \mid f(x_1, \ldots, x_n) = 0 \text{ for all } f \in J \}.$$

Given any subset $S \subset k^n$ define

$$\mathcal{I}(S) = \{ f \in k[x_1, \ldots, x_n] \mid f(x_1, \ldots, x_n) = 0 \text{ for all } (x_1, \ldots, x_n) \in S \}.$$

Note $\mathcal{I}(S)$ is an ideal in the ring $k[x_1, \ldots, x_n]$. We have the properties:

$I \subset J$ implies $\mathcal{V}(I) \supset \mathcal{V}(J)$.

$S \subset T$ implies $\mathcal{I}(S) \supset \mathcal{I}(T)$.

$\mathcal{V}(\sum_\alpha J_\alpha) = \cap_\alpha \mathcal{V}(J_\alpha)$.

$\mathcal{V}(IJ) = \mathcal{V}(I) \cup \mathcal{V}(J)$.

$\mathcal{I}(S \cup T) = \mathcal{I}(S) \cap \mathcal{I}(T)$.

Any ideal J in $k[x_1, x_2, \ldots, x_n]$ is finitely generated. If $g_1, \ldots, g_k$ are generators of J we will write $J = \langle g_1, \ldots, g_k \rangle$. If k is algebraically closed, the Hilbert Nullstellensatz says that if $J \subset k[x_1, \ldots, x_n]$ is an ideal then

$$\mathcal{I}(\mathcal{V}(J)) = \sqrt{J} = \{ f \in k[x_1, \ldots, x_n] \mid f^m \in J \text{ for some } m \}.$$

This is not true for $k = \mathbf{R}$, for example if $J = \langle x^2 + y^2 \rangle \subset \mathbf{R}[x, y]$ then $\mathcal{V}(J) = (0, 0) \in \mathbf{R}^2$, and $\mathfrak{I}(\mathcal{V}(J)) = \langle x, y \rangle$. The real nullstellensatz is more complicated, c.f., [**BCR**].

Definitions: A subset $V \subset k^n$ is called a k-*algebraic set* if it is the set of roots of a collection of polynomials in n-variables. That is, if $V = \mathcal{V}(J)$ for some $J \subset k[x_1, \dots, x_n]$. In this book we will mainly consider the case $k = \mathbf{R}$ and sometimes $k = \mathbf{C}$, i.e., *real* and *complex algebraic sets*. We sometimes also call a k-algebraic set an *affine* k-algebraic set to distinguish it from the projective k-algebraic sets we define later in section 3.

A very useful property of any real algebraic set V is that there is a polynomial p so that $V = p^{-1}(0)$. Just take $p = p_1^2 + \cdots p_k^2$ where $p_1, \dots, p_k$ are the generators of $\mathfrak{I}(V)$.

Definitions: Given any subset $S \subset k^n$, the *Zariski closure* of S, denoted $\mathrm{Cl}_k S$, is defined to be the smallest k-algebraic set containing S, i.e., it is the algebraic set $\mathcal{V}(\mathfrak{I}(S))$. It follows that a set $S \subset k^n$ is algebraic if and only if it equals its Zariski closure. In fact we could put a topology on any algebraic set called the *Zariski topology* where the closed sets are just the algebraic subsets. However, we will soon be specializing to the cases $k = \mathbf{R}$ or $\mathbf{C}$ and unless we explicitly state otherwise we use the usual metric topology on real or complex algebraic sets. But we say a subset of an algebraic set S is *Zariski open* if it is $S - T$ for some algebraic set T.

In some sense, the natural collection of spaces to look at is the collection of Zariski open subsets of algebraic sets. This is because as we see below, these are the natural domains of entire rational functions. Also in the real case, any Zariski open subset is isomorphic to an algebraic set anyway. Thus we are motivated to make the following definitions:

Definitions: A k-*Zopen set* (pronounced 'kay zee open set') is a subset of some k^n which is Zariski open in its Zariski closure. In other words, it is the difference of two k algebraic sets. A k-*Zclosed subset* A of a Zopen set U is a subset of U which is the intersection of U with an algebraic set. In other words, $\mathrm{Cl}_k(A) \cap U = A$. A k-*Zopen subset* V of a k-Zopen set U is a subset $V \subset U$ so that $U - V$ is a k-Zclosed subset. A k-*Zopen neighborhood* of a point x in a Zopen set U is a Zopen subset V of U so that $x \in V$. We usually omit the k- when the field k is clear, thus we refer to Zopen sets, Zopen subsets and Zopen neighborhoods.

Any collection $\mathcal{U}$ of algebraic sets in k^n has a minimal member, that is there exists $V_0 \in \mathcal{U}$ such that for all elements $V \in \mathcal{U}$ either $V = V_0$ or $V \not\subset V_0$. The corresponding result for Zopen sets is not true, consider the collection of Zariski open subsets of an infinite field k, the nonempty ones are just compliments of finite subsets.

Exercise: Show that any collection of Zclosed subsets of a Zopen set X has a minimal member.   ◇

Definitions: Let $V \subset k^n$ and $W \subset k^m$ be any subsets. A function $f\colon V \to W$ is called a *polynomial* if there is a polynomial function $g\colon k^n \to k^m$ such that $g(V) \subset W$ and $f = g|_V$. We say a function $f\colon V \to W$ is *regular* or an *entire rational function* if it is locally the quotient of polynomials, i.e., for every $x \in V$ there is a Zopen neighborhood U of x in k^n and polynomials $p\colon U \to k^m$ and $q\colon U \to k$ so that q is nowhere 0 on $U \cap V$ and $f(y) = p(y)/q(y)$ for all $y \in U \cap V$. Normally we only use the case where V and W are Zopen sets, but we will occasionally use the more general notion when convenient.

It is standard practice in algebraic geometry to talk about rational functions. These are like the entire rational functions above except that they might not be defined everywhere. We will avoid this notion since we are looking at algebraic sets from a topological point of view. Thus we always want to know where functions are defined.

It is a useful fact that for $k = \mathbf{R}$ we may take the local expression of an entire rational function as a quotient of two polynomials to be global, in other words:

Proposition 2.1.1 *If $k = \mathbf{R}$ and $f\colon U \to W$ is a rational function we may find polynomials $p\colon U \to \mathbf{R}^m$ and $q\colon U \to \mathbf{R}$ so that $q(U) \subset \mathbf{R} - 0$ and $f(x) = p(x)/q(x)$ for all $x \in U$.*

Proof: Suppose $U \subset \mathbf{R}^n$. Let $\mathcal{S}$ be the set of algebraic subsets S of $\mathbf{R}^n$ such that there are polynomials $p_S\colon \mathbf{R}^n \to \mathbf{R}^m$ and $q_S\colon \mathbf{R}^n \to \mathbf{R}$ so that $q_S^{-1}(0) = S$ and $f(y) = p_S(y)/q_S(y)$ for all $y \in U - S$. Note that $\mathcal{S}$ is closed under finite intersection since

$$f(y) = (p_T(y)q_T(y) + p_S(y)q_S(y))/(q_T(y)^2 + q_S(y)^2)$$

for all $y \in U - (S \cap T)$ if S and T are in $\mathcal{S}$. But arbitrary intersections of algebraic sets are actually finite, or alternatively, if T is a minimal member of $\mathcal{S}$ we must then have $T = \bigcap_{S \in \mathcal{S}} S$. So it suffices to show that for each $x \in U$ there is an $S \in \mathcal{S}$ so that $x \notin S$. This is almost immediate from the definition of rational function, there is an S and polynomials $p\colon \mathbf{R}^n \to \mathbf{R}^m$ and $q\colon \mathbf{R}^n \to \mathbf{R}$ so that $q^{-1}(0) \subset S$ and $f(y) = p(y)/q(y)$ for all $y \in U - S$. Now take a polynomial r so that $r^{-1}(0) = S$ and let $p_S = rp$ and $q_S = rq$.   □

For $k = \mathbf{C}$ we have no analogue of Proposition 2.1.1. For example, if $S = \mathcal{V}(xy - zw)$ and $T = \mathcal{V}(y, z)$ and $f\colon S - T \to \mathbf{C}$ is x/z if $z \neq 0$ and w/y if $y \neq 0$ we cannot find p and q as above.

Exercise: Show that entire rational functions are continuous in the Zariski topology, i.e., if $f\colon U \to W$ is entire rational and $X \subset W$ is a Zclosed subset then $f^{-1}(X)$ is a Zclosed subset. Show that the converse is not true.   ◇

Definition: If V and W are Zopen sets then we say $f\colon V \to W$ is a *birational isomorphism* if it has an entire rational inverse, i.e., there is an entire rational function $g\colon W \to V$ so that $g \circ f$ and $f \circ g$ are identities. We say V and W are *birationally isomorphic* if there is a birational isomorphism $f\colon V \to W$. This notion should be distinguished from the algebraic geometer's notion of birational equivalence, V and W could be birationally equivalent even though their topologies may be widely different. The idea is that a birational isomorphism is a homeomorphism which preserves all algebraic structure.

Definitions: We say a Zopen set V is *reducible* if $V = V_1 \cup V_2$ for some Zclosed subsets V_1, V_2 such that $V_1 \neq V \neq V_2$, otherwise we say V is *irreducible*.

By standard algebraic geometry, if V is a k algebraic set then V can be written uniquely as $V = \bigcup_{i=1}^{m} V_i$ where each V_i is an irreducible algebraic set and no V_i is contained in another V_j. Each such V_i is called an irreducible component of V. Also V is irreducible if and only if $\mathfrak{I}(V)$ is a prime ideal.

We leave the following Lemma as an exercise:

Lemma 2.1.2 *Let U be a k-Zopen set and let $V = Cl_k(U)$. Let $V_1, \dots, V_m$ be the irreducible components of V. Then each $U_i = U \cap V_i$ is an irreducible Zclosed subset of U and any irreducible Zclosed subset U' of U must be contained in some U_i. Furthermore, no U_i is contained in another U_j. In particular U is irreducible if and only if V is irreducible.*

As a consequence of Lemma 2.1.2, we then know that U is irreducible if and only if $\mathfrak{I}(U)$ is a prime ideal. Also U can be written uniquely as $U = \bigcup_{i=1}^{m} U_i$ where each U_i is an irreducible Zclosed subset and no U_i is contained in another U_j. Each such U_i is called an *irreducible component* of U. The decomposition $U = \bigcup_{i=1}^{m} U_i$ is called the *irreducible decomposition* of U.

The irreducible components of a complex algebraic set are connected (c.f., Corollary 4.16 of [**M**]), but they need not be connected in the real case. For example the real algebraic set $\mathcal{V}\left(y^4 + x^4 - 5x^2 + 4\right)$ is irreducible but topologically it has two components, each of which is diffeomorphic to a circle.

Next we note that birational isomorphisms preserve Zclosed subsets and the irreducible decomposition. We leave the proof of Lemma 2.1.3 as an exercise - the hint is that $f = g^{-1}$.

Lemma 2.1.3 *Let $f\colon V \to W$ be a birational isomorphism.*

    a) *If $Z \subset V$ is a Zclosed subset then $f(Z)$ is a Zclosed subset.*

    b) *Let V_i, $i = 1, \dots, m$ be the irreducible components of V. Let W_i, $i = 1, \dots, n$ be the irreducible components of W. Then $m = n$ and after renumbering, $f(V_i) = W_i$ for all i.*

We alluded above to the following:

Lemma 2.1.4 *Any real Zopen set U is birationally isomorphic to a real algebraic set.*

Proof: Take polynomials p and q so that $p^{-1}(0) = \mathrm{Cl}_{\mathbf{R}}(U) - U$ and $q^{-1}(0) = \mathrm{Cl}_{\mathbf{R}}(U)$. Suppose $U \subset \mathbf{R}^n$. Let

$$V = \{\,(x,t) \in \mathbf{R}^n \times \mathbf{R} \mid q(x) = 0 \ \text{ and } \ tp(x) = 1\,\}.$$

Then $x \mapsto (x, 1/p(x))$ is a birational isomorphism of U to the algebraic set V. $\qquad\square$

Definition: A continuous map $f\colon X \to Y$ between topological spaces is called *proper* if $f^{-1}(K)$ is compact for each compact set $K \subset Y$.

Lemma 2.1.5 *Let $p\colon \mathbf{R}^n \to \mathbf{R}$ be a polynomial with $p^{-1}(0)$ compact. Then for some m the polynomial map $x \mapsto |x|^{2m}p(x)$ is proper.*

Proof: Let $\theta\colon \mathbf{R}^n - 0 \to \mathbf{R}^n - 0$ be the inversion through the unit sphere, $\theta(x) = x/|x|^2$. Let p have degree d. Then after clearing denominators, $q(x) = |x|^{2d}p \circ \theta(x)$ is a polynomial and it vanishes on $0 \cup \theta(p^{-1}(0))$. If $p^{-1}(0)$ is contained in a ball of radius r, then $\theta(p^{-1}(0))$ lies outside a ball of radius $1/r$. So 0 is isolated in $q^{-1}(0)$ and the Lojasiewicz inequality ([**L1**]) then implies that for some e and $c > 0$ we have $|q(x)| \geq c|x|^e$ for all x near 0. But then for x near infinity, $|p(x)| = |x|^{-2d}|q \circ \theta(x)| \geq c|x|^{-2d-e}$ so if $2m > 2d + e$ the result follows. $\qquad\square$

Corollary 2.1.6 *If $V \subset \mathbf{R}^n$ is a compact real algebraic set, then $V = p^{-1}(0)$ for some proper polynomial $p\colon \mathbf{R}^n \to \mathbf{R}$.*

Definitions: Let V be a subset of k^n (which we usually take to be a k-Zopen set). Define $\Gamma(V) = k[x_1,\dots,x_n]/\Im(V)$. There is a natural one to one correspondence between $\Gamma(V)$ and the polynomial maps from V to k. $\Gamma(V)$ is a finitely generated k-algebra. If $V \subset k^n$ and $W \subset k^m$ and $f = (f_1,\dots,f_m)\colon V \to W$ is a polynomial map, then we define $\Gamma(f)\colon \Gamma(W) \to \Gamma(V)$ by $\Gamma(f)(g) = g \circ f$, i.e., $\Gamma(f)(x_i) = f_i(x_1,\dots,x_n)$ for $i = 1, 2, \dots, m$. We also denote $\Gamma(f) = f^*$.

Note that Γ gives a contravariant functor between the categories:

$$\left\{\begin{array}{c} k \text{ algebraic sets,} \\ \text{polynomial morphism} \end{array}\right\} \overset{\Gamma}{\longrightarrow} \left\{\begin{array}{c} \text{finitely generated } k \text{ algebras} \\ k \text{ algebra morphism} \end{array}\right\}$$

The functor Γ is one to one (or more precisely, it is one to one on isomorphism classes of objects). If k is algebraically closed, a simple application of the Nullstellensatz shows that the image of Γ is the finitely generated k-algebras with no nonzero nilpotents (i.e., if I is the ideal of the relations of the generators, then $\sqrt{I} = I$). This correspondence explains the close relationship between the algebraic geometry over algebraically closed fields, and the commutative algebra.

The absence of this type of 'algebraization' functor gives real algebraic sets a different flavor; they are in a sense related more to topology than algebra. However real algebraic sets do enjoy some special algebraic properties such as Proposition 2.1.1 and Lemma 2.1.4. There are many more such properties. We will exploit them in a systematic manner as we go along.

In a similar manner, if U is a subset of some k^n, we may define $\Gamma^r(U)$ to be the ring of entire rational functions $f \colon U \to \mathbf{R}$. By Proposition 2.1.1 we see that for $k = \mathbf{R}$, $\Gamma^r(U)$ is obtained from $\Gamma(U)$ by allowing the inversion of polynomials which are nowhere zero on U. If k is algebraically closed and S is a k-algebraic set, $\Gamma^r(S) = \Gamma(S)$, c.f., Proposition 1.11 of [**M**]. Note that this result fails dramatically for $k = \mathbf{R}$. For example the entire rational function $f \colon \mathbf{R} \to \mathbf{R}$, $f(x) = 1/(1 + x^2)$ is not a polynomial.

If $V \subset U$ then we have an ideal $\mathfrak{I}_U(V)$ in the ring $\Gamma(U)$, namely the ideal of polynomial functions on U which vanish on V. It is naturally isomorphic to $\mathfrak{I}(V)/\mathfrak{I}(U)$. Likewise, we have an ideal $\mathfrak{I}_U^r(V)$ of entire rational functions vanishing on V.

If $f \colon U \to V$ is a polynomial and J is an ideal in $\Gamma(V)$ then we have a pullback ideal $f^*(J)$ in $\Gamma(U)$ where $f^*(J) = \{\, g \circ f \mid g \in J \,\}$. Similarly, if $f \colon U \to V$ is an entire rational function and J is an ideal in $\Gamma^r(V)$, we have a pullback ideal $f^*(J)$ in $\Gamma^r(U)$ where $f^*(J) = \{\, g \circ f \mid g \in J \,\}$.

2. Singularities of Real and Complex Algebraic Sets

From now on in this book, k will refer to either $\mathbf{R}$ or $\mathbf{C}$. We will use many facts from the first chapter of the wonderful book [**M**] for results on complex algebraic sets and use Whitney's seminal paper [**W**] to derive consequences for real algebraic sets. We must exercise some care in the use of [**M**] though since most results are only stated for irreducible complex algebraic sets. This makes perfect sense from the algebraic point of view, but from our geometric point of view it is not natural to restrict attention to irreducible algebraic sets. Doing so would mean we would ignore such questions as how the irreducible components intersect one another and how they together contribute to the topology of the algebraic set.

We will show in this section that any Zopen set has a stratification, i.e., if X is a Zopen set then we can find Zclosed subsets X_i, $i = 0, \ldots, n$ with $X_{i-1} \subset X_i$ so that each $X_i - X_{i-1}$ is either empty or a real or complex i dimensional submanifold of k^m. To do this we introduce the notion of singular and nonsingular points of a Zopen set. There is no general agreement on how to do this, different definitions make sense in different contexts. However, we prefer the definition below since it ensures that the singular set has lower dimension, and thus immediately gives a stratification. (The reader should beware that the 'stratification' on page 6 of [**M**] is not a stratification in the usual sense since the strata are not disjoint).

Definition: Let $V \subset k^n$ be a Zopen set. We say $x \in V$ is *nonsingular of dimension d in V* if there is a neighborhood U of x in k^n, and polynomials $f_i \in \mathfrak{I}(V)$ for $i = 1, \dots, n - d$ such that:

1) $U \cap V = U \cap \bigcap_{i=1}^{n-d} f_i^{-1}(0)$.

2) The gradients $\nabla f_i(x)$ for $i = 1, \dots, n - d$ are linearly independent.

Thus this is the algebraic analogue of V being a smooth d dimensional submanifold of k^n. By the implicit function theorem, a subset $V \subset k^n$ is a smooth (or complex analytic) manifold near x if 1) and 2) above are satisfied with smooth (or holomorphic) f_i.

Exercise: Suppose that 1) and 2) above hold for entire rational functions $f_i \in \Gamma_U^r(V)$. Show that x is nonsingular of dimension d in V. $\diamond$

Exercise: Show that if U is a Zopen subset of V then x is nonsingular of dimension e in U if and only if x is nonsingular of dimension e in V. $\diamond$

Definition: If V is a Zopen set we define $\dim(V)$ or $\dim V$ to be the largest d so that V has a point which is nonsingular of dimension d. Later we will find that all Zopen sets have nonsingular points, hence a dimension, but for now we must act as if $\dim V$ is not always defined. When we wish to emphasize the field k, we sometimes write $\dim_k V$ for $\dim V$.

Definitions: If V is a Zopen set we define $Nonsing(V)$ or $Nonsing\, V$ to be the set of points of V which are nonsingular of dimension $\dim V$. We define $Sing(V)$ or $Sing\, V$ to be its complement, $\text{Sing}\, V = V - \text{Nonsing}\, V$. The points of $\text{Sing}\, V$ are called *singular points* of V and the points of $\text{Nonsing}\, V$ are called the *nonsingular points* of V. We will show below that $\text{Sing}\, V$ is a Zclosed subset of V. We say a Zopen set V is *nonsingular* if $V = \text{Nonsing}\, V$, i.e, if $\text{Sing}\, V = \emptyset$.

Definition: We can associate to every real algebraic set a complex algebraic set by the process of complexification. If $V \subset \mathbf{R}^n$ is a real algebraic set, then by considering the inclusion $\mathbf{R}^n \subset \mathbf{C}^n$ we can assume $V \subset \mathbf{C}^n$. Then the *complexification* $V_{\mathbf{C}}$ of V is defined to be $\text{Cl}_{\mathbf{C}}(V)$. In other words $V_{\mathbf{C}}$ is the smallest complex algebraic set containing V.

If $\mathfrak{I}(V)$ is generated by $f_1, \dots, f_m$, we can consider each f_i to be a complex polynomial $f_{i\mathbf{C}}$, and define $\mathfrak{I}_{\mathbf{C}}$ to be the ideal in $\mathbf{C}[x_1, \dots, x_n]$ generated by $f_{i\mathbf{C}}, \dots, f_{m\mathbf{C}}$. Then $V_{\mathbf{C}} = \mathcal{V}(\mathfrak{I}_{\mathbf{C}})$.

$V_{\mathbf{C}}$ satisfies the following properties, (see [**W**])

1) $V_{\mathbf{C}} \cap \mathbf{R}^n = V$.

2) If $\bar{} : \mathbf{C}^n \to \mathbf{C}^n$ is the complex conjugation map, then $\overline{V}_{\mathbf{C}} = V_{\mathbf{C}}$.

3) If $V = V_1 \cup \cdots \cup V_m$ is the decomposition of V into irreducible components, then $V_{\mathbf{C}} = (V_1)_{\mathbf{C}} \cup \cdots \cup (V_m)_{\mathbf{C}}$ is the decomposition of $V_{\mathbf{C}}$ into irreducible components.

One does not have a canonical complexification of a real Zopen set V. The reason is that there is no smallest complex Zopen set containing V. So there are lots of complexifications. A complexification of V will be a set of the form $\mathrm{Cl}_{\mathbf{C}}(V) - T$ where T is any complex algebraic set which is invariant under complex conjugation so that $T \cap \mathbf{R}^n = \mathrm{Cl}_{\mathbf{R}}(V) - V$. Thus the complexification of a Zopen set is really a germ at V of complex algebraic sets. Whenever we talk below of the complexification of a Zopen set, we will sometimes be a bit sloppy and just pick a convenient representative as above. Note that any complexification satisfies properties 1), 2) and 3) above.

Let $V \subset k^n$ be a Zopen set and $x \in V$. Then $\{ \nabla f(x) \mid f \in \mathfrak{I}(V) \}$ is a vector space where $\nabla f(x)$ is the gradient of f at x. Define

$$
\begin{aligned}
r_x(V) &= \dim\{ \nabla f(x) \mid f \in \mathfrak{I}(V) \}, \\
r(V) &= \max\{ r_x(V) \mid x \in V \}, \\
V^0 &= \{ x \in V \mid r_x(V) = r(V) \}.
\end{aligned}
$$

Notice that if $\mathfrak{I}(V) = \langle f_1, \dots, f_m \rangle$ then $r_x(V)$ is the dimension of the span of $\{\nabla f_i(x),\ i = 1, \dots, m\}$. Also, if $k = \mathbf{R}$ and $x \in V$ then $r_x(V) = r_x(V_{\mathbf{C}})$.

Exercise: If U is any Zopen neighborhood of x in V, show that $r_x(U) = r_x(V)$.

$\diamond$

We could also define r in terms of entire rational functions rather than polynomials since $\{ \nabla f(x) \mid f \in \mathfrak{I}(V) \} = \{ \nabla f(x) \mid f \in \mathfrak{I}^r(V) \}$.

Denote $V^- = V - V^0$, then V^- is a Zclosed subset, since if $\mathfrak{I}(V) = \langle f_1, \dots, f_m \rangle$ and $A(x)$ is the $m \times n$ matrix with i-th row $\nabla f_i(x)$ then V^- is the set of $x \in V$ so that all $r \times r$ minors of $A(x)$ have determinant zero, where $r = r(V)$.

We will show that when V is irreducible, then $V^0 = \mathrm{Nonsing}\,V$ and $V^- = \mathrm{Sing}\,V$. This is not true in general, for example if $I = \langle xy, y(y-1) \rangle$, and $V = \mathcal{V}(I) \subset \mathbf{R}^2$ then $V = (\mathbf{R} \times 0) \cup (0,1)$, $V^0 = (0,1)$ but $\mathrm{Nonsing}\,V = \mathbf{R} \times 0$. Some books on algebraic geometry use V^- and V^0 as definitions for singular and nonsingular points of an algebraic set. As we will see later, this is equivalent to our definition if V is irreducible but otherwise it is different.

Proposition 2.2.1 *If V is a k-Zopen set, then*

    a) $V^0 = V \cap (Cl_k(V))^0$.
    b) $r(Cl_k(V)) = r(V)$.

If $k = \mathbf{R}$ then for any complexification $V_{\mathbf{C}}$ of V,

    c) $V^0 \subset V_{\mathbf{C}}^0$ *and* $V^- \subset V_{\mathbf{C}}^-$.
    d) $r(V) = r(V_{\mathbf{C}})$.

Proof: Suppose $V \subset k^n$. Let $\mathfrak{I}(V) = \langle f_1, \dots, f_m \rangle$ and let $r = r(V)$. Note $\mathfrak{I}(V) = \mathfrak{I}(\mathrm{Cl}_k(V))$. Pick $x \in V^0$ then $r = r_x(V)$. For any $z \in k^n$, let $A(z)$ be the $m \times n$ matrix with i-th row $\nabla f_i(z)$. Let W be the set of all $z \in k^n$ such that all $(r+1) \times (r+1)$ minors of $A(z)$ have determinant 0. Then W is an algebraic set and $W \supset V$ so $W \supset \mathrm{Cl}_k(V)$. Hence $r_y(\mathrm{Cl}_k(V)) \leq r$ for all $y \in \mathrm{Cl}_k(V)$. But $r = r_x(V) = r_x(\mathrm{Cl}_k(V))$ so a) and b) follow.

Now suppose $k = \mathbf{R}$. Recall $\mathfrak{I}(V_{\mathbf{C}}) = \langle f_{1\mathbf{C}}, \dots, f_{m\mathbf{C}} \rangle$. Consequently $r_x(V_{\mathbf{C}}) = r_x(V)$ for all $x \in V$. Let $A_{\mathbf{C}}$ be the complexification of A, i.e., for any $z \in \mathbf{C}^n$, let $A_{\mathbf{C}}(z)$ be the $m \times n$ matrix with i-th row $\nabla f_{i\mathbf{C}}(z)$. Let Z be the set of all $z \in \mathbf{C}^n$ such that all $(r+1) \times (r+1)$ minors of $A(z)$ have determinant 0. Then Z is a complex algebraic set containing V, hence $Z \supset V_{\mathbf{C}}$ by the definition of complexification. So for $z \in V_{\mathbf{C}}, r_z(V_{\mathbf{C}}) \leq r$. But if $x \in V_0, r = r_x(V) = r_x(V_{\mathbf{C}})$ so $x \in V_{\mathbf{C}}^0$ and $r(V_{\mathbf{C}}) = r$. Now if $x \in V^-, r_x(V_{\mathbf{C}}) = r_x(V) < r$ so $x \in V_{\mathbf{C}}^-$. So c) and d) hold also. $\qquad \square$

Proposition 2.2.2 *If W is an irreducible Zclosed subset of an irreducible Zopen set V and $W \neq V$ then $r(W) > r(V)$.*

Proof: By Lemma 2.1.2 and Proposition 2.2.1 we may as well assume that W and V are algebraic sets and $k = \mathbf{C}$. But then this is Proposition 1.14 of [**M**]. $\qquad \square$

Proposition 2.2.3 *Let $V \subset k^n$ be an irreducible k-Zopen set, and $x \in V^0$. Let $f_i \in \mathfrak{I}(V)$ for $i = 1, 2, \dots, r(V)$ be polynomials such that all $\nabla f_i(x)$ are linearly independent. Then for any $g \in \mathfrak{I}(V)$ there are polynomials h_i for $i = 0, 1, \dots, r(V)$ such that $h_0 \cdot g = \sum_{i=1}^{r(V)} h_i \cdot f_i$ and $h_0(x) \neq 0$. In fact, for some Zopen neighborhood U of x in k^n; $f_1, \dots, f_{r(V)}$ generate the ideal $\mathfrak{I}_U^r(V \cap U)$ of $\Gamma^r(U)$.*

Proof: By Lemma 2.1.2 and Proposition 2.2.1 we may as well assume that V is an algebraic set. When $k = \mathbf{C}$ this is Corollary 1.20 of [**M**]. The case $k = \mathbf{R}$ follows from $k = \mathbf{C}$ as follows. We know that $x \in V^0 \subset V_{\mathbf{C}}^0$ and $r = r(V) = r(V_{\mathbf{C}})$. By the case $k = \mathbf{C}$ there are $h_i \in \mathbf{C}[x_1, \dots, x_n]$, $i = 0, 1, \dots, r$ with $h_0(x) \neq 0$ and $h_0 \cdot g_{\mathbf{C}} = \sum h_i \cdot f_{i\mathbf{C}}$. Let $h_i = u_i + \sqrt{-1}v_i$ for $u_i, v_i \in \mathbf{R}[x_1, \dots, x_n]$. After multiplying by $\sqrt{-1}$ if necessary, we may assume $u_0(x) \neq 0$. Then $u_0 \cdot g = \sum u_i \cdot f_i$.

To see that $f_1, \dots, f_{r(V)}$ generate the ideal $\mathfrak{I}_U^r(V \cap U)$, let $g_1, \dots, g_m$ be generators of $\mathfrak{I}(V)$ and pick polynomials h_{ji} so $h_{i0}g_i = \sum_{j=1}^{r(V)} h_{ij} \cdot f_j$ and $h_{i0}(x) \neq 0$. Let $h = \prod_{i=1}^m h_{i0}$. Let $U = k^n - h^{-1}(0)$. Let $U' = \{(x, t) \in k^n \times k \mid th(x) = 1\}$ and let $\pi: U' \to U$ be projection, note it is a birational isomorphism.

If $k = \mathbf{R}$, then Proposition 2.1.1 implies $f_1, \dots, f_{r(V)}$ generate the ideal $\mathfrak{I}_U^r(V \cap U)$. If $k = \mathbf{C}$, take any entire rational function $\varphi: U \to \mathbf{C}$. Then $\varphi\pi$ is a entire rational function on the complex algebraic set U' and hence is a polynomial by Proposition 1.11 of [**M**]. So $\varphi = p/h^b$ for some polynomial p and some b. It then follows that $f_1, \dots, f_{r(V)}$ generate the ideal $\mathfrak{I}_U^r(V \cap U)$. $\qquad \square$

The following Lemma is only an intermediate result because Proposition 2.2.6 below is much stronger.

Lemma 2.2.4 *(intermediate results) Let $V \subset k^n$ be a k-Zopen set, then:*

a) *If $x \in V$ is nonsingular of dimension e, then $r_x(V) \geq n - e$ and there is an irreducible component S of V so that $x \in S$ and x is nonsingular of dimension e in S.*

b) *If $\operatorname{Nonsing} V \neq \emptyset$, then there is an irreducible component S of V with $\operatorname{Nonsing} S \neq \emptyset$ and $\dim S \geq \dim V$.*

Proof: First we will prove a). Let U be a neighborhood of x, and $f_i \in \mathfrak{I}(V)$, $i = 1, 2, \ldots, n - e$ so that $U \cap V = U \cap \bigcap_{i=1}^{n-e} f_i^{-1}(0)$ and $\nabla f_i(x)$ are linearly independent. Let $S_1, \ldots, S_m$ be the irreducible components of V which contain x. We will see in Proposition 2.2.6 below that $m = 1$, but for now we pretend otherwise. By the implicit function theorem we can shrink U so that $U \cap V$ is a connected analytic manifold. Suppose $U \cap S_i \neq U \cap V$ for all $i = 1, \ldots, m$. Then for each i there is a $g_i \in \mathfrak{I}(U \cap S_i)$ with $g_i \notin \mathfrak{I}(U \cap V)$. But g_i is an analytic function on a connected analytic manifold $U \cap V$; hence if g_i were zero on an open subset of $U \cap V$, it would be zero on $U \cap V$. So each $U \cap g_i^{-1}(0)$ is nowhere dense in $U \cap V$. Therefore each $U \cap S_i$ is nowhere dense in $U \cap V$. So $U \cap V \neq \bigcup (U \cap S_i) = U \cap (\bigcup S_i) = U \cap V$, contradiction. Hence one of S_i, which we call S, must satisfy $U \cap S = U \cap V$. Therefore $x \in S$ is nonsingular of dimension e in S. To see that $r_x(V) \geq n - e$, note that $\{ \nabla f(x) \mid f \in \mathfrak{I}(V) \} \supset$ the span of $\{ \nabla f_1(x), \ldots, \nabla f_{n-e}(x) \}$ which has dimension $n - e$.

To see b), let $e = \dim V$ and pick $x \in V$ nonsingular of dimension e. By a), we know x is nonsingular of dimension e in some irreducible component S of V. Then $\dim S \geq e = \dim V$ by definition. (Actually Proposition 2.2.5 below shows that $\dim S = \dim V$). $\qquad\square$

Proposition 2.2.5 *If $U \subset k^n$ is an irreducible k-Zopen set and $V = Cl_k(U)$, then*

a) *$\operatorname{Nonsing} U$ is nonempty.*

b) *$\dim V = \dim U = n - r(U)$.*

c) *$\operatorname{Nonsing} U = U \cap \operatorname{Nonsing} V = U^0$.*

d) *$\operatorname{Sing} U = U \cap \operatorname{Sing} V = U^-$.*

e) *If $W \subset U$ is an irreducible Zclosed subset with $W \neq U$, then $\dim W < \dim U$.*

Proof: First we will prove a). Recall $r(U) = r(V)$ by Proposition 2.2.1b. Pick $x \in U^0$, then there are $f_j \in \mathfrak{I}(U)$, $j = 1, 2, \ldots, r(U)$ so that $\nabla f_j(x)$ are linearly independent. By Proposition 2.2.3 there is a Zopen neighborhood O of x in k^n so that $\mathfrak{I}_O^r(U \cap O)$ is generated by $f_1, \ldots, f_{r(U)}$. Hence $O \cap U = O \cap \bigcap f_j^{-1}(0)$.

So x is nonsingular of dimension $n - r\left(U\right)$, hence $\dim U \geq n - r\left(U\right)$, in particular Nonsing $U \neq \emptyset$.

We now prove b). Let $\mathcal{S}$ be the set of all irreducible algebraic sets, let $\mathcal{A} = \left\{ S \in \mathcal{S} \mid \dim S > n - r\left(S\right) \right\}$. Suppose $\mathcal{A}$ is nonempty. Let S be a minimal member of $\mathcal{A}$. We have Nonsing $S \cap S^0 = \emptyset$ since if $x \in S^0$ then x is nonsingular of dimension $n - r\left(S\right)$ by above. So Nonsing $S \subset S^-$. But Nonsing S is open in S so if $x \in$ Nonsing S we know that x is nonsingular of dimension $\dim S$ in S^-. So Nonsing $S^- \neq \emptyset$ and $\dim S^- \geq \dim S$. By Lemma 2.2.4 there is an irreducible component T of S^- (in particular $T \subsetneq S$) such that $\dim T \geq \dim S^-$. This along with Proposition 2.2.2 gives, $\dim T \geq \dim S^- \geq \dim S > n - r\left(S\right) > n - r\left(T\right)$ which contradicts the minimality of S. So for every irreducible algebraic set W, $\dim W = n - r\left(W\right)$. In particular, V is irreducible by Lemma 2.1.2 so $\dim V = n - r(V) = n - r(U)$. Now $n - r(U) = \dim V \geq \dim U \geq n - r(U)$ so b) follows.

We showed above that $U^0 \subset$ Nonsing U, but then c) and d) are consequences of b) and Propositions 2.2.1a and 2.2.4a. Also e) follows from b) and Proposition 2.2.2. $\qquad\square$

In case V is not irreducible the following Proposition describes the relation between the nonsingular points of the irreducible components of V and the nonsingular points of V. This is proven by a different route in [**P**].

Proposition 2.2.6 Let V be a k-Zopen set. Then Nonsing V is not empty. Also $x \in V$ is nonsingular of dimension e if and only if x is contained in exactly one irreducible component S of V, $x \in$ Nonsing $S = S^0$ and $\dim S = e$.

Proof: Let $V = S_0 \cup S_1 \cup \cdots \cup S_m$ be the irreducible decomposition of V, so each S_i is irreducible and no S_i is contained in any other. Suppose $x \in$ Nonsing $S_0 - \bigcup_{i=1}^m S_i$ and $\dim S_0 = e$. Suppose $V \subset k^n$. Then there is an open neighborhood U of x in k^n, and polynomials $f_j \in \mathcal{I}\left(S_0\right)$, $j = 1, \dots, n - e$ such that $U \cap S_0 = \bigcap_{j=1}^{n-e} f_j^{-1}\left(0\right) \cap U$ and $\nabla f_j\left(x\right)$ are linearly independent. By shrinking U we can assume that $U \cap S_0 = U \cap V$. Since $x \notin \bigcup_{i=1}^m S_i$ there is a polynomial $h \in \mathcal{I}\left(\bigcup_{i=1}^m S_i\right)$ with $h\left(x\right) \neq 0$. We may assume $h\left(U\right) \subset k - \{0\}$. Then

$$U \cap V = U \cap S_0 = \bigcap_{j=1}^{n-e} \left(hf_j\right)^{-1}\left(0\right)$$

and the vectors

$$\nabla\left(hf_j\right)\left(x\right) = h\left(x\right) \nabla f_j\left(x\right)$$

are linearly independent and

$$hf_j \in \mathcal{I}\left(S_0\right) \cap \mathcal{I}\left(\bigcup_{i=1}^n S_i\right) = \mathcal{I}\left(V\right).$$

Hence x is nonsingular of dimension e in V.

Conversely assume x is nonsingular of dimension e in V. By Lemma 2.2.4 there exists an irreducible component S of V with $x \in S$ and x is nonsingular of dimension e in S. In particular $\dim S \geq e$, and by Proposition 2.2.5 and Lemma 2.2.4a, $\dim S = n - r(S) \leq n - (n - e) = e$. Hence $\dim S = e$.

Let S' be any other irreducible component of V with $x \in S'$. Pick $f_i \in \mathfrak{I}(V) \subset \mathfrak{I}(S)$, $i = 1, \dots, n - e$ such that $\nabla f_i(x)$ are linearly independent. Then if $g \in \mathfrak{I}(S)$, by Proposition 2.2.3 there are polynomials h_i, $i = 0, \dots, n - e$ such that $h_0 \cdot g = \sum_{i=1}^{n-e} h_i f_i$ and $h_0(x) \neq 0$. Then since $\mathfrak{I}(V) \subset \mathfrak{I}(S')$, we know that $h_0 \cdot g \in \mathfrak{I}(S')$. We have $S' = \left(S' \cap h_0^{-1}(0)\right) \cup \left(S' \cap g^{-1}(0)\right)$ and $x \in S' - S' \cap h_0^{-1}(0)$. Then by irreducibility $S' = S' \cap g^{-1}(0)$, i.e., $g \in \mathfrak{I}(S')$. So we have shown $\mathfrak{I}(S) \subset \mathfrak{I}(S')$, so $S' \subset S$. This is a contradiction since by definition, no irreducible component is contained in any other. □

Corollary 2.2.7 *If V is a Zopen set and $V = S_0 \cup S_1 \cup \cdots \cup S_k$ is its decomposition into irreducible components then $\dim V = \max\{\dim S_i\}$ and*

$$\text{Nonsing } V = \bigcup_{\dim S_i = \dim V} \left(\text{Nonsing } S_i - \bigcup_{i \neq j} S_j\right).$$

Exercise: Show that $\dim(\text{Sing } V) < \dim V$.   ◇

We also get the following corollary since any irreducible component of V is either contained in W or $V - W$.

Corollary 2.2.8 *Let V be an algebraic set and let $W \subset V$ be an algebraic subset so that $W \subset \text{Nonsing } V$ and $\dim V = \dim W$. Then $V - W$ is an algebraic set. Furthermore, $\text{Sing}(V - W) = \text{Sing } V$ unless $W = \text{Nonsing } V$.*

We also get the following corollary of Propositions 2.2.2 and 2.2.6.

Lemma 2.2.9 *Every proper algebraic subset S of an irreducible algebraic set V must have $\dim S < \dim V$.*

The decomposition into singular and nonsingular points is preserved under birational isomorphism, although it is not obvious under our definition.

Lemma 2.2.10 *Let $f : V \to W$ be a birational isomorphism. If $x \in V$ is nonsingular of dimension d in V then $f(x)$ is nonsingular of dimension d in W. In particular, $f(\text{Sing } V) = \text{Sing } W$.*

Proof: By Proposition 2.2.6 and 2.1.3b, we may as well assume V and W are irreducible. Now the Lemma for $k = \mathbf{C}$ follows from 1.9 and 1.13 of [**M**] along with Proposition 2.2.5. But by Proposition 2.2.1 the $k = \mathbf{C}$ case implies the $k = \mathbf{R}$ case.   □

We now get a version of Proposition 2.2.3 for the reducible case.

Proposition 2.2.11 *Let $V \subset k^n$ be a Zopen set. Pick $x \in \text{Nonsing}\, V$ and let $f_i \in \mathfrak{I}(V)$ for $i = 1, \ldots, n - \dim V$ be such that all $\nabla f_i(x)$ are linearly independent. Then for any $g \in \mathfrak{I}(V)$ there are polynomials h_i for $i = 0, \ldots, n - \dim V$ such that:*

    a) $h_0(x) \neq 0$.
    b) $h_0 g = \sum h_i f_i$.

In fact, for some Zopen neighborhood U of x in k^n; $f_1, \ldots, f_{r(V)}$ generate the ideal $\mathfrak{I}_U^r(V \cap U)$ of $\Gamma^r(U)$.

Proof: Let S be the irreducible component of V which contains x. By Proposition 2.2.5 and 2.2.6 $\dim S = \dim V = n - r(S)$ and $x \in S^0$. Hence Proposition 2.2.3 implies our result. $\qquad\qquad\qquad\qquad\qquad\qquad\qquad\qquad\qquad\qquad\qquad\qquad$ $\square$

The following Lemma gives a more convenient way of testing for nonsingularity of subsets.

Lemma 2.2.12 *Let U and V be Zopen sets with $U \subset V$. Suppose $x \in U$ is nonsingular of dimension d in V. Then x is nonsingular of dimension e in U if and only if there are a Zopen neighborhood O of x in V and an entire rational function $f \colon O \to k^{d-e}$ so that*

    a) $O \cap U = O \cap f^{-1}(0)$.
    b) *The map on tangent spaces $df \colon TO_x \to Tk_x^{d-e}$ has rank $d - e$.*

Proof: Suppose $V \subset k^n$. Choose $g_i \in \mathfrak{I}(V)$ and a neighborhood W of x in k^n so that $W \cap V = W \cap \bigcap_{i=1}^{n-d} g_i^{-1}(0)$ and $\nabla g_i(x)$, $i = 1, \ldots, n - d$ are linearly independent. Note that TO_x is the set of vectors perpendicular to all ∇g_i.

Suppose first that O and f satisfying a) and b) exist. Let f_i be the i-th coordinate of f. Then after perhaps shrinking O a bit we may write each f_i as the quotient of two polynomials $f_i = p_i/q_i$ where q_i is nowhere 0 on O. Pick a polynomial $r \in \mathfrak{I}(V - O)$ so that $r(x) \neq 0$. But then setting $g_{n-d+i} = r \cdot p_i$, $i = 1, \ldots, d - e$ and shrinking W so that $W \cap V \subset O - r^{-1}(0)$, we see that $W \cap U = W \cap \bigcap_{i=1}^{n-e} g_i^{-1}(0)$ and $\nabla g_i(x)$, $i = 1, \ldots, n-e$ are linearly independent. So x is nonsingular of dimension e in U.

Now suppose x is nonsingular of dimension e in U. We wish to find f and O satisfying a) and b). We know that $r_x(U) = n - e$ so we may pick $g_i \in \mathfrak{I}(U)$, $i = n - d + 1, \ldots, n - e$ so that $\nabla g_i(x)$, $i = 1, \ldots, n - e$ are linearly independent. By Proposition 2.2.11, there is a Zopen neighborhood O' of x in k^n so that $\mathfrak{I}_{O'}^r(U \cap O')$ is generated by g_i, $i = 1, \ldots, n - e$. Let $O = V \cap O'$ and $f = (g_{1+n-d}, \ldots, g_{n-e})$. Note if $x \in O \cap f^{-1}(0) - U$ there is a $g \in \mathfrak{I}(U)$ so that $g(x) \neq 0$. But since $g \in \mathfrak{I}(U)$ we know $g(x) = \sum h_i(x) g_i(x) = 0$ and we have a contradiction. $\qquad\qquad\qquad\qquad\qquad\qquad\qquad\qquad\qquad\qquad\qquad\qquad$ $\square$

Next we get an algebraic version of transversality.

Definition: Recall that if $f \colon M \to N$ is a smooth map between smooth manifolds and $L \subset N$ is a smooth submanifold of N then we say f is *transverse* to L at a point $x \in M$ if either $f(x) \notin L$ or $df(TM_x) + TL_{f(x)} = TN_{f(x)}$, i.e., the normal bundle of L at $f(x)$ is generated by images of tangent vectors to M. We write $f \pitchfork L$ at x if f is transverse to L at x, and say $f \pitchfork L$ if $f \pitchfork L$ at all $x \in M$.

One consequence of transversality at x is that near x, $f^{-1}(L)$ is a smooth submanifold of M with codimension equal to the codimension of L. The following Lemma gives the algebraic version of this result.

Lemma 2.2.13 *Suppose $f \colon V \to W$ is an entire rational function between Zopen sets and $U \subset W$ is a Zclosed subset. Suppose x is nonsingular of dimension a in V, $f(x)$ is nonsingular of dimension b in W, $f(x)$ is nonsingular of dimension c in U and $f \pitchfork U$ at x. Then x is nonsingular of dimension $a - b + c$ in $f^{-1}(U)$.*

Proof: Let $Z = f^{-1}(U)$. By Lemma 2.2.12 we may choose a Zopen neighborhood O of $f(x)$ in W and an entire rational function $g \colon O \to k^{b-c}$ so that dg has rank $b - c$ at $f(x)$ and $U \cap O = g^{-1}(0) \cap O$. Let $h = g \circ f$ and $O' = f^{-1}(O)$. Then $Z \cap O' = h^{-1}(0) \cap O'$ and dh has rank $b - c$ at x by transversality. Hence by Lemma 2.2.12, x is nonsingular of dimension $a - b + c$ in $f^{-1}(U)$.   □

We also have transversality for maps.

Definition: If $f \colon M \to N$ and $g \colon L \to N$ are two smooth maps and $(x, y) \in M \times L$ then we say $f \pitchfork g$ at (x, y) if either $f(x) \neq g(y)$ or $f(x) = g(y) = z$ and $df(TM_x) + dg(TL_y) = TN_z$. We say $f \pitchfork g$ if $f \pitchfork g$ at all $(x, y) \in M \times L$.

Lemma 2.2.14 *Suppose $f \colon V \to W$ and $g \colon U \to W$ are entire rational functions between Zopen sets. Suppose x is nonsingular of dimension a in V, $f(x)$ is nonsingular of dimension b in W, y is nonsingular of dimension c in U, $f(x) = g(y)$ and $f \pitchfork g$ at (x, y). Let*

$$Z = \{ (v, u) \in V \times U \mid f(v) = g(u) \}$$

be the pullback of f and g. Then (x, y) is nonsingular of dimension $a - b + c$ in Z.

Proof: The map $(f, g) \colon V \times U \to W \times W$ is transverse to the diagonal $D = \{ (w, w) \in W \times W \}$ at (x, y) and $Z = (f, g)^{-1}(D)$. But D is nonsingular of dimension b at $(f(x), f(x))$, $W \times W$ is nonsingular of dimension $2b$ at $(f(x), f(x))$ and $V \times U$ is nonsingular of dimension $a + c$ at (x, y). So the result follows from Lemma 2.2.13.   □

The complexification of a nonsingular real algebraic set need not be nonsingular, and the complexification of a regular function need not be regular. However the following shows that up to isomorphism they are.

Lemma 2.2.15 *Let $\rho\colon W \to V$ be an entire rational function from a nonsingular real Zopen set. Then there is a nonsingular real algebraic set W' and a birational isomorphism $\eta\colon W \to W'$ such that the complexification $W'_{\mathbf{C}}$ is nonsingular and $\rho \circ \eta^{-1}\colon W' \to V$ extends to an entire rational function from $W'_{\mathbf{C}}$ to $V_{\mathbf{C}}$.*

Proof: Suppose $W \subset \mathbf{R}^m$ and $V \subset \mathbf{R}^k$. Let $X = \mathrm{Cl}_{\mathbf{R}}(W) - W$. By Proposition 2.1.1 there are polynomials $p\colon \mathbf{R}^m \to \mathbf{R}^k$ and $q\colon \mathbf{R}^m \to \mathbf{R}$ so that $q^{-1}(0) \cap W$ is empty and $\rho = p/q|_W$. After multiplying p and q by a polynomial r with $r^{-1}(0) = X$ we may assume $X \subset q^{-1}(0)$. Let $p_{\mathbf{C}}$ and $q_{\mathbf{C}}$ be the complexifications of p and q. Let $Y = \mathrm{Cl}_{\mathbf{C}}(W)$. Pick a polynomial h with real coefficients so that $\mathrm{Sing}\, Y \subset h^{-1}(0)$ and $h^{-1}(0) \cap W$ is empty. For example, h could be the sum of squares of real and imaginary parts of generators of $\mathcal{I}(\mathrm{Sing}\, Y)$. Let

$$
\begin{aligned}
W' &= \{\, (x,t) \in \mathrm{Cl}_{\mathbf{R}}(W) \times \mathbf{R} \mid th(x)q(x) = 1 \,\} \\
&= \{\, (x,t) \in W \times \mathbf{R} \mid t = 1/h(x)q(x) \,\} \quad \text{and} \\
\eta(x) &= (x, 1/(h(x)q_{\mathbf{C}}(x)))
\end{aligned}
$$

Then

$$
W'_{\mathbf{C}} = \{\, (x,t) \in Y \times \mathbf{C} \mid th(x)q_{\mathbf{C}}(x) = 1 \,\}
$$

is nonsingular since it is isomorphic to

$$
Y - (h^{-1}(0) \cup q_{\mathbf{C}}^{-1}(0)) \subset \mathrm{Nonsing}\, Y.
$$

Also $\rho\eta^{-1}(x,t) = th(x)p(x)$ which extends to the polynomial map $th(x)p_{\mathbf{C}}(x)$.
$\square$

Zopen sets have a canonical stratification induced by their singularities. If X is a Zopen set we may define $X_0 = X$, $X_1 = \mathrm{Sing}\, X$ and $X_i = \mathrm{Sing}\, X_{i-1}$. Since $\dim X_{i+1} < \dim X_i$, this is a well defined finite process and gives a stratification of X. Furthermore by enlarging the X_i's we can make this stratification satisfy the Whitney conditions. These are some niceness conditions which in particular imply that this is a conelike stratification (c.f., [**Wa**]).

If $f\colon V \to W$ is an entire rational function between Zopen sets then $f(V)$ need not be a Zopen set. For example, the projection of a circle in $\mathbf{R}^2$ to the x axis is an interval. However $f(V)$ is something called a semialgebraic set if $k = \mathbf{R}$ or a constructible set if $k = \mathbf{C}$. A *semialgebraic set* is a finite union of finite intersections of sets of the form $\{\, x \in \mathbf{R}^n \mid p(x) > 0 \,\}$ or $\{\, x \in \mathbf{R}^n \mid p(x) = 0 \,\}$ where p is a polynomial. A *constructible set* is a finite union of finite intersections of sets of the form $\{\, x \in \mathbf{C}^n \mid p(x) \neq 0 \,\}$ or $\{\, x \in \mathbf{C}^n \mid p(x) = 0 \,\}$ where p is a polynomial. By the Tarski-Seidenberg theorem, the image of a semialgebraic or constructible set under an entire rational function is semialgebraic or constructible [**BCR**]. The only other properties we need are that semialgebraic and constructible sets are triangulable and thus have a dimension and this dimension equals the dimension of the Zariski closure [**L2**].

The following Lemma allows us to convert any algebraic stratification of a real algebraic set into the canonical singular stratification. It is a useful device to enlarge singularities of a real algebraic set to fit a given requirement, provided we are willing to change it by a homeomorphism.

Proposition 2.2.16 *Let Y be any real Zopen set and let*

$$Y_0 \subset Y_1 \subset \cdots \subset Y_n = Y$$

be Zclosed subsets of Y so that $Y_{i-1} \supset \operatorname{Sing} Y_i$ and $\dim Y_{i-1} < \dim Y_i$ for all $i = 1, \ldots, n$. Then there is a real algebraic set Z and an entire rational function $h \colon Z \to Y$ so that h is a homeomorphism and

$$h^{-1}(Y_{i-1}) = \operatorname{Sing}\left(h^{-1}(Y_i)\right)$$

for $i = 1, \ldots, n$.

Proof: Pick the largest k so that there is a real algebraic set Z' and an entire rational function $h' \colon Z' \to Y$ so that h' is a homeomorphism, $h'^{-1}(Y_{i-1}) = \operatorname{Sing}(h'^{-1}(Y_i))$ for $i = 1, \ldots, k$ and $h'^{-1}(Y_{i-1}) \supset \operatorname{Sing}(h'^{-1}(Y_i))$ for $i = 1, \ldots, n$. Note you can do this for $k = 0$ by Lemmas 2.1.4 and 2.2.10. So $k \geq 0$.

If $k = n$ we are done so suppose $k < n$. We will obtain a contradiction and thus prove the Lemma. Suppose $Z' \subset \mathbf{R}^m$. Let $W = h'^{-1}(Y_k)$ and let $p \colon \mathbf{R}^m \to \mathbf{R}$ be a polynomial so that $W = p^{-1}(0)$. Let $d \colon \mathbf{R}^m \to [0, \infty)$ be the function $d(x) =$ the distance from x to W. Pick a compact set $K \subset W$ which contains an open subset of Nonsing X for every irreducible component X of W. By the Lojasiewicz inequality [**L1**] we know that $|p(x)| \geq c \cdot d(x)^b$ for all x in some neighborhood U of K and some $c > 0$ and some integer $b > 0$.

Let $S = \{(y, t) \in \mathbf{R}^m \times \mathbf{R} \mid t^{2b+1} = p(y)^2\}$. Projection to $\mathbf{R}^m$ induces an entire rational function $g \colon S \to \mathbf{R}^m$ which is homeomorphism. Let $Z'' = g^{-1}(Z') = Z' \times \mathbf{R} \cap S$ and let $h'' \colon Z'' \to Y$ be the entire rational function $h' \circ g|$. Since $g|_{W \times 0}$ is a birational isomorphism to W, we know that $h''^{-1}(Y_{i-1}) = \operatorname{Sing}(h''^{-1}(Y_i))$, $i = 1, \ldots, k$. It is easy to check directly from the definition that $g^{-1}(T)$ is nonsingular for any nonsingular Zopen set $T \subset \mathbf{R}^m - W$. Hence for $i > k$ we have

$$\operatorname{Sing}(h''^{-1}(Y_i)) \subset g^{-1}\big(\operatorname{Sing}(h'^{-1}(Y_i)) \cup W\big) \subset g^{-1}h'^{-1}(Y_{i-1})$$

So to obtain a contradiction we only need show that $W \times 0 = h''^{-1}(Y_k) = \operatorname{Sing}(h''^{-1}(Y_{k+1}))$. The idea of why this is so is that S bends in a sharp cusp normal to W. A precise proof is as follows.

Let $V = h''^{-1}(Y_{k+1})$. We know $W \times 0 \supset \operatorname{Sing} V$ so if $W \times 0 \neq \operatorname{Sing} V$ there is an irreducible component X of $W \times 0$ so that $X \cap \operatorname{Sing} V \neq X$. Then $\dim(\operatorname{Sing} X \cup (X \cap \operatorname{Sing} V)) < \dim X$ so since $X \subset V$ we may pick some x in Nonsing $X \cap$ Nonsing $V \cap K$. Since $\dim X < \dim V$ and Nonsing $X \cap$ Nonsing V is a submanifold of Nonsing V, we can find an analytic arc $\alpha \colon [-\epsilon, \epsilon] \to$ Nonsing V so that $\alpha(0) = x$ and $\alpha'(0)$ is nonzero and perpendicular to the tangent space of Nonsing X at x. Let $\alpha(t) = (\alpha_1(t), \alpha_2(t))$ where $\alpha_1(t) \in \mathbf{R}^m$ and $\alpha_2(t) \in \mathbf{R}$. Now $\alpha_2(t) \geq 0$ since

$$\alpha_2(t)^{2b+1} = (p\alpha_1(t))^2 \geq 0.$$

So $\alpha_2'(0) = 0$, hence $\alpha_1'(0) \neq 0$. But

$$d(\alpha_1(t)) \geq |\alpha_1'(0)| \cdot |t|/2$$

for small t so

$$|\alpha_2(t)| = |p \circ \alpha_1(t)|^{2/(2b+1)} \geq c^{2/(2b+1)} d(\alpha_1(t))^{2b/(2b+1)} \geq c'|t|$$

for some constant $c' > 0$ and small t. This contradicts the analyticity of $\alpha_2(t)$ since $\alpha_2(t) \approx at^e$ for small t where $e = ord(\alpha_2) \geq 2$. $\qquad\square$

3. Projective Algebraic Sets

Definition: The multiplicative group $k - 0$ acts on $k^{n+1} - (0, \ldots, 0)$ by multiplication. Define

$$
\begin{aligned}
k\mathbf{P}^n &= k^{n+1} - (0, \ldots, 0)/k - \{0\} \\
&= k^{n+1} - (0, \ldots, 0)/x \sim \lambda x \quad \text{for} \quad \lambda \in k - \{0\}.
\end{aligned}
$$

We denote a point in $k\mathbf{P}^n$ by an $(n+1)$-tuple $[x_0, \ldots, x_n]$. Of course these coordinates are not unique since we are free to multiply all coordinates by any nonzero λ.

For example for $k = \mathbf{R}$ we have $\mathbf{R}\mathbf{P}^n$ which can also be identified by $S^n/x \sim -x$, the quotient of the sphere by the antipodal map. We can think of $k\mathbf{P}^n$ as the space of lines in k^{n+1} where $[x_0, \ldots, x_n]$ corresponds to the line $\{(tx_0, \ldots, tx_n) \in k^{n+1} \mid t \in k\}$.

There are coordinate charts $\theta_i \colon k^n \to k\mathbf{P}^n$ for $i = 0, \ldots, n$ given by

$$\theta_i(x_1, \ldots, x_n) = [x_1, \ldots, x_i, 1, x_{i+1}, \ldots, x_n].$$

These charts cover $k\mathbf{P}^n$ and give it the structure of a smooth or complex manifold.

Definitions: Any homogeneous polynomial $f \in k[x_0, \ldots, x_n]$ describes a subset of $k\mathbf{P}^n$ as follows

$$\hat{V}(f) = \{[x_0, \ldots, x_n] \in k\mathbf{P}^n \mid f(x_0, \ldots, x_n) = 0\}.$$

Since $f(\lambda(x_0, \ldots, x_n)) = \lambda^d f(x_0, \ldots, x_n)$ for some d, the above expression $f(x_0, \ldots, x_n) = 0$ makes sense. Hence $\hat{V}(f)$ is well defined. In general if J is a set of homogeneous polynomials in $k[x_0, \ldots, x_n]$ we have a subset

$$\hat{V}(J) = \{[x_0, \ldots, x_n] \in k\mathbf{P}^n \mid f(x_0, \ldots, x_n) = 0 \text{ for all } f \in J\}.$$

Such a $\hat{\mathcal{V}}(J)$ is called a *projective algebraic set*. In a similar manner to the affine case we define projective Zopen sets, Zclosed subsets, Zopen subsets and Zopen neighborhoods. If we need to distinguish projective Zopen sets from our previous notion of Zopen sets in k^n, we will call the Zopen sets of k^n *affine*.

Definition: If $S \subset k\mathbf{P}^n$ and $T \subset k\mathbf{P}^m$ we say $f \colon S \to T$ is an *entire rational function* if for each $x \in S$ there are homogeneous polynomials p_i, $i = 0, \ldots, m$ all with the same degree and a Zopen neighborhood U of x in $k\mathbf{P}^n$ so that $U \cap \hat{\mathcal{V}}(p_j) = \emptyset$ for some j and $f(y) = [p_0(y), \ldots, p_m(y)]$ for all $y \in U \cap S$. Note this is well defined since the p_i's all have the same degree.

Note the chart θ_j pulls back projective algebraic sets to algebraic sets. If $\hat{\mathcal{V}}(f_1, \ldots, f_m) \subset k\mathbf{P}^n$ is a projective algebraic set and

$$\hat{f}_i (x_1, \ldots, x_n) = f_i (x_1, \ldots, x_{j-1}, 1, x_j, \ldots, x_n)$$

then

$$\theta_j^{-1} \left(\hat{\mathcal{V}}(f_1, \ldots, f_m) \right) = \mathcal{V} \left(\hat{f}_1, \ldots, \hat{f}_m \right).$$

Definitions: We can also associate to any algebraic set V a projective algebraic set $\mathcal{P}(V)$ as follows. If $f \in k[x_1, \ldots, x_n]$ we can write $f = \sum_{i=0}^d f_i$ where each f_i is homogeneous of degree i. Let

$$\tilde{f}(x_0, x_1, \ldots, x_n) = \sum_{i=0}^d x_0^{d-i} f_i (x_1, \ldots, x_n)$$

Then $\tilde{f}$ is homogeneous of degree d. It is called the *homogenization* of f. In particular when $x_0 \neq 0$, $\tilde{f}(x_0, \ldots, x_n) = x_0^d f(x_1/x_0, x_2/x_0, \ldots, x_n/x_0)$. If $S \subset k^n$ and J is the homogenization of the polynomials in $\mathcal{I}(S)$, then J defines a projective algebraic set $\mathcal{P}(S)$. In particular $S = \theta_0^{-1}(\mathcal{P}(S))$. If an algebraic set $S \subset k^n$ has the property $\theta_0(S) = \mathcal{P}(S)$ we call S *projectively closed*.

Definition: A polynomial $f \in \mathbf{R}[x_1, \ldots, x_n]$ is called *overt* if its homogenization $\tilde{f}(x_0, \ldots, x_n)$ has no zeroes with $x_0 = 0$ except for $x = 0$. Equivalently, $f_d(x) = 0$ implies $x = 0$.

Exercise: For $k = \mathbf{R}$, the algebraic set $\mathcal{V}(x^2 + y^2 - 1)$ is projectively closed but $\mathcal{V}(x^2 + y^4 - 1)$ is not.   ◇

Exercise: The only projectively closed complex algebraic sets have dimension 0.   ◇

Exercise: The following are equivalent for a real algebraic set S:
   a) S is projectively closed.
   b) $\mathcal{I}(S)$ contains an overt polynomial.
   c) $S = p^{-1}(0)$ for some overt polynomial.   ◇

Exercise: A function $f\colon S \to T$ is an entire rational function between affine subsets if and only if $\theta_i f \theta_j^{-1}\colon \theta_j(S) \to \theta_i(T)$ is an entire rational function between projective subsets for some or any j and i.   ◇

We can define singular and nonsingular points of projective Zopen sets. A point x in a projective Zopen set V is nonsingular of dimension d in V if and only if $\theta_j^{-1}(x)$ is nonsingular of dimension d in $\theta_j^{-1}(V)$ for some (and hence all) j with x in the image of θ_j.

Projective algebraic sets behave like algebraic sets and there is a natural $1-1$ correspondence

$$\left\{ \begin{array}{l} \text{Projective algebraic} \\ \text{sets } S \text{ in } k\mathbf{P}^n \end{array} \right\} \longleftrightarrow \left\{ \begin{array}{l} \text{Nonempty algebraic sets } S' \text{ in } k^{n+1} \\ \text{such that } \lambda S' \subset S' \text{ for all } \lambda \in k \end{array} \right\}$$

To see this correspondence let S be a projective algebraic set given by the homogeneous polynomials J then $\mathcal{V}(J) \subset k^{n+1}$ gives an algebraic set S' invariant under multiplication by elements of k. Conversely let $S' \subset k^{n+1}$ be an algebraic set with $\lambda S' \subset S'$ for all $\lambda \in k$. Let $\mathcal{I}(S') = \langle f_1, \ldots, f_m \rangle$ and $f_i = \sum_{j=0}^{d_i} f_{ij}$ where each f_{ij} is homogeneous of degree j. Then for $x \in S'$, $0 = f_i(\lambda x) = \sum_{j=0}^{d_i} f_{ij}(\lambda x) = \sum_{j=0}^{d_i} \lambda^j f_{ij}(x)$. Since this equation holds for each i and $\lambda \in k$ we must have each $f_{ij}(x) = 0$. Therefore $\mathcal{I}(S') = \langle f_{11}, \ldots, f_{m,d_m} \rangle$ is generated by homogeneous polynomials and defines a projective algebraic set S in $k\mathbf{P}^n$.

Our main objects of interest will still be affine Zopen sets, since they are easier to deal with and by Proposition 2.4.1 below any real projective Zopen set is birationally isomorphic to an affine Zopen set. But projective Zopen sets are often useful objects. Since $k\mathbf{P}^n$ is compact, they give a way of compactifying algebraic sets. By means of the embedding θ above, we can identify any affine Zopen set with a projective Zopen set.

One ingredient which seems to be missing is cartesian product. One could of course talk about algebraic subsets of cartesian products of projective spaces, but there is no need to do so. There is a canonical embedding called the Segre embedding $\sigma\colon k\mathbf{P}^n \times k\mathbf{P}^m \to k\mathbf{P}^{nm+n+m}$ defined by $\sigma([x_0, \ldots, x_n], [y_0, \ldots, y_m] = [x_0 y_0, \ldots, x_n y_m]$. Its image is a projective algebraic set and σ is a birational isomorphism with respect to the sensible algebraic structure on $k\mathbf{P}^n \times k\mathbf{P}^m$ (c.f. Proposition 2.12 of [**M**]). Hence from now on we allow ourselves to take cartesian products of projective Zopen sets; without saying so explicitly we always identify them with a projective Zopen set via the Segre embedding.

As an application of projective algebraic sets we present the following.

Proposition 2.3.1 *Let $f\colon V \to W$ be an entire rational function between affine or projective Zopen sets. Then there is a Zclosed subset X of W with $\dim X < \dim W$ so that $f|\colon V - f^{-1}(X) \to W - X$ is a locally trivial fibration.*

Proof: The affine case follows from the projective case via the chart θ_0 above. So we only need prove the projective case. By replacing W by Nonsing W and V by $f^{-1}(\text{Nonsing } W)$, we may as well assume W is nonsingular. Suppose $V \subset k\mathbf{P}^n$. By replacing V by the graph of f we can assume that $V \subset W \times k\mathbf{P}^n$ and f is induced by the projection. Let V' be the Zariski closure of V in $W \times k\mathbf{P}^n$ and let $V'' = V' - V$. Take an algebraic Whitney stratification of V' compatible with V''. In particular we have k algebraic subsets $\{V_i\}$ such that Sing $V_i \subset V_{i-1} \subset V_i$ and $V_i - V_{i-1}$ are the strata, c.f., [**Wa**]. Furthermore, any connected component of a stratum is contained in either V'' or V. If dim $W = m$ let

$$Y_i = \{\ y \in V_i - V_{i-1} \mid d\pi_y \text{ has rank} < m\ \}$$

where $\pi\colon V' \to W$ is the induced projection. Now $\bigcup \pi(Y_i)$ is a constructible or semialgebraic set and it has measure zero by Sard's Theorem, hence it and its Zariski closure have smaller dimension than W. So if we let $X = \text{Cl}_k(\bigcup \pi(Y_i))$, then dim $X <$ dim W. By Thom's first isotopy lemma (c.f., [**GWPL**] p.58), $\pi\colon (V' - V'') \bigcap (W - X) \times k\mathbf{P}^n \to W - X$ is a locally trivial fibration. $\square$

Since irreducible complex Zopen sets are connected, we see that if W above is irreducible, then all the sets $f^{-1}(x)$ for $x \in W - X$ are homeomorphic. In the real case we do not get such a strong result, but Proposition 2.3.1 does imply that there is a natural way to associate a notion of degree to entire rational functions between real Zopen sets as follows.

Proposition 2.3.2 *Let V and W be real affine or projective Zopen sets with W irreducible. Let $f\colon V \to W$ be an entire rational function. Then there is a $\delta(f) \in \mathbf{Z}/2\mathbf{Z}$ and a Zclosed subset X of W with dim $X <$ dim W, such that for each $z \in W - X$ the Euler characteristic of $f^{-1}(z)$ satisfies $\chi(f^{-1}(z)) \equiv \delta(f)$ mod 2. We call $\delta(f)$ the degree of f.*

Proof: Just as in 2.3.1, the projective case implies the affine case so we will just do the projective case. We will use various unspecified generalizations of affine notions to projective notions. If the reader feels uncomfortable with this, just think of this as a proof in the affine case and use Proposition 2.4.1 below to get the projective case (which we never use anyway). Let $V_{\mathbf{C}}$ and $W_{\mathbf{C}}$ and $f\colon V_{\mathbf{C}} \to W_{\mathbf{C}}$ be complexifications of V, W and f. By Proposition 2.3.1, there is a Zclosed subset $Y \subset W_{\mathbf{C}}$ so that

$$\dim_{\mathbf{C}}(Y) < \dim_{\mathbf{C}}(W_{\mathbf{C}}) = \dim_{\mathbf{R}}(W)$$

and

$$f_{\mathbf{C}}|\colon V_{\mathbf{C}} - f_{\mathbf{C}}^{-1}(Y) \to W_{\mathbf{C}} - Y$$

is a locally trivial fibration. Let $X = Y \cap W =$ the real points of Y. Note $X_{\mathbf{C}} \subset Y$ so

$$\dim X = \dim_{\mathbf{C}}(X_{\mathbf{C}}) \leq \dim_{\mathbf{C}}(Y) < \dim W.$$

Also for any $x \in W$ we know that $f^{-1}(x)$ is the fixed point set of the complex conjugation involution on $f_{\mathbf{C}}^{-1}(x)$. Hence

$$\chi(f^{-1}(x)) = \chi(f_{\mathbf{C}}^{-1}(x)) \bmod 2.$$

But $W_{\mathbf{C}} - Y$ is connected so all the sets $f_{\mathbf{C}}^{-1}(x)$ for $x \in W - X$ are homeomorphic. So if we set $\delta(f) = \chi(f^{-1}(x))$ for some $x \in W - X$ we obtain our result.   □

If we let W be $\mathrm{Cl}_{\mathbf{R}} f(V)$ then $\delta(f) = 1$ so we must have $\mathrm{Cl}_{\mathbf{R}} f(V) - f(V) \subset X$. Consequently:

Corollary 2.3.3 *Suppose V is an irreducible real Zopen set and suppose $f \colon V \to \mathbf{R}^m$ is an entire rational function such that $\chi\left(f^{-1}(x)\right)$ is odd for a dense set of points $x \in f(V)$. Then $\dim\left(\mathrm{Cl}_{\mathbf{R}} f(V) - f(V)\right) < \dim f(V)$.*

The above Corollary 2.3.3 was originally proven by Benedetti and Tognoli in [**BT**] under the special assumption that f is an embedding. Proposition 2.3.2 is from [**AK7**].

4. Grassmannians

Definitions: Let $G_k(n,m)$ denote the Grassmannian of m-planes in k^n. Let $\mathfrak{M}_k(n)$ denote the set of $(n \times n)$-matrices with coefficients in k. When $k = \mathbf{R}$ there is a natural identification with projection matrices

$$G_{\mathbf{R}}(n,m) = \{\, L \in \mathfrak{M}_{\mathbf{R}}(n) \mid L^2 = L,\ L = L^t,\ \mathrm{trace}(L) = m \,\}$$

given by m-plane $\mapsto$ the matrix of orthogonal projection onto that plane. When $k = \mathbf{R}$ we will denote $G_{\mathbf{R}}(n,m)$ by $G(n,m)$. The above identification makes $G(n,m)$ a real algebraic set in $\mathbf{R}^{n^2}$.

Exercise: $G(n,m)$ is nonsingular. Hint: Show it is homogeneous, that there is a birational automorphism taking any point to any other point.   ◇

If we define:

$$
\begin{aligned}
E(n,m) &= \{\, (L,y) \in G(n,m) \times \mathbf{R}^n \mid Ly = y \,\} \\
E^*(n,m) &= \{\, (L,y,t) \in E(n,m) \times \mathbf{R} \mid |y|^2 + t^2 = t \,\}
\end{aligned}
$$

then the natural projections

$$
\begin{aligned}
E(n,m) &\ \to\ G(n,m) \\
E^*(n,m) &\ \to\ G(n,m)
\end{aligned}
$$

are the universal $\mathbf{R}^n$-bundle and the corresponding n sphere bundle obtained by compactifying each fiber of $E(n,m)$.

Exercise: Show that $E(n,m)$ and $E^*(n,m)$ are nonsingular algebraic sets.

◇

There is a natural entire rational diffeomorphism

$$\mu \colon \mathbf{RP}^n \to G\left(n+1,1\right)$$

given by $\mu[x_0,\ldots,x_n] = \left(x_i x_j / \sum_{\ell=0}^n x_\ell^2\right)$.

When $k = \mathbf{R}$ projective algebraic sets and algebraic sets coincide:

Proposition 2.4.1 *A set $S \subset \mathbf{RP}^n$ is a projective algebraic set if and only if $\mu\left(S\right) \subset G\left(n+1,1\right)$ is an algebraic set in $\mathbf{R}^{(n+1)^2}$.*

Proof: Let $S = \hat{\mathcal{V}}\left(f_1,\ldots,f_m\right) \subset \mathbf{RP}^n$ where each f_i is a homogeneous polynomial of degree d_i. Then we claim that there are polynomials $p_i \colon \mathbf{R}^n \to \mathbf{R}$ such that $p_i\mu\left(x\right) = |x|^{-2d_i} f_i^2\left(x\right)$ for all $x \in \mathbf{RP}^n$. The existence of p_i is best understood by examples. For example, when $f_i\left(x_0,x_1,x_2\right) = x_0^3 - x_1 x_2^2$ then

$$
\begin{aligned}
|x|^{-6} f_i^2\left(x\right) &= |x|^{-6}\left(x_0^6 - 2x_0^3 x_1 x_2^2 + x_1^2 x_2^4\right) \\
&= p_i \circ \mu\left(x\right)
\end{aligned}
$$

where $p_i\left(L\right) = L_{00}^3 - 2L_{00}L_{01}L_{22} + L_{11}L_{22}^2$.

Then $\mu\left(S\right) = V\left(p_1,\ldots,p_m\right) \cap G\left(n+1,1\right)$.

Conversely suppose $S = \mathcal{V}\left(p_1,\ldots,p_m\right) \cap G\left(n+1,1\right)$, and $p_i = \sum_{j=0}^{d_i} p_{ij}$ with p_{ij} homogeneous of degree j. Then the polynomials

$$
\begin{aligned}
f_i\left(x_0,\ldots,x_n\right) &= |x|^{2d_i} p_i \circ \mu[x_0,\ldots,x_n] \\
&= \sum_{j=0}^{d_i} |x|^{2(d_i-j)} p_{ij}\left(x_0^2, x_0 x_1,\ldots,x_{n-1} x_n, x_n^2\right)
\end{aligned}
$$

are homogeneous of degree $2d_i$. Hence $\hat{\mathcal{V}}\left(f_1,\ldots,f_m\right) = \mu^{-1}\left(S\right)$.   □

Corollary 2.4.2 *A subset $S \subset \mathbf{RP}^k$ is a projective Zopen set if and only if $\mu\left(S\right) \subset G\left(k+1,1\right)$ is a Zopen set in $\mathbf{R}^{(k+1)^2}$.*

Let $V \subset \mathbf{R}^n$ be a Zopen set with $m = \dim V$. Then we have the natural maps classifying the tangent and normal bundles to Nonsing V:

$\alpha \colon$ Nonsing $V \to G\left(n,m\right)$ where $\alpha\left(x\right) =$ the tangent plane to V at x
$\beta \colon$ Nonsing $V \to G\left(n,n-m\right)$ where $\beta\left(x\right) =$ the normal plane to V at x

Proposition 2.4.3 *α and β are entire rational functions.*

Proof: Let I denote the $n \times n$ identity matrix. Then $\alpha(x) = I - \beta(x)$ so it suffices to prove $\beta(x)$ is entire rational.

Pick any $y \in$ Nonsing$V$. Pick $f_1,\ldots,f_{n-m} \in \mathcal{I}(V)$ so that the gradients ∇f_i are linearly independent at y. Let $A(x)$ be the $n \times (n-m)$ matrix whose i-th column is $\nabla f_i(x)$ and let $A^t(x)$ be its transpose. Then by standard linear algebra we know that

$$\beta(x) = A(x)\left(A^t(x)A(x)\right)^{-1} A^t(x)$$

as long as $A(x)$ has rank $n - m$. But by Cramer's rule,

$$\left(A^t(x)A(x)\right)^{-1} = P(x)/q(x)$$

for some polynomial matrix $P(x)$ where $q(x) = \det\left(A^t(x)A(x)\right)$. But $A(x)$ has rank $n - m$ if and only if $q(x) \neq 0$ so

$$\beta(x) = A(x)P(x)A^t(x)/q(x)$$

for all $x \in V - q^{-1}(0)$. Note that $q(y) \neq 0$, so $\beta(x)$ is an entire rational function.
□

Corollary 2.4.4 *Let $L \subset V \subset \mathbf{R}^n$ be nonsingular Zopen sets. Then there is an entire rational function $\gamma\colon L \to G(n, k)$ where $k = \dim V - \dim L$, such that $\gamma(x)$ is the subspace of the tangent plane to V at x which is perpendicular to the tangent space of L at x. In other words, γ classifies the normal bundle of L in V.*

Proof: Let r and s be the codimensions of L and V respectively in $\mathbf{R}^n$. Let $\beta_L\colon L \to G(n, r)$ and $\beta_V\colon V \to G(n, s)$ be the entire rational functions of Proposition 2.4.3 classifying the normal bundles of L and V in $\mathbf{R}^n$. Note $\beta_L\beta_V = \beta_V$ so $\beta_V\beta_L = \beta_V$ since β_V and β_L are symmetric. Then $\gamma\colon L \to G(n, k)$ is given by $\gamma = \beta_L - \beta_V$, since

$$. \ (\beta_L - \beta_V)^2 = \beta_L^2 - \beta_L\beta_V - \beta_V\beta_L + \beta_V^2 = \beta_L - \beta_V - \beta_V + \beta_V = \beta_V - \beta_L$$

and the image of $\beta_L - \beta_V$ is the required subspace. See Figure II.4.1.   □

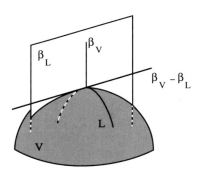

FIGURE II.4.1. Projecting to the normal bundle of L in V

5. Blowing Up

Definitions: Let M be a real or complex manifold and let N be a proper submanifold of M. We will construct a new manifold $\mathfrak{B}(M, N)$ called the *blowup of M along N*. We will also construct a proper map $\pi(M, N)\colon \mathfrak{B}(M, N) \to M$. We call N the *center* of the blowup and call $\pi(M, N)$ the *blowup map* or *blowup projection*.

Let $\rho\colon E \to N$ be the projective normal bundle of N in M, i.e., E is the space of lines in the normal bundle of N. So if N has codimension n, ρ will be a smooth fibre bundle with fibre $k\mathbf{P}^{n-1}$, real or complex projective space. As a point set $\mathfrak{B}(M, N)$ is $M - N$ union E. We then put a natural manifold structure on this space. The map $\pi(M, N)$ is the identity on $M - N$ and is the bundle projection to N on E.

Let us start with an example. Suppose $M = k^m$ and $N = k^{m-n} = \{\, x \in k^m \mid x_1 = x_2 = \cdots = x_n = 0 \,\}$. Let $\eta\colon L \to k\mathbf{P}^{n-1}$ be the canonical line bundle over $k\mathbf{P}^{n-1}$, so

$$L = \{\, (x, \lambda) \in k^n \times k\mathbf{P}^{n-1} \mid x \text{ is a point in the line } \lambda \,\}.$$

and $\eta(x, \lambda) = \lambda$. Note that there is a canonical isomorphism $\nu\colon L - k\mathbf{P}^{n-1} \to k^n - 0$ where $\nu(x, \lambda) = x$. Then $\mathfrak{B}(k^m, k^{m-n}) = L \times k^{m-n}$ and $\pi(k^m, k^{m-n})$ is the map sending $((x, \lambda), y)$ to (x, y). We are identifying $0 \times k\mathbf{P}^{n-1} \times k^{m-n}$ with the projective normal bundle of k^{m-n} in the obvious way.

The smooth structure on $\mathfrak{B}(M, N)$ for general M and N is obtained locally from the above construction. In particular, suppose $\theta\colon U \to M$ is an embedding onto an open subset $V \subset M$ where U is an open subset of k^m and $\theta^{-1}(N) = U \cap k^{m-n}$. Let π denote $\pi(k^m, k^{m-n})$ and let $U' \subset \mathfrak{B}(k^m, k^{m-n})$ be $\pi^{-1}(U)$. We have a bijection of sets $\zeta\colon U' \to \pi(M, N)^{-1}(V)$ given by $\zeta(z) = \theta\pi(z) \in M - N$ for $z \in U' - \pi^{-1}(k^{m-n})$ and $\zeta(z) =$ the line through $d\theta(v)$ if $z \in \pi^{-1}(k^{m-n})$ is the line through a normal vector v. We now put the smooth structure on $\mathfrak{B}(M, N)$ so that all such maps ζ are open embeddings.

Exercise: This construction is well defined, i.e., different embeddings θ give rise to the same smooth structure where they overlap.   ◇

There is a global description of the smooth structure on $\mathfrak{B}(M, N)$ in the real case, or in the complex case if N has a tubular neighborhood. Let $\mu\colon T \to N$ be the normal bundle of N and let $\eta\colon L \to E$ be the canonical line bundle over its projectivization E, so

$$L = \{\, (x, \lambda) \in T \times E \mid x \text{ is a point in the line } \lambda \,\}.$$

Identify E and N with their 0-sections in L and T. Then there is a canonical isomorphism $\nu\colon L - E \to T - N$ induced by projection to the first factor. Let $\psi\colon T \to M$ be a tubular neighborhood of N. Then $\mathfrak{B}(M, N)$ is the manifold obtained by gluing L and $M - N$ together via the embedding $\psi\nu\colon L - E \to M - N$. We have a natural smooth projection $\pi\,(M, N)\colon \mathfrak{B}\,(M, N) \to M$ which is the identity on $M - N$ and is ρ on E. In the complex case one may not be able to do the above construction exactly, since a tubular neighborhood may not exist.

This is all less complicated than it sounds. For example, let us find $\mathfrak{B}\,(\mathbf{R}^3, 0)$. Then E is just $\mathbf{RP}^2$ and we can identify L with $\mathbf{RP}^3 - [1, 0, 0, 0]$. The maps η and ν are given by $\eta[w, x, y, z] = [x, y, z]$ and $\nu[w, x, y, z] = (x, y, z)\,w/\left(x^2 + y^2 + z^2\right)$. In this case, ψ can be onto, so we just get $\mathfrak{B}\,(\mathbf{R}^3, 0) = L$ and $\pi\,(\mathbf{R}^3, 0) = \psi\nu$.

Another description of $\mathfrak{B}\,(\mathbf{R}^3, 0)$ is obtained by describing its charts. The manifold $E = \mathbf{RP}^2$ has three charts $[1, y, z], [x, 1, z]$ and $[x, y, 1]$ and of course the bundle L is trivial over each chart. This gives us three charts for $L = \mathfrak{B}\,(\mathbf{R}^3, 0)$ and the map $\pi\,(\mathbf{R}^3, 0)$ is given by $(x, xy, xz)\,, (xy, y, yz)$ and (xz, yz, z) respectively on the three charts. We will find this concrete description of the blowup useful so we will state it generally.

Let $k = \mathbf{R}$ or $\mathbf{C}$ and let $\xi_{in}\colon k^m \to k^m$ denote the map whose j-th coordinate is:

$$\xi_{in}\,(x_1, x_2, \dots, x_m)_j = \begin{cases} x_j x_i & \text{if } i \neq j \leq n \\ x_j & \text{if } j = i \text{ or } j > n \end{cases}$$

Then the following describes exactly the local topology of a blowup:

Lemma 2.5.1 *Let N be a proper submanifold of the real or complex manifold M. Let $\theta\colon U \to M$ be an embedding onto an open set V in M where $U \subset k^m$ is open and $\theta^{-1}\,(N) = \{\,(x_1, x_2, \dots, x_m) \in U \mid x_i = 0 \text{ for all } i \leq n\,\}$. Then $\pi\,(M, N)^{-1}\,(V)$ is covered by n charts $\varphi_i\colon U_i \to \mathfrak{B}\,(M, N)$ where $U_i = \xi_{in}^{-1}\,(U)$ and*

$$\theta^{-1} \circ \pi\,(M, N) \circ \varphi_i = \xi_{in}|_{U_i}.$$

Proof: From the local definition of the smooth structure on $\mathfrak{B}(M, N)$, it suffices to do this for the case $M = k^m$, $N = k^{m-n}$, $U = M$ and $\theta = $ the identity. Define $\tau_i\colon k^m \to k^n - 0$ by

$$\tau_i(x) = (x_1, \dots, x_{i-1}, 1, x_{i+1}, \dots, x_n)$$

and let $\psi\colon k^n - 0 \to k\mathbf{P}^{n-1}$ be the map $\psi(x) = [x_1, \dots, x_n]$. Define

$$\varphi_i(x) = ((x_i\tau_i(x), \psi\tau_i(x)), (x_{n+1}, \dots, x_m)) \in L \times k^{m-n} = \mathfrak{B}(k^m, k^{m-n}).$$

Then $\varphi_i^{-1}\colon V_i \to k^m$ is defined by

$$\varphi_i^{-1}((x, [y_1, \dots, y_n]), x_{n+1}, \dots, x_m) = (y_1/y_i, \dots, y_i, \dots, y_n/y_i, x_{n+1}, \dots, x_m)$$

where $V_i = \text{Image}\,\varphi_i = \{\,((x, y), z) \in L \times k^{m-n} \mid y_i \neq 0\,\}$. $\square$

Exercise: If $N \subset M$ has codimension one, then $\mathfrak{B}(M, N) = M$ and $\pi(M, N) =$ identity. ⬦

Exercise: If N is a point in M and M is real then $\mathfrak{B}(M, N) \approx M \# \mathbf{RP}^m$, a connected sum of M with $\mathbf{RP}^m$. ⬦

Exercise: If $N \subset M$ has trivial normal bundle then

$$\mathfrak{B}(M, N) = (M - U(N)) \bigcup_{\partial} N \times \mathbf{RP}^m_0$$

where $U(N) \approx N \times \mathrm{Int}(B^m)$ is a tubular neighborhood of N in M and $\mathbf{RP}^m_0$ is $\mathbf{RP}^m - \mathrm{Int}(B^m)$. ⬦

We want to define a notion of blowing up in the algebraic category in such a way that it coincides with our previous notion in the case of nonsingular algebraic sets.

Definitions: Let $S \subset k^n$ be a Zopen set and let $J \subset \Gamma^r(S)$ be an ideal. For example, if $T \subset S$ is a Zclosed subset we can take $J = \mathfrak{I}^r_S(T)$. We define the blowup $\pi(S, J): \mathfrak{B}(S, J) \to S$ as follows. Let $\langle f_1, \ldots, f_m \rangle$ be generators of J. Then $\mathfrak{B}(S, J)$ is the smallest Zclosed subset of $S \times k\mathbf{P}^{m-1}$ which contains

$$A = \{ (x, [f_1(x), \ldots, f_m(x)]) \mid x \in S - \mathcal{V}(J) \}.$$

We define $\pi(S, J)$ to be the restriction of the projection map $S \times k\mathbf{P}^{m-1} \to S$.

In the case $k = \mathbf{R}$ we can alternately define $\mathfrak{B}(S, J)$ to be $\mathrm{Cl}_{\mathbf{R}} C \cap S \times G(m, 1)$ where

$$C = \{ (x, \mu([f_1(x) \ldots f_m(x)])) \mid x \in S - \mathcal{V}(J) \}$$

and where $\mu: \mathbf{RP}^{m-1} \to G(m, 1)$ is the isomorphism introduced in section 4. These two definitions are equivalent by Corollary 2.4.2 since the map $\mathrm{id} \times \mu$ carries A to C. The main point is that in the case $k = \mathbf{R}$ we may assume the blowup is an affine Zopen set.

The blowup $\pi(S, J): \mathfrak{B}(S, J) \to S$ is independent of the generators of J up to birational isomorphism of $\mathfrak{B}(S, J)$ commuting with $\pi(S, J)$. To see this we need only consider the case of adding one generator since any generating set may be obtained from any other by a sequence of such additions and deletions. So let our generating set be $\langle f_1, \ldots, f_m \rangle$ as above and pick any $f \in J$. Pick $u_i \in \Gamma^r(S)$ so that $f = \sum u_i f_i$. Now let

$$
\begin{aligned}
A_1 &= \{ (x, [f_1(x), \ldots, f_m(x)]) \mid x \in S - \mathcal{V}(J) \}, \\
A_2 &= \{ (x, [f_1(x), \ldots, f_m(x), f(x)]) \mid x \in S - \mathcal{V}(J) \} \quad \text{and} \\
A_3 &= \{ (x, [y_1, \ldots, y_{m+1}]) \mid x \in S \text{ and } y_{m+1} = \sum u_i(x) y_i \}.
\end{aligned}
$$

Let B_1 and B_2 be the Zariski closures of A_1 and A_2 in $S \times k\mathbf{P}^{m-1}$ and $S \times k\mathbf{P}^m$. Then A_3 is a Zclosed subset of $S \times k\mathbf{P}^m$ and $A_2 \subset A_3$, so $B_2 \subset A_3$.

Define $\varphi\colon S \times k\mathbf{P}^{m-1} \to A_3$ and $\theta\colon A_3 \to S \times k\mathbf{P}^{m-1}$ by $\varphi(x, [y_1 \ldots y_m]) = (x, [y_1 \ldots y_m, \sum u_i(x) y_i])$ and $\theta = \varphi^{-1}$. The maps φ and θ are entire rational functions and hence birational isomorphisms. Since $\varphi(A_1) = A_2$ and φ and θ take Zclosed subsets to Zclosed subsets (by the analogue of Lemma 2.1.3a) we must have $\varphi(B_1) = B_2$. Thus φ restricts to a birational isomorphism of B_1 to B_2 which commutes with projection to S. As an exercise, the reader can fill in the details.

The following lemmas are easy exercises.

Lemma 2.5.2 *If M is a Zopen set and $J \subset \Gamma^r(M)$ is an ideal and T is any irreducible component of the blowup $\mathfrak{B}(M, J)$ then $T \not\subset \pi(M, J)^{-1}(\mathcal{V}(J))$.*

Lemma 2.5.3 *If M is a Zopen set and $J \subset \Gamma^r(M)$ is an ideal then the map*

$$\pi(M, J)\,|\colon \mathfrak{B}(M, J) - \pi(M, J)^{-1}(\mathcal{V}(J)) \to M - \mathcal{V}(J)$$

is a birational isomorphism.

Exercise: Show that if $U \subset V$ is a Zopen subset then there is a unique birational isomorphism $h\colon \mathfrak{B}(U, J) \to \pi(V, J)^{-1}(U)$ so that $\pi(V, J) \circ h = \pi(U, J)$. $\diamond$

Definitions: We can do the same thing in the smooth category. Note that on a smooth manifold M we have a ring $C^\infty(M)$ of smooth real valued functions on M. If $K \subset M$ we have an ideal $\mathfrak{I}_M^\infty(K)$ of smooth functions vanishing on K. (We sometimes write $\mathfrak{I}^\infty(K)$ instead of $\mathfrak{I}_M^\infty(K)$ when M is clear from context). Likewise if J is an ideal in $C^\infty(M)$ we have a set

$$\mathcal{V}(J) = \{\, x \in M \mid f(x) = 0 \text{ for all } f \in J \,\}.$$

If $f\colon N \to M$ is smooth and J is an ideal in $C^\infty(M)$ we have a pullback ideal $f^*(J)$ in $C^\infty(N)$ defined just as in the algebraic case.

Note that $f^*(J)$ is finitely generated if J is. Not all ideals in $C^\infty(M)$ are finitely generated though, for example $\bigcap_{\epsilon > 0} \mathfrak{I}_\mathbf{R}^\infty([0, \epsilon])$ in $C^\infty(\mathbf{R})$.

Definition: If J is finitely generated and $K \subset M$ we may define a blowup

$$\pi(K, J) : \mathfrak{B}(K, J) \to K$$

just as in the algebraic case. Pick generators f_i, $i = 0, \ldots, k$ for J. We let $\mathfrak{B}(K, J)$ be the closure of $\{\, (x, [f_0(x) \ldots f_k(x)]) \in K \times \mathbf{RP}^k \mid x \in K - \mathcal{V}(J) \,\}$ in $K \times \mathbf{RP}^k$ and let $\pi(K, J)$ be induced by projection to the first factor.

Exercise: Show that this blowup is independent of the generators chosen. $\diamond$

Exercise: Show that if $L \subset M$ is a smooth proper submanifold then $\mathfrak{I}_M^\infty(L)$ is finitely generated and $\mathfrak{B}(M, \mathfrak{I}_M^\infty(L))$ is a smooth submanifold of $M \times \mathbf{RP}^k$. $\diamond$

We have a smooth analogue of Proposition 2.2.11.

Lemma 2.5.4 *Let $L \subset M$ be a closed smooth submanifold of a smooth manifold M. Pick $p \in L$ and pick $f_i \in \mathfrak{I}_M^\infty(L)$, $i = 1, \ldots, n$ where n is the codimension of L so that the gradients $\nabla f_i(p)$ are linearly independent. Then there is a neighborhood V of p in M so that for any $g \in \mathfrak{I}_M^\infty(L)$ there are smooth $u_i \colon V \to \mathbf{R}$ so that $g(x) = \sum_{i=1}^n u_i(x) f_i(x)$ for all $x \in V$.*

Proof: By the inverse function theorem, we may pick a coordinate chart $\theta \colon U \to M$ so that $\theta^{-1}(L) = \{ x \in U \mid x_i = 0 \ \text{for all} \ i \le n \}$ and $f_i \theta(x) = x_i$ for all $x \in U$ and $i = 1, \ldots, n$ where U is some ball in $\mathbf{R}^m$. Then $g\theta(x) = 0$ if $x_i = 0$ for all $i \le n$. Hence

$$g\theta(x) = \int_0^1 d/dt \left(g\theta(tx_1, \ldots, tx_n, x_{n+1}, \ldots, x_m) \right) dt = \sum_{i=1}^n x_i g_i(x)$$

where $g_i(x) = \int_0^1 \partial g\theta / \partial x_i (tx_1, \ldots, tx_n, x_{n+1}, \ldots, x_m) \, dt$. Now just let $V = \theta(U)$ and $u_i = g_i \theta^{-1}$.   □

It would be very confusing if all these different types of blowups we have defined were different. Fortunately, for smooth blowups this is not the case, as the following Lemma shows. As a consequence of Lemma 2.5.5 below, if M and L are nonsingular algebraic sets we can always set $\mathfrak{B}(M, L) = \mathfrak{B}(M, \mathfrak{I}(L))$. We will always do so, thus $\mathfrak{B}(M, L)$ will always be given an algebraic structure. This saves us from having to distinguish notationally between topological and algebraic blowups.

Lemma 2.5.5 *If M is a smooth manifold and $L \subset M$ is a smooth proper submanifold then there is a unique diffeomorphism*

$$\mu \colon \mathfrak{B}(M, L) \to \mathfrak{B}(M, \mathfrak{I}_M^\infty(L))$$

so that

$$\pi(M, L) = \pi(M, \mathfrak{I}_M^\infty(L)) \circ \mu.$$

If M is a nonsingular Zopen set and L is a nonsingular Zclosed subset of M then $\mathfrak{B}(M, \mathfrak{I}^r(L))$ is a nonsingular Zopen set and there is a unique diffeomorphism

$$\eta \colon \mathfrak{B}(M, L) \to \mathfrak{B}(M, \mathfrak{I}_M^r(L))$$

so that

$$\pi(M, L) = \pi(M, \mathfrak{I}_M^r(L)) \circ \eta.$$

Proof: We will just sketch the proof and leave the details as an exercise for the reader. First we need to show that $\mathfrak{B}(M, \mathfrak{I}_M^r(L))$ is nonsingular in the algebraic case. This follows from Proposition 2.2.11.

Now we prove existence and uniqueness of η and μ simultaneously. For convenience we call them both μ. Let J be $\mathfrak{I}_M^r(L)$ or $\mathfrak{I}_M^\infty(L)$ as the case may be. Note that μ restricted to $\pi(M, L)^{-1}(M - L)$ must be $\pi(M, J)^{-1} \circ \pi(M, L)|$. But then the fact that $\pi\left(M,L\right)^{-1}\left(M-L\right)$ is dense in $\mathfrak{B}\left(M,L\right)$ gives uniqueness of μ, once we have shown existence. Furthermore, we need only prove that μ is a local diffeomorphism since μ embeds the dense subset $\pi\left(M,L\right)^{-1}\left(M-L\right)$ onto a dense subset.

So pick $p \in L$. If f_i, $i = 1,\dots,b$ are the generators of J used to define $\mathfrak{B}\left(M,J\right)$, we may after reordering assume that $\nabla f_i\left(p\right)$, $i = 1,\dots,n$ are linearly independent where n is the codimension of L in M. Then by Proposition 2.2.11 or Lemma 2.5.4 we know that there is a neighborhood V of p in M and smooth (or entire rational) functions $u_{ji}\colon V \to k$ so that $f_j\left(x\right) = u_{j1}\left(x\right)f_1\left(x\right) + \cdots + u_{jn}\left(x\right)f_n\left(x\right)$ for all $x \in V$ and $j = n+1,\dots,b$. We may also suppose $V \cap L = V \cap \bigcap_{i=1}^{n} f_i^{-1}\left(0\right)$. By the inverse function theorem we may pick a coordinate chart $\theta\colon U \to M$ so that $f_i\theta\left(x\right) = x_i$ for all $i = 1,\dots,n$, U is open in k^m and $p \in \theta\left(U\right) \subset V$. We have a diffeomorphism

$$\theta'\colon \mathfrak{B}\left(U,\langle x_1,\dots,x_n\rangle\right) \to \pi\left(M,J\right)^{-1}\left(\theta\left(U\right)\right)$$

given by

$$\theta'\left(x,[y_1,\dots,y_n]\right) = \left(\theta\left(x\right),[y_1,\dots,y_n,y'_{n+1},\dots,y'_b]\right)$$

where $y'_j = \sum_{i=1}^{n} u_{ji}\left(\theta\left(x\right)\right)\cdot y_i$. On the other hand, Lemma 2.5.1 gives a diffeomorphism

$$\theta''\colon \mathfrak{B}\left(U,\langle x_1,\dots,x_n\rangle\right) \to \pi\left(M,L\right)^{-1}\left(\theta\left(U\right)\right)$$

given by $\theta''\left(x,[y_1,\dots,y_n]\right) =$

$$\varphi_i\left(y_1/y_i,\dots,y_{i-1}/y_i,x_i,y_{i+1}/y_i,\dots,y_n/y_i,x_{n+1},\dots,x_m\right)$$

where i is chosen so that $y_i \neq 0$. Then μ is locally $\theta' \circ \theta''^{-1}$ which is a diffeomorphism. Hence μ is globally a diffeomorphism. □

Definition: An ideal J of functions is *locally principal* if each point has a neighborhood so that if we restrict the ideal to the neighborhood, we get a principal ideal (i.e., an ideal with one nonzero generator).

For example if L is nonsingular and codimension one then $\mathfrak{I}\left(L\right)$ is locally principal. In fact, if L is any codimension one Zclosed subset of a nonsingular Zopen set V then $\mathfrak{I}^r_V\left(L\right)$ is locally principal, c.f., Corollary 1.30 of [**M**].

Lemma 2.5.6 *The map* $\pi\left(M,J\right)\colon \mathfrak{B}\left(M,J\right) \to M$ *is a diffeomorphism (or a birational isomorphism in the algebraic case) if and only if J is locally principal.*

Proof: We leave this as an exercise. Note that disallowing $\langle 0\rangle$ as a principal ideal is essential. □

Lemma 2.5.7 *Suppose M and N are smooth manifolds or nonsingular Zopen sets and J is a finitely generated ideal in $C^\infty(M)$ or $\Gamma^r(M)$ and $f \colon N \to M$ is smooth or an entire rational function. Then there is a unique smooth map $\mu \colon \mathfrak{B}(N, f^*(J)) \to \mathfrak{B}(M, J)$ so that*

$$\pi(M, J) \circ \mu = f \circ \pi(N, f^*(J)).$$

The map μ is an entire rational function in the algebraic case.

Proof: This is an easy exercise. Just write down the definition of the blowup using the obvious generators for the pullback ideal. $\qquad\square$

We will often want to lift a map to the blowup. The following Lemma tells when we can do so.

Lemma 2.5.8 *Suppose M and N are smooth manifolds or nonsingular Zopen sets and $f \colon N \to M$ is a smooth map (or entire rational function in the algebraic case) and J is a finitely generated ideal in $C^\infty(M)$ or $\Gamma^r(M)$. Suppose also that $f^{-1}(\mathcal{V}(J))$ is nowhere dense in N. Then the ideal $f^*(J)$ is locally principal if and only if there is a smooth map (or entire rational function) $g \colon N \to \mathfrak{B}(M, J)$ so that $\pi(M, J) \circ g = f$.*

Proof: If $f^*(J)$ is locally principal, then Lemmas 2.5.6 and 2.5.7 imply the existence of g. On the other hand, it is an easy exercise to see that $\pi(M, J)^*(J)$ is locally principal. Hence if g exists, $f^*(J) = g^*\big(\pi(M, J)^*(J)\big)$ is locally principal also. $\qquad\square$

Exercise: Where was nowhere density used in the above proof? $\qquad\diamond$

Definition: We say a smooth map $f \colon N \to M$ *hits* a submanifold $L \subset M$ *cleanly* if $f^{-1}(L)$ is a submanifold and df injects the normal bundle of $f^{-1}(L)$ at each point into the normal bundle of L.

For example, if f is transverse to L or if $f(N) \subset L$ then f hits L cleanly.

Lemma 2.5.9 *Suppose M and N are smooth manifolds or nonsingular Zopen sets, L is a smooth submanifold or nonsingular Zclosed subset of M and the smooth map (or entire rational function) $f \colon N \to M$ hits L cleanly. Then there is a unique smooth map $\mu \colon \mathfrak{B}\big(N, f^{-1}(L)\big) \to \mathfrak{B}(M, L)$ compatible with projections. The map μ is an entire rational function in the algebraic case.*

$$
\begin{array}{ccc}
\mathfrak{B}\big(N, f^{-1}(L)\big) & \xrightarrow{\ \mu\ } & \mathfrak{B}(M, L) \\
\downarrow \pi\big(N, f^{-1}(L)\big) & & \downarrow \pi(M, L) \\
N & \xrightarrow{\ f\ } & M
\end{array}
$$

Proof: Notice that $f^*(\mathfrak{I}_M^\infty(L)) = \mathfrak{I}_N^\infty\big(f^{-1}(L)\big)$ by Lemma 2.5.4 and a partition of unity argument. So Lemma 2.5.7 implies this result in the smooth case.

In the algebraic case, we know by Proposition 2.2.11 that for each $x \in N$ there is a Zariski open neighborhood U of x in N such that $(f|_U)^* (\mathfrak{I}_M^r (L)) = \left(\mathfrak{I}_U^r \left(f^{-1}(L) \cap U \right) \right)$. By Lemma 2.5.7 this means μ is locally rational, hence entire rational. $\square$

Definition: Suppose that $\pi (M, J) : \mathfrak{B} (M, J) \to M$ is a blowup map and suppose that $N \subset M$ is a submanifold. Then the *strict transform* of N is the closure in $\mathfrak{B} (M, J)$ of the subset $\pi (M, J)^{-1} (N - \mathcal{V}(J))$ (or the Zariski closure in the algebraic case). We will also refer to a strict transform as a *strict preimage*.

It is immediate from the definition that the strict transform of N is $\mathfrak{B} (N, J|_N)$. The point is that there is a natural embedding of the blowup of a subset into the blowup of the whole. We will often use this in the case where N is smooth, $J = \mathfrak{I}_M^\infty (L)$ or $\mathfrak{I}_M^r (L)$ and N hits L cleanly. Then $J|_N = \mathfrak{I}_N^\infty (N \cap L)$ or $\mathfrak{I}_N^r (N \cap L)$ so the strict transform of N is just $\mathfrak{B} (N, N \cap L)$.

Lemma 2.5.10 *Let Y be a smooth manifold or nonsingular Zopen set and let W be a smooth proper submanifold or nonsingular Zclosed subset of Y and let L be a smooth proper submanifold or nonsingular Zclosed subset of W. Let $X = \mathfrak{B}(W, L)$ and $Z = \mathfrak{B}(Y, L)$. Then there is a unique smooth map or entire rational function $h \colon \mathfrak{B}(Z, X) \to \mathfrak{B}(Y, W)$ compatible with the blowup projections, i.e., the following commutes:*

$$
\begin{array}{ccc}
\mathfrak{B} (\mathfrak{B}(Y, L), \mathfrak{B}(W, L)) & \xrightarrow{\ h\ } & \mathfrak{B} (Y, W) \\
\downarrow \pi(Z, X) & & \downarrow \pi(Y, W) \\
\mathfrak{B}(Y, L) & \xrightarrow[\pi(Y, L)]{} & Y
\end{array}
$$

Proof: If h exists, it must be unique since it equals $\pi(Y, W)^{-1} \circ \rho \circ \pi(Z, X)$ on the dense subset $\pi(Z, X)^{-1} \rho^{-1}(Y - W)$. To prove existence, it suffices by Lemma 2.5.8 to prove that $\pi(Z, X)^* \rho^* (\mathfrak{I}_Y^r (W))$ is locally principal. We leave this as an exercise. $\square$

Exercise: Show that 2.5.10 remains true if we assume $L \pitchfork W$ instead of $L \subset W$. The natural conjecture at this point is that it is true if L intersects W cleanly, but show that this is not true. $\diamond$

The alert reader will have noticed that whereas the blowup of an affine real Zopen set is affine, the blowup of an affine complex Zopen set is usually not affine. This leads to a problem, since we will usually want to do several blowups in succession. What we could do is define blowups of projective Zopen sets. The reader interested in doing so could try to do this as a rather long exercise or look at how it is done in some standard algebraic geometry text. (One idea is to glue together the blowups of a covering by Zopen sets.) However, for the purposes of this book the reader has the option of not worrying about this. We will define something we call a semiblowup which has the advantage of being affine. It deletes certain parts of the blowup, but for the cases we will be interested in the deleted parts will be far away from the real points so we don't need them. We should emphasize that this is only an option, the proofs in this book can use either these semiblowups or the traditional blowups.

The idea of semiblowups is that we pretend we are in the real case and try to make the blowup affine as we do in the real case. We will almost succeed, except that some points will be missing. To be precise we let $G(m, 1)_{\mathbf{C}}$ be the complexification of $G(m, 1)$. We have a birational isomorphism $\mu_{\mathbf{C}} \colon \mathbf{CP}^{m-1} - Z \to G(m, 1)_{\mathbf{C}}$ where Z is some projective algebraic subset with no real points. In particular, $\mu_{\mathbf{C}}([z_1 \dots z_n])$ is the complex $m \times m$ matrix with (i, j)-th coordinate $z_i z_j / \sum z_k^2$ and $Z = \{ [z_1 \dots z_n] \mid \sum z_i^2 = 0 \}$. If $J = \langle f_1, \dots, f_m \rangle$ then the *semiblowup* $\mathfrak{B}^-(M, J)$ is the Zariski closure of

$$\{ (z, \mu_{\mathbf{C}}([f_1(z) \dots f_m(z)])) \mid z \in M - \mathcal{V}(J) \quad \text{and} \quad \sum f_i(z)^2 \neq 0 \}.$$

Of course this depends on the generators chosen. Anyway, this semiblowup is birationally isomorphic to a Zopen subset of the blowup. However it is immediate that $\mathfrak{B}^-(M_{\mathbf{C}}, J_{\mathbf{C}}) = \mathfrak{B}(M, J)_{\mathbf{C}}$ for real M and J. Since all the complex algebraic sets we look at will be complexifications of real algebraic sets these semiblowups will be good enough, in fact the above complexification property makes them preferable.

Hironaka's celebrated resolution theorem in [**H**] (also see [**BM**]) says that by a sequence of blowups an algebraic set can be made nonsingular , in particular:

Theorem 2.5.11 ([H]) *Let X be a Zopen set (real or complex), then there is a smooth Zopen set $\tilde{X}$ and an entire rational function $\pi \colon \tilde{X} \to X$, such that π is a composition of blowing up projections along smooth centers of dimension less than $\dim X$, i.e., $\tilde{X}$ is obtained from X by a sequence of blowing up operations.*

The following Lemma would be vacuous for $k = \mathbf{C}$ since $\mathrm{Cl}_{\mathbf{C}}(X) = \mathrm{Cl}(X)$.

Lemma 2.5.12 *Let X be a real Zopen set and let $X' = \mathrm{Cl}_{\mathbf{R}}(X)$. Then*

$$X' - \mathrm{Cl}(X) \subset \mathrm{Sing}\, X'.$$

If J is an ideal in $\Gamma^r(X')$ and $\mathcal{V}(J) \subset X' - \mathrm{Cl}(X)$ then

$$\mathrm{Sing}(\mathfrak{B}(X', J)) = \pi(X', J)^{-1}(\mathrm{Sing}\, X').$$

Proof: Suppose $x \in \mathrm{Nonsing}\, X' - \mathrm{Cl}(X)$. Let T be the irreducible component of X' which contains x. By Corollary 2.2.7 we know that $x \in \mathrm{Nonsing}\, T$. Now $X' - X$ is a real algebraic set so $T - X$ is a Zclosed subset of T. Hence $\dim(T - X) < \dim T$ by Lemma 2.2.9. But since $x \notin \mathrm{Cl}(X)$ we know that $T - X$ contains an open subset of $\mathrm{Nonsing}\, T$ and thus $\dim(T - X) = \dim T$, a contradiction. So $\mathrm{Nonsing}\, X' - \mathrm{Cl}(X) = \emptyset$, i.e., $X' - \mathrm{Cl}(X) \subset \mathrm{Sing}\, X'$.

Let $Z = \mathfrak{B}(X', J)$ and $\pi = \pi(X', J)$. The second part now follows from the first part since if $Y = \pi^{-1}(X)$ then $\mathrm{Cl}_{\mathbf{R}}(Y) = Z$ and thus

$$\pi^{-1}(\mathcal{V}(J)) \subset Z - \pi^{-1}(\mathrm{Cl}(X)) \subset Z - \mathrm{Cl}(Y) \subset \mathrm{Sing}\, Z$$

but π restricts to a birational isomorphism over $X' - \mathcal{V}(J)$ so by Lemma 2.2.10, $\mathrm{Sing}\, Z - \pi^{-1}(\mathcal{V}(J)) = \pi^{-1}(\mathrm{Sing}\, X' - \mathcal{V}(J))$. □

Recall that $V \subset \mathbf{R}^n$ is projectively closed if $\theta(V)$ is an algebraic subset of $\mathbf{RP}^n$ where $\theta(x) = [1, x_1, \ldots, x_n]$. As opposed to other properties of algebraic sets we have discussed, being projectively closed is not an algebraic invariant. In fact the following Theorem implies that any compact real algebraic set is birationally isomorphic to a projectively closed algebraic set.

Theorem 2.5.13 *Let W be real Zopen set and let V be a compact Zclosed subset of W. Then there is a Zopen set Y and a birational isomorphism $\varphi \colon W \to Y$ so that $\varphi(V)$ is a projectively closed real algebraic set.*

Proof: By induction on dimension we may as well assume that $\mathrm{Sing}\, V$ is a projectively closed real algebraic set. Suppose $W \subset \mathbf{R}^n$. Let $\mu \colon \mathbf{RP}^n \to G(n+1, 1)$ be the isomorphism of section 4. Let $\theta \colon \mathbf{R}^n \to \mathbf{RP}^n$ be the embedding $\theta(x_1, \ldots, x_n) = [1, x_1, \ldots, x_n]$. Let W' be the Zariski closure of $\mu\theta(W)$. Let P be the algebraic set $W' - \mu\theta(W)$. Let V' be the Zariski closure of $\mu\theta(V)$, let $V'' = V' \cap P = V' - \mu\theta(V)$ and let $S = \mu\theta(\mathrm{Sing}\, V)$. Then S is an algebraic set by Proposition 2.4.1. Note that $\mathrm{Sing}\, V' = V'' \cup S$ by Lemmas 2.5.12 and 2.2.10. Notice that W', V', V'' and S are all projectively closed since $G(n+1, 1)$ is contained in the unit sphere in $\mathbf{R}^{(n+1)^2}$.

By Theorem 2.5.11 we may blow up $V' - S$ to make it nonsingular. In particular, there are real Zclosed subsets $L_i \subset V_i \subset Z_i$, $i = 0, \ldots, m$ and entire rational functions $\rho_i \colon Z_i \to Z_{i-1}$ with $Z_0 = G(n+1, 1) - S$, $V_0 = V' - S$ so that:

1) $Z_i = \mathfrak{B}(Z_{i-1}, L_{i-1})$, $V_i = \mathfrak{B}(V_{i-1}, L_{i-1})$ and $\rho_i = \pi(Z_{i-1}, L_{i-1})$.
2) L_i is a smooth Zclosed subset of V_i.
3) $\rho_1 \circ \cdots \circ \rho_i(L_i) \subset \mathrm{Sing}\, V' - S = V''$.
4) $\mathrm{Sing}\, V_m = \emptyset$.

Let $X_0 = G(n+1, 1)$ and $U_0 = V'$. Since $L_0 \subset V''$ we know $\mathrm{Cl}_{\mathbf{R}}(L_0) \subset V''$, but then L_0 is an algebraic set since it is a Zclosed subset of $V_0 = V' - S$, which means that $\mathrm{Cl}_{\mathbf{R}}(L_0) - L_0 \subset S$. Thus $\mathrm{Cl}_{\mathbf{R}}(L_0) - L_0 \subset S \cap V'' = \emptyset$.

Let $X_1 = \mathfrak{B}(X_0, L_0)$, $\pi_1 = \pi(X_0, L_0)$ and let $U_1 \subset X_1$ be the strict transform of U_0. We have a canonical inclusion $Z_1 \subset X_1$ so that ρ_1 is the restriction of π_1 and $V_1 \subset U_1$. Then $L_1 \subset \pi_1^{-1}(V'')$ which is disjoint from $X_1 - Z_1 = \pi_1^{-1}(S)$, So L_1 is an algebraic set. Let $X_2 = \mathfrak{B}(X_1, L_1)$ etc., and we obtain real algebraic sets $L_i \subset U_i \subset X_i$, $i = 0, \ldots, m$ and entire rational functions $\pi_i \colon X_i \to X_{i-1}$ so that:

a) $X_i = \mathfrak{B}(X_{i-1}, L_{i-1})$, $U_i = \mathfrak{B}(U_{i-1}, L_{i-1})$ and $\pi_i = \pi(X_{i-1}, L_{i-1})$.

b) $\pi_1 \circ \cdots \circ \pi_i (L_i) \subset V''$.

c) $\operatorname{Sing} U_m = \pi_m^{-1} \circ \cdots \circ \pi_1^{-1}(S)$.

Note that if we take the usual model of the blowup, each X_i is contained in a product of $G(n_i, 1)$'s and hence is projectively closed. Thus U_m is projectively closed. Let $\pi \colon X_m \to X_0$ be the composition of the π_i's. Then b) implies that $\pi| \colon X_m - \pi^{-1}(V'') \to X_0 - V''$ is a birational isomorphism.

We will let $Y = \pi^{-1}\mu\theta(W)$ and $\varphi = \pi^{-1}\mu\theta|_W$. Then it suffices to show that $\varphi(V) = U_m$, i.e., $\pi^{-1}(V'') \cap U_m$ is empty.

By repeated applications of Lemma 2.5.12 we know that $\operatorname{Sing} V_i = V_i \cap \pi_i^{-1} \circ \cdots \circ \pi_1^{-1}(V'')$. Setting $i = m$ we then get $\emptyset = \operatorname{Sing} V_m = V_m \cap \pi^{-1}(V'')$ so $\pi^{-1}(V'') \cap U_m = \pi^{-1}(V'') \cap (U_m - V_m) = \pi^{-1}(V'') \cap \pi^{-1}(S) = \emptyset$. $\square$

Exercise: Why do we need to first make $\operatorname{Sing} V$ projectively closed to make the above proof work? $\diamond$

Corollary 2.5.14 *Suppose X is a compact real Zopen set. Then X is birationally isomorphic to a projectively closed algebraic set.*

Proof: This is just a corollary of Theorem 2.5.13 with $V = W = X$. $\square$

6. Blowing Down

We call the opposite of an algebraic blowing up operation, an algebraic blowing down. In the real algebraic category there are many other ways of blowing down Zopen sets. For example we can crush down Zclosed subsets by taking their algebraic quotient spaces as in the following proposition, which generalizes [**AK1**] and [**AK2**].

Proposition 2.6.1 ([AK15]) *Suppose V and U are real Zopen sets and W is a Zclosed subset of U. Suppose V is compact and $p \colon V \to U$ is an entire rational function. Let X be the quotient space obtained from the disjoint union of V and W after identifying each $v \in p^{-1}(W)$ with $p(v) \in W$. Then there are a real Zopen set Y, entire rational functions $f \colon V \to Y$ and $g \colon W \to Y$ and a homeomorphism $h \colon Y \to X$ so that*

a) *hf and hg are just the quotient maps $V \subset V \cup W \to X$ and $W \subset V \cup W \to X$.*

b) *$W' = g(W)$ is a Zclosed subset of Y.*

c) *$g \colon W \to W'$ is a birational isomorphism.*

d) *If $\dim W < \dim(V - p^{-1}(W))$ then $\operatorname{Sing} Y \subset W' \cup f(\operatorname{Sing} V)$.*

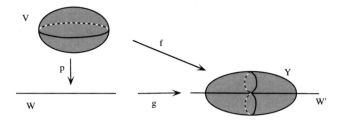

FIGURE II.6.1. Algebraic blowing down

Proof: By Corollary 2.5.14 we may as well assume that V is a projectively closed algebraic set. Suppose $V \subset \mathbf{R}^m$. After a translation, we may assume that $0 \notin V$. Pick polynomials $q \colon \mathbf{R}^m \to \mathbf{R}$ and $r \colon U \to \mathbf{R}$ so that q is overt, $V = q^{-1}(0)$ and $W = r^{-1}(0)$. By Proposition 2.1.1 we may write $p = t/u$ for polynomials t and u so that u is nowhere 0 on V. Extend p to all of $\mathbf{R}^m$ by setting $p = tu/(u^2 + q^{2b})$ where b is chosen so that the degree of $u^2 + q^{2b}$ is greater than the degree of tu. Let the degree of q be d. Define $s \colon \mathbf{R}^m \times U \to \mathbf{R}$ by

$$s(x, y) = r(y)^{2d} \left((y - p(x/r(y)))^2 + q(x/r(y))^2 \right).$$

Then after clearing denominators we see that $s(x, y)$ is an entire rational function.

Let $Y = s^{-1}(0)$. Define $f \colon V \to \mathbf{R}^m \times U$ by $f(v) = (v \cdot rp(v), p(v))$ and $g \colon W \to \mathbf{R}^m \times U$ by $g(w) = (0, w)$ and let $W' = g(W) = 0 \times W$.

Assertion 2.6.1.1 $Y = f(V) \cup W'$.

Proof: Notice that if $y \in W$ then $r(y) = 0$ and so $s(x, y) = q_d(x)^2$ where q_d is the homogeneous part of $q(x)$ of degree d. Hence $Y \cap (\mathbf{R}^m \times W) = 0 \times W = W'$ since overtness of q guarantees that $q_d(x) = 0$ implies $x = 0$. Now suppose $y \notin U - W$, then $r(y) \neq 0$. So if $s(x, y) = 0$ for some x, we have $y = p(x/r(y))$ and $q(x/r(y)) = 0$ so $x/r(y) \in V$. But then $(x, y) = f(x/r(y))$. So $Y - \mathbf{R}^m \times W \subset f(V - p^{-1}(W))$. Conversely, if $v \in V$ then

$$
\begin{aligned}
sf(v) &= s(v \cdot rp(v), p(v)) \\
&= rp(v)^{2d} \left((p(v) - p(v))^2 + q(v)^2 \right) = 0
\end{aligned}
$$

since $q(v) = 0$, so $f(V) \subset Y$ and so $Y = f(V) \cup W'$. □

Assertion 2.6.1.2 *There is a homeomorphism $h \colon Y \to X$ so that hf and hg are the quotient maps.*

Proof: Suppose that $f(v) = g(w)$ for some $v \in V$ and $w \in W$. Then $w = p(v)$, so $w \sim v$. Likewise, if $g(w) = g(w')$ then $w = w'$. Finally, if $f(v) = f(v')$ then $p(v) = p(v')$ so $v \sim v'$ if $p(v) \in W$. But we also have $v \cdot rp(v) = v' \cdot rp(v)$ so if $p(v) \notin W$ then $rp(v) \neq 0$ so $v = v'$.

Conversely, if $w \sim v$ for $w \in W$ and $v \in V$ then $p(v) = w$ so $f(v) = (0, w) = g(w)$. So h exists by the universal property of the quotient topology.   □

Assertion 2.6.1.3 $W' = Y \cap 0 \times U$, so W' is a Zclosed subset of Y.

Proof: If $(0, y) \in Y$ then

$$0 = s(0, y) = r(y)^{2d} \left((y - p(0))^2 + q(0)^2 \right).$$

But $q(0) \neq 0$ since $0 \notin V$ so $r(y) = 0$ and hence $y \in W$.   □

So we only need check that $\operatorname{Sing} Y \subset W' \cup f(\operatorname{Sing} V)$. The entire rational function $f' \colon Y - W' \to V - p^{-1}(W)$ defined by $f'(x, y) = x/r(y)$ is the inverse of $f|$, so by Lemma 2.2.10, $\operatorname{Sing}(Y - W') = f(\operatorname{Sing}(V - p^{-1}(W)) \subset f(\operatorname{Sing} V)$ and $\dim(Y - W') = \dim(V - p^{-1}(W))$. So d) follows.   □

A quick proof of Proposition 2.6.1 when W is a point can be seen from the following since then $X = V/p^{-1}(W) =$ the one point compactification of $V - p^{-1}(W)$.

Lemma 2.6.2 *The one point compactification of a real Zopen set V is homeomorphic to an algebraic set.*

Proof: Since V is birationally isomorphic to an algebraic set, we may as well assume V is an algebraic subset of some $\mathbf{R}^n$. We can assume $0 \notin V$. Let $f \colon \mathbf{R}^n \to \mathbf{R}$ be a polynomial of degree d such that $f^{-1}(0) = V$. Define $F(x) = |x|^{2d} f\left(x/|x|^2\right)$, then after clearing denominators F becomes a polynomial. But $F^{-1}(0)$ is homeomorphic to the one point compactification of V, since $x \mapsto x/|x|^2$ is an inversion through the origin.   □

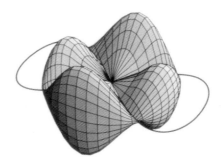

FIGURE II.6.2. Squishing a sphere to a figure 8

In Proposition 2.6.1 if $U \subset \mathbf{R}^n$, $V \subset U \times \mathbf{R}^k$, and p is induced by projection $p(u, z) = u$ then, as in [**AK2**] and [**AK6**], for the equation of Y we can alternatively use:

$$r(u)^d q(u, z/r(u)) = 0$$

Here, $q(u,z)$ and $r(u)$ are polynomials with $q^{-1}(0) = V$ and $r^{-1}(0) = W$, d is the degree of q, and q has the property that $q_d^{-1}(0) \cap (0 \times \mathbf{R}^k) = \{0\}$ (notice that this is a weaker assumption than that q be overt). In this case, Y is topologically the crushing of V to W, i.e., it is obtained by identifying of the points of V which lie over W with W. For example, when $q = x^2 + y^2 + z^2 - 2$ and $r = x^2 - y^3$ we get a sphere folded over a cusp as in the example in Chapter I, Figure I.4.3. By replacing r by $r = y^4 - 4y^2 + 4x^2$ we get a sphere folded over a figure eight as in Figure II.6.2 above.

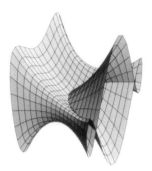

FIGURE II.6.3. Squishing a hyperboloid to a parabola

Similarly, by taking $q = x^2 - y^2 + z^2 - 1$ and $r = y - x^2$ we obtain a hyperboloid folded over a parabola as in Figure II.6.3.

7. Algebraic Homology

Definitions: Let V be a subset of some $\mathbf{R}^n$. We let $H_i^A(V)$ denote the subgroup of $H_i(V; \mathbf{Z}/2\mathbf{Z})$ consisting of classes $g_*([S])$ where $g\colon S \to V$ is an entire rational function, S is a compact nonsingular i-dimensional real algebraic set and $[S]$ is the fundamental class of S in $H_i(S; \mathbf{Z}/2\mathbf{Z})$. By replacing S with the graph of g we can always assume that $S \subset V \times \mathbf{R}^n$ for some n and g is induced by the projection. We call the elements of $H_*^A(V)$ *algebraic homology classes*.

Exercise: Show that by the resolution theorem 2.5.11 the nonsingularity assumption on S is unnecessary. (In particular, a compact real algebraic set has a fundamental class – a fact which can also be seen using [**Su1**] and the triangulability of real algebraic sets [**L2**].) $\diamond$

Exercise: Suppose $f\colon V \to W$ is an entire rational function. Show that $f_*(H_*^A(V)) \subset H_*^A(W)$. $\diamond$

If V is a real Zopen set then we also know that $H_i^A(V)$ is the subgroup of $H_i(V; \mathbf{Z}/2\mathbf{Z})$ generated by Zclosed subsets T of V with $V \cap \mathrm{Cl}(\mathrm{Nonsing}\, T)$ compact. To see this, suppose $g\colon S \to V$ is an entire rational function where S is a nonsingular irreducible compact real algebraic set and $g_*([S]) \neq 0$. Then

g is generically odd to one, otherwise we would have $g_*([S]) = 0$. Hence by Corollary 2.3.3, if we take the Zariski closure $T = \mathrm{Cl_R} g(S) \cap V$ of $g(S)$ in V, then $T - g(S)$ has smaller dimension than $g(S)$. Consequently, the fundamental class of T represents $g_*([S])$.

Definition: We say that V has *totally algebraic homology* if all its homology classes are algebraic, i.e., $H_*^A(V) = H_*(V; \mathbf{Z}/2\mathbf{Z})$.

For example the Grassmannian $G(m, n)$ has totally algebraic homology, since its homology is generated by the Schubert cycles which are easily seen to be algebraic (c.f., [**AK1**]). This last fact along with Corollary 2.4.4 implies that, if $V \subset W$ are nonsingular Zopen sets then the Poincare duals of the Stiefel-Whitney classes of the the normal bundle as well as the tangent bundle of V are algebraic.

A very useful property of algebraic sets with totally algebraic homology is that their bordism groups are generated by algebraic sets. Recall $\mathfrak{N}_*(V)$ is the unoriented bordism group of V, generated by the bordism classes of maps from closed smooth manifolds $f \colon M \to V$.

Lemma 2.7.1 *A nonsingular real Zopen set V has totally algebraic homology if and only if every element of $\mathfrak{N}_*(V)$ is represented by some $\pi \colon W \to V$ where W is a nonsingular algebraic set contained in $V \times \mathbf{R}^n$ for some n and π is induced by projection.*

Proof: By [**Th**], the evaluation map $ev \colon \mathfrak{N}_*(V) \to H_*(V; \mathbf{Z}/2\mathbf{Z})$ given by $ev(f \colon M \to V) = f_*[M]$ is onto, so one direction is proven.

Assume that V is totally algebraic. By [**Th**] and [**CF**], $\mathfrak{N}_*(V)$ is generated by projections

$$\pi_{ij} \colon Y_i \times Z_j \to Z_j \overset{\alpha_j}{\to} V$$

where $\{Y_i\}$ generate $\mathfrak{N}_*$ (point) and $\{ev(\alpha_j)\}$ generate $H_*(V; \mathbf{Z}/2\mathbf{Z})$. By assumption we can take Z_j to be nonsingular algebraic subsets of $V \times \mathbf{R}^m$ for some m. By [**Mi**] we can assume that Y_i are nonsingular algebraic sets in some $\mathbf{R}^n$. Then $Z_j \times Y_i \subset V \times \mathbf{R}^m \times \mathbf{R}^n$ and the projections $\pi_{ij} \colon Z_j \times Y_i \to V$ generate $\mathfrak{N}_*(V)$. By translating we can make all $Z_j \times Y_i$ disjoint, since the group operation is the disjoint union we are done. $\qquad\square$

Definition: Suppose V and W are nonsingular Zopen sets. A *rational diffeomorphism* is an entire rational function $f \colon V \to W$ which is also a diffeomorphism.

Notice that we do not require f^{-1} to be rational so f might not be a birational isomorphism. For example, we could take V to be the algebraic set $\mathcal{V}(y^3 - x^2 - 1) \subset \mathbf{R}^2$ and let $f \colon V \to \mathbf{R}$ be projection to the x axis. Then f is a rational diffeomorphism but its inverse is not rational since $\sqrt[3]{x^2 + 1}$ is not an entire rational function.

The following exercise shows that the effect of a rational diffeomorphism is to increase the number of entire rational functions on a Zopen set without changing the diffeomorphism type.

Exercise: Show that if $f\colon V \to W$ is a rational diffeomorphism, then $\Gamma^r(f)$ injects $\Gamma^r(W)$ into $\Gamma^r(V)$ but $\Gamma^r(f)(\Gamma^r(W)) \neq \Gamma^r(V)$ unless f^{-1} is entire rational.   ◇

If $f\colon V \to W$ is a rational diffeomorphism and W has totally algebraic homology, then V has totally algebraic homology. This is a consequence of the next Lemma.

Lemma 2.7.2 *If* $f\colon V \to W$ *is a rational diffeomorphism then*

$$f_*\left(H_*^A(V)\right) = H_*^A(W)$$

Proof: We leave this as an exercise. Use the pullback Zopen set

$$T = \{\,(v,s) \in V \times S \mid f(v) = g(s)\,\}$$

$$
\begin{array}{ccc}
T & \longrightarrow & V \\
\downarrow & & \downarrow \\
S & \longrightarrow & W
\end{array}
$$

Note that the map $T \to S$ is a rational diffeomorphism so $T \to V$ represents f_*^{-1} of $S \to W$.   □

There are many examples of nonsingular algebraic sets V with the property $H_*^A(V) \neq H_*(V; \mathbf{Z}/2\mathbf{Z})$. Hence not every V has totally algebraic homology. For example:

Theorem 2.7.3 ([AK7]) *For any* $n \geq 3$ *there exists a connected nonsingular real algebraic set* V *of dimension* n *such that* $H_k^A(V) \neq H_k(V; \mathbf{Z}/2\mathbf{Z})$ *for all* $k = 2, 3, \ldots, n-1$

Proof: The proof of this theorem uses a result from Section 8. For completeness we choose not to delay this proof to Section 8. Let V_1 be any closed smooth manifold of dimension n which contains two disjointly embedded circles with trivial normal bundles $S_1 \sqcup S_2 \subset V_1$ such that the inclusion induces an injection $H_1(S_1 \sqcup S_2; \mathbf{Z}/2\mathbf{Z}) \to H_1(V_1; \mathbf{Z}/2\mathbf{Z})$. Let $X \subset \mathbf{R}^m$ be an irreducible algebraic set diffeomorphic to $S_1 \sqcup S_2$. For example,

$$X = \mathcal{V}\left(\langle x_2^4 + \left(x_1^2 - 1\right)\left(x_1^2 - 4\right), x_3, \ldots, x_m \rangle\right)$$

in $\mathbf{R}^m$. If $m > 2n$ we can extend the embedding of $S_1 \cup S_2$ to an embedding of V_1 in $\mathbf{R}^m$. By Theorem 2.8.4 we can, after perhaps increasing m, isotop V_1 to a nonsingular algebraic set V_2 keeping X fixed, i.e., $X \subset V_2$. Let V be the algebraic blow up $V = \mathcal{B}(V_2, X) \to V_2$ and let $\pi = \pi(V_2, X)$. Write $X = X_1 \cup X_2$, where X_i are the components of X. Then, $\pi^{-1}(X) \approx X \times \mathbf{RP}^{n-2}$.

Let $\gamma_j \in H_j(V; \mathbf{Z}/2\mathbf{Z})$ be the homology classes corresponding to $X_1 \times \mathbf{RP}^{j-1}$, $j = 1, \dots, n-1$. We claim that none of γ_j for $j = 2, \dots, n-1$ are algebraic.

Suppose γ_j were algebraic for some $j \geq 2$, then we could find a nonsingular algebraic set $Y_j \subset V \times \mathbf{R}^s$ for s large, representing this class. By using Theorem 2.8.4 we could isotop Y_j in $V \times \mathbf{R}^s$ so that it is transverse to $\pi^{-1}(X) \times \mathbf{R}^s$, see

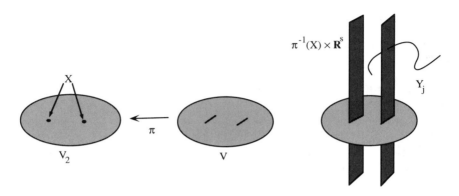

FIGURE II.7.1. V and Y_j

Figure II.7.1. Then the algebraic set $Y_{j-1} = Y_j \cap \pi^{-1}(X) \times \mathbf{R}^s$ would represent γ_{j-1}. This is because

$$
\begin{aligned}
D^{-1}(\gamma_{j-1}) &= D^{-1}(\gamma_j) \cup D^{-1}([\pi^{-1}(X_1)]) \\
&= D^{-1}(\gamma_j) \cup D^{-1}([\pi^{-1}(X)]) \\
&= D^{-1}(\rho_*[Y_j]) \cup D^{-1}([\pi^{-1}(X)]) \\
&= D^{-1}(\rho_*[Y_j \cap \pi^{-1}(X) \times \mathbf{R}^s])
\end{aligned}
$$

where D is the Poincare duality map, and $\rho: Y_j \to V$ is the induced projection.

By repeating this process we eventually obtain a nonsingular algebraic set $Y_1 \subset X \times \mathbf{R}^s \subset V \times \mathbf{R}^s$ representing γ_1. Since $\pi_*(\gamma_1) = [X_1]$, we have $\pi_* \rho_*([Y_1]) = [X_1]$. But then $\rho^{-1} \pi^{-1}(x)$ is an odd number of points for generic $x \in X_1$, and an even number of points for generic $x \in X_2$. This contradicts Corollary 2.3.3. So none of the classes γ_j are algebraic.

Note that if one did not care about the connectedness of V, one could just take $V = X \times \mathbf{RP}^{n-1}$ and also get $H_1^A(V) \neq H_1(V, \mathbf{Z}/2\mathbf{Z})$. Also this proof would be a bit simpler since it would not be necessary to use Theorem 2.8.4 to construct V. Then this result just follows from the fact that the algebraic cohomology is closed under cup product.   □

The existence of non-algebraic homology classes in nonsingular algebraic sets was also independently established by several other authors (c.f., [BT] and [BCR]). In [BD1] an example of a smooth manifold M along with a homology class α of M is given, so that α can not be algebraic in any nonsingular algebraic set V which is diffeomorphic to M. On the other hand it is amusing to note that:

Theorem 2.7.4 ([AK11]) *Every closed smooth manifold is homeomorphic to a real algebraic set with totally algebraic homology.*

Hence the totally algebraic set in the conclusion of Theorem 2.7.4 must sometimes have singularities.

8. Making Smooth Objects Algebraic

Suppose $M \subset \mathbf{R}^n$ is a smooth compact manifold which you wish to make diffeomorphic to a nonsingular Zopen set. The naive approach to doing this would be to take a tubular neighborhood U of M and take a map $f \colon U \to E(n, k)$ to the Grassmannian where k is the codimension of M, so that f is transverse to $G(n, k)$ and $f^{-1}(G(n, k)) = M$. Then you approximate f by a polynomial p. Transversality guarantees that $p^{-1}(0)$ has a component near M which is diffeomorphic to M. There are two problems with this approach however. One problem is that the image of p might not be contained in $E(n, k)$, just in the ambient Euclidean space. The second problem is that you have no control over the part of $p^{-1}(0)$ outside of U.

The first problem is addressed in Theorem 2.8.3 below. Basically, it says that the approximating polynomial p can have image in $E(n, k)$ at the expense of adding more variables to the domain. In Nash's original paper on the subject [**N**], he did this in a different way by exploiting particular properties of $E(n, k)$ but we choose a more general approach which is more useful.

The second problem was solved in [**To1**]. We address it in Theorem 2.8.4 below. The solution requires that more care be taken when approximating f. In particular, if f is already a polynomial on a certain nonsingular algebraic subset L we ask that the approximation p satisfy $p|_L = f|_L$. The question of when you can do this leads to the following definition.

Definition: We call a Zopen set $L \subset \mathbf{R}^n$ *nice* if for every $x \in L$ there is a neighborhood U of x in $\mathbf{R}^n$ so that $\mathfrak{I}_U^\infty(U \cap L) = C^\infty(U)\mathfrak{I}(L)$. In other words, locally any smooth function vanishing on L is a linear combination of polynomials vanishing on L.

Tognoli has given a precise characterization of such sets [**To3**]. However, in this book we only need a few special cases of nice Zopen sets obtained from the following exercises. (Hint: Use Lemma 2.5.4.)

Exercise: Nonsingular Zopen sets are nice. ◇

Exercise: The disjoint union of nice real algebraic sets is nice. ◇

Exercise: The union of nonsingular codimension one Zclosed subsets of a nonsingular Zopen set is nice.   ◇

   The importance of nice algebraic sets is that any smooth function vanishing on a nice Zopen set L can be approximated near a compact set by a polynomial still vanishing on L, as we shall see in Lemma 2.8.1.

   But not all Zopen sets are nice, for example $V = \{\,(x,y) \mid y^2 = x^3 - x^2\,\}$ fails to be nice at the isolated point $(0,0)$ because, for example, x vanishes on $V \cap U$ for a neighborhood U of $(0,0)$. For that reason we make the following definition.

Definition: Let $L \subset \mathbf{R}^n$ be a Zopen set and let $P \subset \mathbf{R}^n$ be a set so $L \subset P$. Then we say (P,L) is an *approximable pair* if for each $x \in L$ there is an open neighborhood U of x in $\mathbf{R}^n$ so that $\mathfrak{I}_U^\infty(U \cap P) \subset C^\infty(U)\mathfrak{I}(L)$. In other words, locally any smooth function vanishing on P is a linear combination of polynomials vanishing on L.

Exercise: If L or P is nice, then (P,L) is an approximable pair.   ◇

Exercise: Suppose for each $x \in L$ there is a Zopen neighborhood U_x of x in $\mathbf{R}^n$ and a neighborhood V_x of x in U_x and a nice Zopen set K_x so that $L \cap U_x \subset K_x$ and $K_x \cap V_x \subset P$. Then (P,L) is an approximable pair.   ◇

Lemma 2.8.1 *Suppose L is a Zclosed subset of a Zopen set V, $P \subset V$, (P,L) is an approximable pair, $K \subset V$ is compact and $f\colon (V,P) \to (\mathbf{R},0)$ is a smooth function.*

   a) *Then f can be arbitrarily closely C^∞-approximated on K by polynomials $\varphi\colon (V,L) \to (\mathbf{R},0)$. In other words, for any $\epsilon > 0$ and any integer k we may approximate f by a polynomial $\varphi\colon (V,L) \to (\mathbf{R},0)$ so that for any partial derivative D of order $\le k$ and any $x \in V \cap K$ we have $|D\varphi(x) - Df(x)| \le \epsilon$.*

   b) *Suppose f vanishes on $V - K$. Then for any $\epsilon > 0$ and any integers k and m we may approximate f by an entire rational function $\varphi\colon (V,L) \to (\mathbf{R},0)$ so that for any partial derivative D of order $\le k$ and any $x \in V$ we have $|D\varphi(x) - Df(x)| \le \epsilon(1 + |x|^2)^{-m}$.*

Proof: Let $\mathfrak{I}(L) = \langle p_1,\dots,p_k\rangle$. Suppose $V \subset \mathbf{R}^n$. Extend f in any way to a smooth function $f\colon (\mathbf{R}^n, P) \to (\mathbf{R},0)$.

Assertion 2.8.1.1 *For each $x \in V$ there are a neighborhood U of x in V and smooth functions $u_i\colon U \to \mathbf{R}$ for $i = 1,\dots,k$ so that $f(y) = \sum_{i=1}^{k} u_i(y)p_i(y)$ for all $y \in U$.*

Proof: If $x \in L$ this is immediate since (P,L) is an approximable pair. If $x \in V - L$ then $p_i(x) \ne 0$ for some i. So set $U = V - p_i^{-1}(0)$, $u_i = f/p_i$ and $u_j = 0$ for $j \ne i$.   □

Assertion 2.8.1.2 *There are smooth functions* $u_i\colon V \to \mathbf{R}$ *for* $i = 1, \ldots, k$ *so that* $f(y) = \sum_{i=1}^{k} u_i(y)p_i(y)$ *for all* $y \in V$.

Proof: Piece together the local functions u_i in Assertion 2.8.1.1 with a partition of unity. □

Now we may prove a) by approximating u_i on K by polynomials q_i and setting $\varphi = \sum_{i=1}^{k} q_i p_i$.

Now suppose that f vanishes on $V - K$. Pick any $\epsilon > 0$, k and m. Pick a polynomial $r\colon \mathbf{R}^n \to \mathbf{R}$ so that $\mathrm{Cl}_{\mathbf{R}}(V) - V = r^{-1}(0)$. Define

$$\mu\colon \mathbf{R}^n - r^{-1}(0) \to \mathbf{R}^{n+1}$$

by $\mu(x) = (x, 1/r(x))$. Let $\theta\colon \mathbf{R}^{n+1} - 0 \to \mathbf{R}^{n+1} - 0$ be inversion through the unit sphere, $\theta(x) = x/|x|^2$. Note θ^2 is the identity.

Assertion 2.8.1.3 $V' = \theta\mu(V) \cup 0$ *and* $L' = \theta\mu(L) \cup 0$ *are compact real algebraic sets.*

Proof: First, $\mu(V)$ is a real algebraic set since

$$\mu(V) = \{\, (x,t) \in \mathrm{Cl}_{\mathbf{R}}(V) \times \mathbf{R} \mid t\,r(x) = 1 \,\}.$$

Likewise $\mu(L)$ is a real algebraic set. Then V' and L' are real algebraic sets by the proof of Lemma 2.6.2. They are compact since $\mu(V)$ is closed and misses 0, hence the inversion $\theta\mu(V)$ is compact. □

Now $\theta\mu(K)$ is compact and misses 0 so there is a compact neighborhood Q of 0 in $\mathbf{R}^{n+1}$ so that $Q \cap \theta\mu(K) = \emptyset$. Let $P' = Q \cup \theta\mu(P)$.

Assertion 2.8.1.4 (P', L') *is an approximable pair.*

Proof: Pick any $y \in L'$. If $y = 0$ then $\mathfrak{I}_U^\infty(U \cap P') = \langle 0 \rangle \subset C^\infty(U)\mathfrak{I}(L')$ for small enough U. So suppose $y \neq 0$. Then $y = \theta\mu(x)$ for some $x \in L$. Pick an open neighborhood U of x in $\mathbf{R}^n$ so that $\mathfrak{I}_U^\infty(U \cap P) \subset C^\infty(U)\mathfrak{I}(L)$. The image of $\theta\mu$ is a manifold T so there is a neighborhood U' of y in $\mathbf{R}^{n+1}$ so that the restriction $\theta\mu|_U$ maps U diffeomorphically onto $T \cap U'$. Suppose now that $h\colon (U', U' \cap P') \to (\mathbf{R}, 0)$ is smooth. Then $h \circ \theta \circ \mu|_U \in \mathfrak{I}_U^\infty(U \cap P)$ so we may find smooth $v_i\colon U \to \mathbf{R}$ so that $h\theta\mu(z) = \sum_{i=1}^{k} v_i(z)p_i(z)$ for all $z \in U$.

Now if $\pi\colon \mathbf{R}^{n+1} \to \mathbf{R}^n$ is projection, we have $\pi\theta(\theta\mu(z)) = \pi\mu(z) = z$ so for all $u \in \mu(U)$ we have

$$h\theta(u) = \sum_{i=1}^{k} v_i(\pi(u))p_i(\pi(u)).$$

Note that $p_i\pi \in \mathfrak{I}(\mu(L))$. Let $h'\colon \theta(U') \to \mathbf{R}$ be defined by

$$h'(u) = h(u) - \sum_{i=1}^{k} v_i(\pi\theta(u))p_i(\pi\theta(u)).$$

Then h' vanishes on the manifold $\mu(U)$ which is given by the equation $tr(x) = 1$. Hence by Lemma 2.5.4, we know

$$h'(u) \in \langle tr(x) - 1 \rangle \subset C^\infty(\theta(U'))\mathfrak{I}(\mu(L)).$$

So $h \circ \theta \in C^\infty(\theta(U'))\mathfrak{I}(\mu(L))$ which implies that $h \in C^\infty(U')\mathfrak{I}(\theta\mu(L))$ since θ is a birational isomorphism. But $C^\infty(U')\mathfrak{I}(\theta\mu(L)) = C^\infty(U')\mathfrak{I}(L')$ as long as we make sure $0 \notin U'$, so we are done.   □

Pick a very large integer b. How big we pick it depends on how big m and k are. We can define a smooth function $g\colon V \to \mathbf{R}$ by $g(x) = (1 + |x|^2)^b f(x)$ and define $h\colon (V', L') \to (\mathbf{R}, 0)$ by by $h(x) = g \circ \mu^{-1} \circ \theta^{-1}(x)$ on $V' - 0$ and $h(0) = 0$. Note that h is smooth since g vanishes outside K. By part a) of this lemma, h can be approximated by a polynomial $\psi\colon (V', L') \to (\mathbf{R}, 0)$. Let $\varphi(x) = \psi \circ \theta \circ \mu(x)(1 + |x|^2)^{-b}$. Then φ approximates f, and if b is chosen large enough and ψ is close enough, the approximation φ will be as close as we wanted. The idea is that any partial derivative of $\psi \circ \theta \circ \mu - g$ can be written in terms of partials of $\psi - h$ (which are small) and partials of θ and ψ which are polynomials and hence bounded by $|x|$ to some power.   □

The next theorem has a long history. Originally Seifert showed that any compact codimension one submanifold of $\mathbf{R}^n$ is isotopic to a nonsingular real algebraic subset. His ideas are easily extended to show that any null homologous compact codimension one submanifold of a nonsingular real algebraic set is isotopic to a nonsingular real algebraic set. This (and in particular its relative version) is all we really need for the results in this book. However it can be generalized. In [**AK4**] and [**AK7**], we showed that a compact codimension one smooth submanifold of a nonsingular real algebraic set is isotopic to a nonsingular algebraic subset if it is homologous to a union of nonsingular algebraic subsets. Bochnak, Kucharz and Shiota showed the nonsingularity of the algebraic sets was not needed [**BKS**]. The following proof is based on the ideas in [**BKS**]. But first a definition.

Definition: Let V be a real Zopen set of dimension n. We define $AH_{n-1}(V)$ to be the subgroup of $H_{n-1}(\mathrm{Nonsing}\,V; \mathbf{Z}/2\mathbf{Z})$ generated by compact codimension one Zclosed subsets W of V so that $W \subset \mathrm{Nonsing}\,V$.

Hence, if V is compact and nonsingular then $AH_{n-1}(V) = H_{n-1}^A(V)$. In fact the compactness is not necessary, if V is nonsingular then $AH_{n-1}(V) = H_{n-1}^A(V)$. This is because by Theorem 2.5.11 and Lemma 2.6.2, V is birationally isomorphic to a Zopen subset of a compact nonsingular Zopen set V'. Any class in $H_{n-1}^A(V)$ is represented by Zclosed subset W of V with $\mathrm{Cl}(\mathrm{Nonsing}\,W)$ compact. The Zariski closure W' of W in V' represents some homology class supported in V. By [**Th**], there is a compact smooth submanifold M in V homologous to W'. By Theorem 2.8.2 below, there is a nonsingular Zclosed subset X of V' ϵ-isotopic to M. Then $X \subset V$ is homologous to W and represents an element of $AH_{n-1}(V)$.

Theorem 2.8.2 *Suppose V is an n dimensional real Zopen set and M is a smooth closed codimension one submanifold of Nonsing V. Then M is ϵ- isotopic to a nonsingular Zclosed subset of V if and only if $[M] \in AH_{n-1}(V)$.*

Furthermore, a relative version is true. Suppose $L \subset P \subset M$ where L is a Zclosed subset of V and (P, L) is an approximable pair. Then M is ϵ- isotopic to a nonsingular Zclosed subset of V fixing L if and only if $[M] \in AH_{n-1}(V)$.

Proof: We prove the relative version which of course implies the nonrelative version by setting $L = P = \emptyset$. One way is trivial since isotopy implies homology. Let $V' = $ Nonsing V. Pick a compact codimension one Zclosed subset W of V so $W \subset V'$ and W is homologous to M in V'. We may assume all irreducible components of W are codimension 1 by just throwing away all smaller dimensional irreducible components. Let $J = \mathcal{I}_V^r(W)$. Note that J is locally principal by Corollary 1.30 of [**M**] applied to the complexification $W_{\mathbf{C}}$ of W. Let $V'' = \mathfrak{B}(V, J)$ and let $\pi\colon V'' \to V$ be $\pi(V, J)$. Now by Lemma 2.5.6, there is an entire rational function $\mu\colon V \to V''$ so that $\pi\mu$ is the identity. Taking the standard model of the blowup, this means that if J is generated by $f_1, \dots, f_n \in \Gamma^r(V)$ then there is an entire rational function $\theta\colon V \to G(n,1)$ so that $\mu(x) = (x, \theta(x))$. In particular, for $x \in V - W$, $\theta(x)$ is the matrix of projection to the line through $f(x)$ where $f(x) = (f_1(x), \dots, f_n(x))$. Now θ induces a line bundle from the canonical bundle $E(n,1)$ over $G(n,1)$, namely $\rho\colon E \to V$ where $E = \{\, (x, y) \in V \times \mathbf{R}^n \mid \theta(x)(y) = y \,\}$ and $\rho(x, y) = x$. This bundle has a section $\sigma\colon V \to E$ given by $\sigma(x) = (x, f(x))$. Notice that W is the set of zeroes of σ.

Assertion 2.8.2.1 *There is a section $\eta\colon V \to E$ and a compact N so that η is transverse to the zero section $V \times 0$ and $\eta^{-1}(V \times 0) = M$ and $\sigma(x)/\eta(x) > 0$ for all $x \in V - N$.*

Proof: Since M is homologous to W, there is a compact set $N \subset V$ so that $M \cup W$ is the frontier of N. (For example, to get N you can after ϵ-isotoping M, triangulate V so that M and W are subpolyhedra. Then let N be W union the simplices in the homology from M to W.) Now pick a smooth function $f\colon V - W \to \mathbf{R}$ so that $f^{-1}((-\infty, 0]) = N - W$, $f^{-1}(0) = M - W$ and for any point $x \in V$ there is a neighborhood U of x in V and smooth functions $p\colon U \to \mathbf{R}$ and $q\colon U \to \mathbf{R}$ so that $f(x) = p(x)/q(x)$ for all $x \in U - W$ and so that p is a generator of $\mathcal{I}_U^\infty(M \cap U)$ if $x \in M$ and so q is a generator of $\mathcal{I}_U^r(W \cap U)$ if $x \in W$. We can certainly construct such an f locally, but then we may piece together with a partition of unity to get f globally. Now we just let $\eta(x) = f(x)\sigma(x)$ and it satisfies the required properties.

The reader might feel more comfortable first perturbing σ slightly to a σ' transverse to the zero section so that $W' = {\sigma'}^{-1}(V \times 0)$ is transverse to M.

Then W is homologous to W' and the above argument can be done with W' replacing W.   □

Pick η and N as in Assertion 2.8.2.1. The section η is given by $\eta(x) = (x, g(x))$ for some smooth $g: V \to \mathbf{R}^n$. By Lemma 2.1.6 we may pick a proper polynomial $r: V \to \mathbf{R}$ so that $L = r^{-1}(0)$. Pick a large compact neighborhood N' of N in V. We want N' to be so large that if b is the maximum of $r(x)^2$ for $x \in N$, then $4b$ is the minimum of $r(x)^2$ on $V - N'$. Let $N'' = \{ x \in V \mid r(x)^2 \le 2b \}$, then $N \subset N'' \subset N'$.

By Lemma 2.8.1 we may approximate g on N' by a polynomial $p: V \to \mathbf{R}^n$ so that $p|_L = 0$. Then for $x \in N'$, $\theta(x)(p(x)) \approx \theta(x)(g(x)) = g(x)$ so the section $x \mapsto (x, \theta(x)(p(x)))$ is rational and approximates η on N', although it may be wildly different outside N'.

But for large enough m, the map

$$\lambda(x) = \big(x,\, \theta(x)\,(p(x)) + (r(x)^2/(3b))^m \sigma(x)\big)$$

is a section which is close to η near N'' and has no zeroes on $V - N''$. By transversality, the zeroes X of λ are a manifold isotopic to M. By Lemma 2.2.13, X is nonsingular. Also $L \subset X$ so X is ϵ-isotopic to M fixing L.   □

There is also a one dimensional version of this theorem [**AK10**], it says that a smooth curve C in a nonsingular Zopen set V is ϵ-isotopic to a nonsingular Zclosed subset if and only if the homology class $[C]$ lies in the subgroup $AH_1(V)$ of $H_1(V; \mathbf{Z}/2\mathbf{Z})$, generated by nonsingular algebraic curves in V; (W. Kucharz [**Ku**] had independently proved this in the special case when V is orientable).

Before proving the next theorem, let us review a few facts about smooth submanifolds of $\mathbf{R}^n$. Let M be a smooth m-dimensional submanifold of $\mathbf{R}^n$ and let $\beta: M \to G(n, n - m)$ be the canonical map classifying its normal bundle, so $\beta(x)$ is the matrix of orthogonal projection to the subspace of vectors perpendicular to the tangent space of M at x. We may write the normal bundle of M as $E = \{ (x, y) \in M \times \mathbf{R}^n \mid \beta(x)y = y \}$. We have a map $\theta: E \to \mathbf{R}^n$ given by $\theta(x, y) = x + y$. Then the Jacobian of θ is nonsingular at all points of the zero section $M \times 0$ of E, hence by the inverse function theorem and a little elementary topology there is a neighborhood E' of $M \times 0$ in E so that θ maps E' diffeomorphically onto a neighborhood U of M in $\mathbf{R}^n$. (This is just the existence theorem for tubular neighborhoods.) Note that we have a retraction $\rho: U \to M$ given by $\rho(z) = x$ if $z = \theta(x, y)$ for some $(x, y) \in E'$. Actually, if U is chosen small enough, this is just the closest point map.

The following Theorem 2.8.3 is our workhorse theorem for making smooth objects algebraic. Roughly it says that after wiggling U a bit in $V \times \mathbf{R}^k$, we may approximate a smooth function $f: V \to W$ by an entire rational function. Furthermore, this entire rational function equals f on nice sets where f was already entire rational. For the sake of clarity we precede it with an easy version:

Theorem 2.8.3 ([AK1] Easier Version) *Let L, V and W be real algebraic sets with L nice and W nonsingular. Let $U \subset \text{Nonsing}\, V$ be an open neighborhood of L which has compact closure. Let $f\colon V \to W$ be a smooth function so that $f|_L = u$ for some entire rational function $u\colon L \to W$. Then there exists an algebraic set $Z \subset V \times \mathbf{R}^k$ for some k, an open set $Z_0 \subset \text{Nonsing}\, Z$ and an entire rational function $p\colon Z \to W$ such that*

   a) *$L \times 0 \subset Z_0$ and $p|_{L \times 0} = u$.*
   b) *The projection $\pi\colon V \times \mathbf{R}^k \to V$ induces a diffeomorphism from Z_0 to U.*
   c) *$p|_{Z_0}$ approximates $f \circ \pi|_{Z_0}$.*

Proof: Suppose $V \subset \mathbf{R}^n$ and $W \subset \mathbf{R}^m$, then by Proposition 2.4.3 we can find an entire rational function $\delta\colon W \to G(m, m - \dim W) \subset \mathbf{R}^{m^2}$ such that $\delta(x)$ is the normal plane to W at x. Since L is nice, $f|_U$ can be approximated by an entire rational function $g\colon U \to \mathbf{R}^m$ with $g|_L = u$. Define

$$Z = \{\, (x,y) \in \mathbf{R}^n \times \mathbf{R}^m \mid g(x) + y \in W \ \text{ and }\ \delta(g(x) + y)y = y \,\}$$

and define $p\colon Z \to \mathbf{R}^m$ by $p(x,y) = g(x) + y$.

Clearly Z is an algebraic set. Since for arbitrarily small $\epsilon > 0$ we have $|f(x) - g(x)| < \epsilon$ for all $x \in U$, and $f(x) \in W$, there is a unique closest point $w(x)$ on W to $g(x)$. Let $y(x) = w(x) - g(x)$ be the vector from $g(x)$ to $w(x)$. Hence $y(x)$ is perpendicular to W at $w(x) = g(x) + y(x)$, so $y(x)$ is the unique 'small' solution of the equations

$$\left\{ \begin{array}{l} g(x) + y \in W \\ \delta(g(x) + y)y = y \end{array} \right\} = \left\{ \begin{array}{l} g(x) + y \in W \\ y \text{ is } \perp \text{ to } W \text{ at } g(x) + y \end{array} \right\}$$

Let $Z_0 = \{\, (x,y) \in Z \mid x \in U,\ |y| < \epsilon \,\}$, hence if $(x,y) \in Z_0$ then $y = y(x)$, and in particular $p(x,y) = g(x) + y(x) = w(x)$. $\qquad\square$

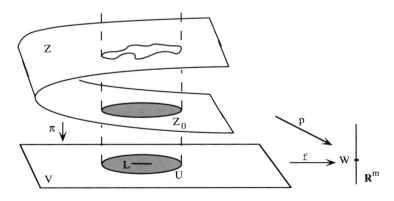

FIGURE II.8.1. Making a map algebraic

We now state and prove the more general version of Theorem 2.8.3. Normally the following general case is applied when $L = P \subset T$ and L is nice.

Theorem 2.8.3 *Let L, V and W be real Zopen sets with L a Zclosed subset of V and suppose $P \subset V$ is such that (P, L) is an approximable pair. Let $T \subset \text{Nonsing}\, V$ be compact and let $f \colon U \to \text{Nonsing}\, W$ be a smooth function from an open neighborhood U of T in $\text{Nonsing}\, V$ so that $f|_{P \cap U} = u$ for some entire rational function $u \colon P \cap U \to \mathbf{R}$.*

Then there exists a Zclosed subset $Z \subset V \times \mathbf{R}^k$ for some k, an open set $Z_0 \subset \text{Nonsing}\, Z$, a neighborhood U' of T in U and an entire rational function $p \colon Z \to W$ such that

  a) *$(L \cap U') \times 0 \subset Z_0$ and $p(x, 0) = u(x)$ for all $x \in L \cap U'$.*
  b) *The projection $\pi \colon V \times \mathbf{R}^k \to V$ induces a diffeomorphism from Z_0 to U', and in fact Z_0 is ϵ-isotopic to $U' \times 0$ fixing $(L \cap U') \times 0$.*
  c) *$p|_{Z_0}$ approximates $f \circ \pi|_{Z_0}$.*

Proof: Suppose $V \subset \mathbf{R}^n$ and $W \subset \mathbf{R}^m$.

Assertion 2.8.3.1 *There is an entire rational function $g \colon V' \to \mathbf{R}^m$ from a Zopen set $V' \subset V$ and a neighborhood U' of T in U so that $U' \subset V'$, $g|_{U'}$ approximates $f|_{U'}$ and $f|_{L \cap U'} = g|$.*

Proof: By Proposition 2.1.1, we may assume u is given as the quotient of two polynomials $u = s/t$ so that t is nowhere 0 on $P \cap U$. Pick a polynomial p so that $V \cap p^{-1}(0) = L$. Let $u' = st/(t^2 + \epsilon p^2)$ for some small $\epsilon > 0$. Note that u' is defined on all of U, in fact it is defined on $V' = V - t^{-1}(0) \cap L$. There is a smooth $f' \colon V \to \mathbf{R}^m$ approximating f on a neighborhood U' of T so that $f'|_{P \cap U} = u'|$. By changing f' outside a neighborhood of T we may as well assume that $f' - u'$ has compact support and $f'|_P = u'|_P$. By Lemma 2.8.1 we may choose an entire rational function $r \colon (V', L) \to (\mathbf{R}^m, 0)$ approximating $f' - u'$ on V'. Then $g = r + u'$ is an entire rational function approximating f on U' and it equals f on $L \cap U$. □

Let $W' = \text{Nonsing}\, W$. Then by Proposition 2.4.3 we can find an entire rational function $\beta \colon W' \to G(m, \dim W) \subset \mathbf{R}^{m^2}$ such that $\beta(x)$ is the tangent plane to W' at x. Let $E = \{ (w, y) \in W' \times \mathbf{R}^m \mid \beta(w)y = 0 \}$ be the normal bundle of W'. There is a neighborhood E' of $W' \times 0$ in E so that the map $\theta(w, y) = w + y$ maps E' diffeomorphically onto a neighborhood O of W'. Let $\rho \colon O \to W'$ be the retraction so that $\rho(z) = w$ if $z = \theta(w, y)$ for some $(w, y) \in E'$. If we choose g close enough to f then we may assume $g(U') \subset O$. Define

$$
\begin{aligned}
B' &= \{ (v, w, y) \in V' \times E \mid y = g(v) - w \} \\
   &= \{ (v, w, y) \in V' \times W' \times \mathbf{R}^m \mid g(v) - w = y \perp W' \text{ at } w \} \\
Z' &= \{ (v, y) \in V' \times \mathbf{R}^m \mid (v, g(v) - y, y) \in B' \} \\
Z &= \text{Cl}_{\mathbf{R}}(Z') \cap V \times \mathbf{R}^m
\end{aligned}
$$

and set $p(v, y) = g(v) - y$.

Now if $(v, y) \in Z - Z'$ then either $v \in V - V'$ or $g(v) - y \in \text{Cl}_{\mathbf{R}}(W) - W'$. In particular, there are no such (v, y)'s for $v \in U'$ and y near 0. Let $B_0 = B' \cap U' \times E'$. We have a map $\varphi \colon U' \to Z$ given by $\varphi(v) = (v, g(v) - \rho g(v))$. Since g approximates f, we know $g(v) - \rho g(v) \sim 0$. So we only need to show that $Z_0 = \varphi(U')$ is an open subset of Nonsing Z.

We have a birational isomorphism $h \colon B' \to Z'$ where $h(v, w, y) = (v, y)$. For $v \in U'$, we know that $h^{-1}\varphi(v) = (v, \rho g(v), g(v) - \rho g(v))$. Since $g(v) \in O$, we know $h^{-1}\varphi(v) \in B_0$.

But then since $h^{-1}\varphi$ is an embedding, it suffices to show that all points of B_0 are nonsingular of dimension equal to $\dim V$. But this follows from Lemma 2.2.14 since B_0 is the pullback of the maps $(w, y) \mapsto w + y$ and $g|_{U'}$ which are transverse since the first map is an embedding onto O. $\square$

We can improve Theorem 2.8.3 provided the bordism class of f is algebraic. For example, Proposition 2.7.1 says this will be true if W is nonsingular and has totally algebraic homology. Again for clarity we precede it with an easy version:

Theorem 2.8.4 ([AK1] Easier Version) *Let $L \subset \mathbf{R}^n$ and W be algebraic sets such that L is nice and W is nonsingular. Suppose we have a closed smooth manifold $M \subset \mathbf{R}^n$ and a smooth map $f \colon M \to W$ so that $L \subset M$, $f|_L = u$ is an entire rational function and the germ of M at L is a germ of a nonsingular algebraic set, and the bordism class (M, f) in $\mathfrak{N}_*(W)$ is algebraic. Then there is a nonsingular algebraic set $V \subset \mathbf{R}^n \times \mathbf{R}^m$ and a diffeomorphism $g \colon V \to M$ and an entire rational function $p \colon V \to W$ such that*

a) $L \times 0 \subset V$.

b) $g|_{L \times 0} =$ *the identity.*

c) $p|_{L \times 0} = f \circ g|_{L \times 0}$ *and p approximates $f \circ g$.*

Proof: To say (M, f) is algebraic means there is a bordism $F_0 \colon X_0 \to W$ from f to an entire rational function $r \colon Y \to W$ where Y is a nonsingular algebraic set. We double this cobordism and get a closed smooth manifold X and a smooth map $F \colon X \to W$ with $M \cup Y \subset X$, M homologous to Y, $F|_M = f$ and $F|_Y = r$, see Figure II.8.2. We can assume $X \subset \mathbf{R}^n \times \mathbf{R}^k$ so that L corresponds to $L \times 0$ and Y corresponds to $0 \times Y$. Let $c = n + k - \dim X$ and let $E = E(n + k, c) \to G = G(n + k, c)$ be the universal bundle over the Grassmannian. Let U be an open tubular neighborhood of X, and $\varphi \colon U \to E$ be the normal bundle map. φ is transverse to G with $\varphi^{-1}(G) = X$, and by Proposition 2.4.3 we can assume that φ is entire rational on $L \times 0 \cup 0 \times Y$. Extend F to U and apply Proposition 2.8.3 to $F \times \varphi \colon U \to W \times E$. We get an algebraic set $Z \subset \mathbf{R}^n \times \mathbf{R}^k \times \mathbf{R}^b$ for some b, and an open set $Z_0 \subset \text{Nonsing } Z$ and an entire rational function $q \colon Z \to W \times E$ such that

1) $L \times 0 \times 0 \cup 0 \times Y \times 0 \subset Z_0$

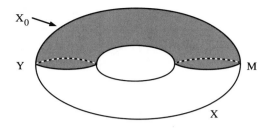

FIGURE II.8.2. The doubled cobordism

2) $q = F \times \varphi$ on $L \times 0 \times 0 \cup 0 \times Y \times 0$
3) $\pi|: Z_0 \to U$ is a diffeomorphism
4) $q|_{Z_0}$ approximates $(F \times \varphi) \circ \pi|_{Z_0}$

By 3), $Z_0 = \{ (x, \theta(x)) \mid x \in U \}$ for some smooth function $\theta(x)$. Let $\tilde{q}(x) = q(x, \theta(x))$. Then $\tilde{q}$ is close to $F \times \varphi$, hence by transversality $\tilde{q}^{-1}(W \times G)$ is isotopic to $(F \times \varphi)^{-1}(W \times G) = X$. Since $q = \tilde{q} \circ \pi$ on Z_0, we have

$$q^{-1}(W \times G) \cap Z_0 = (\pi|_{Z_0})^{-1}(\tilde{q}^{-1}(W \times G)) \approx X.$$

Therefore $q^{-1}(W \times G)$ has a nonsingular component X' which is isotopic to X, and $L \times 0 \times 0 \cup 0 \times Y \times 0 \subset X'$.

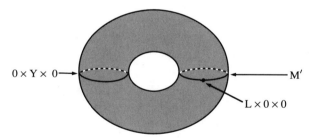

FIGURE II.8.3. The doubled cobordism made algebraic

Under this isotopy, M corresponds to a codimension one smooth submanifold M' of the algebraic set $q^{-1}(W \times G)$ which is homologous to $0 \times Y \times 0$. Hence by Theorem 2.8.2 it is ϵ-isotopic to a nonsingular algebraic subset V fixing $L \times 0 \times 0$. In particular $V \subset X'$. We now let p be $q|_V$ followed by the projection $W \times G \to W$. $\square$

We now give the more complete version of Theorem 2.8.4.

Theorem 2.8.4 *Let* $L \subset S$ *and* W *be real Zopen sets such that* L *is a Zclosed subset of* S, *and* W *is nonsingular. Suppose we have a closed smooth manifold* $M \subset \operatorname{Nonsing} S$, *a set* $P \subset M$ *and a smooth map* $f\colon M \to W$ *so that* (P, L) *is an approximable pair, the bordism class of* $f \times$ *inclusion in* $\mathfrak{N}_*(W \times \operatorname{Nonsing} S)$

is algebraic and $(f, \alpha)|_P$ *is an entire rational function where* $\alpha \colon M \to G(n, k)$ *is the map classifying the normal bundle of* M *in* S. *(i.e., if* $S \subset \mathbf{R}^n$ *then* $\alpha(x)$ *is the space of vectors tangent to* S *and perpendicular to* M *at* x.)

Then there is a nonsingular Zclosed subset V of $S \times \mathbf{R}^m$, a diffeomorphism $g \colon V \to M$ and an entire rational function $p \colon V \to W$ such that

    a) $L \times 0 \subset V$.

    b) $g(x, 0) = x$ and $p(x, 0) = f(x)$ for all $x \in L$.

    c) p approximates $f \circ g$.

Proof: Pick a bordism $F_0 \colon X_0 \to W$ from $f \times$ inclusion to an entire rational function $r \colon Y \to W \times \text{Nonsing}\, S$ where Y is a compact nonsingular algebraic set. We double this cobordism and get a closed smooth manifold X and a smooth map $F_1 \colon X \to W \times \text{Nonsing}\, S$ with $M \cup Y \subset X$, M is homologous to Y in X, $F_1|_M = f \times$ inclusion and $F_1|_Y = r$. By replacing Y with the graph of r we may as well assume $Y \subset \text{Nonsing}\, S \times \mathbf{R}^k$ for some k. Let $F \colon X \to W$ be the composition of F' and the projection $W \times \text{Nonsing}\, S \to W$.

We can assume, after perhaps enlarging k, that $X \subset S \times \mathbf{R}^k \times \mathbf{R}$ so that M corresponds to $M \times 0 \times 0$ and Y corresponds to $Y \times 0$ and a neighborhood of $M \cup Y$ in X is $M \times 0 \times [-1, 1] \cup Y \times [-1, 1]$. Suppose $S \subset \mathbf{R}^n$. Let $c = \dim S + 1 + k - \dim X$ and let $E = E(n+k+1, c) \to G = G(n+k+1, c)$ be the universal bundle over the Grassmannian. Let U be an open tubular neighborhood of X and let $\varphi \colon U \to E$ be the map classifying the normal bundle of X in $S \times \mathbf{R}^k \times \mathbf{R}$. In other words $\varphi(x) = (L, y)$ where, if z is the closest point to x on X, then L is the space of vectors in the tangent space to $S \times \mathbf{R}^k \times \mathbf{R}$ at z which are perpendicular to the tangent space to X at z; and y is the projection of $x - z$ to L.

Note φ is transverse to G with $\varphi^{-1}(G) = X$. Also using Corollary 2.4.4 we see that (F, φ) is the restriction of an entire rational function on $P \times 0 \times 0 \cup Y \times 0$. Extend F to U and apply Proposition 2.8.3 to $F \times \varphi \colon U \to W \times E$. We get an algebraic set $Z \subset S \times \mathbf{R}^k \times \mathbf{R} \times \mathbf{R}^b$ for some b, and an open set $Z_0 \subset \text{Nonsing}\, Z$ and an entire rational function $q \colon Z \to W \times E$ such that

    1) $L \times 0 \times 0 \times 0 \cup Y \times 0 \times 0 \subset Z_0$.

    2) $q = F \times \varphi$ on $L \times 0 \times 0 \times 0 \cup Y \times 0 \times 0$.

    3) The projection $\pi \colon S \times \mathbf{R}^k \times \mathbf{R} \times \mathbf{R}^b \to S \times \mathbf{R}^k \times \mathbf{R}$ induces a diffeomorphism $\pi| \colon Z_0 \to U$ and in fact, $Z_0 = \{ (x, \theta(x)) \mid x \in U \}$ for some small smooth function $\theta(x)$.

    4) $q|_{Z_0}$ approximates $(F \times \varphi) \circ \pi|_{Z_0}$

Let $\tilde{q}(x) = q(x, \theta(x))$. Now $\tilde{q}$ is close to $F \times \varphi$, hence by transversality $X'' = \tilde{q}^{-1}(W \times G)$ is isotopic to $(F \times \varphi)^{-1}(W \times G) = X$. Note that X is transverse to $S \times \mathbf{R}^k \times 0$ and the intersection is $M \times 0 \times 0 \cup Y \times 0$. Hence X'' is transverse to $S \times \mathbf{R}^k \times 0$ also and their intersection is $M'' \cup Y \times 0$ for some submanifold $M'' \subset X''$ and there is a small isotopy taking X to X'' which carries $M \times 0 \times 0$ to M'' and fixes $L \times 0 \times 0 \cup Y \times 0$.

Since $q = \tilde{q} \circ \pi$ on Z_0,

$$q^{-1}(W \times G) \cap Z_0 = (\pi|_{Z_0})^{-1}(\tilde{q}^{-1}(W \times G)) = (\pi|_{Z_0})^{-1}(X'')$$

Therefore $q^{-1}(W \times G)$ has a nonsingular union of connected components X' which is isotopic to X, and $L \times 0 \times 0 \times 0 \cup Y \times 0 \times 0 \subset X'$. Also the Zopen set $X \cap S \times \mathbf{R}^k \times 0 \times \mathbf{R}^b \cap Z$ has a nonsingular union of connected components M' where $M' = (\pi|_{Z_0})^{-1}(M'')$.

The codimension one smooth submanifold M' of the algebraic set $q^{-1}(W \times G)$ is homologous to $Y \times 0 \times 0$. Also $(M', L \times 0 \times 0 \times 0)$ is an approximable pair since M' is a component of $\mathrm{Nonsing}(\mathrm{Cl}_{\mathbf{R}}(M'))$ which is nice. So by Theorem 2.8.2, M' is ϵ-isotopic to a nonsingular Zclosed subset V of $q^{-1}(W \times G)$, via an isotopy fixing $L \times 0 \times 0 \times 0$. In particular $V \subset X'$. We now let p be $q|_V$ composed with the projection $W \times G \to W$. $\square$

Theorem 2.8.4 in fact implies the following (by taking W to be a point and $L = P = \emptyset$).

Corollary 2.8.5 *A closed smooth submanifold M of a nonsingular real algebraic set S can be ϵ-isotoped to a nonsingular algebraic subset of $S \times \mathbf{R}^k$ for some k provided the bordism class of the inclusion $M \hookrightarrow S$ is algebraic.*

The bordism condition is always satisfied if W has totally algebraic homology, and in general it can not be avoided (c.f., Theorem 2.7.3). In Corollary 2.8.5 by letting $S = \mathbf{R}^n$ we obtain a theorem of Tognoli which is an improvement of a previous result of Nash [**N**].

Corollary 2.8.6 ([To1]) *Every closed smooth manifold is diffeomorphic to a nonsingular real algebraic set.*

The next result gives a way of converting the image of an entire rational embedding into subvariety. The image of an algebraic set under a rational embedding does not have to be an algebraic set. The most we can say about it is that its Zariski closure may only contain lower dimensional extra pieces (Corollary 2.3.3). Proposition 2.8.7 says that the image of a rational embedding can be made an algebraic subvariety if we are willing to relax the ambient algebraic set by a rational diffeomorphism.

Proposition 2.8.7 ([AK8]) *Let V, W be nonsingular real Zopen sets with V compact, and let $p\colon V \to W$ be an entire rational function which is an embedding. Then there are nonsingular algebraic sets $V' \subset W'$ and a rational diffeomorphism $r\colon W' \to W$ and a birational isomorphism $q\colon V \to V'$ such that $r \circ q = p$.*

Proof: We may as well assume V and W are algebraic sets, since they are birationally isomorphic to real algebraic sets. Suppose $V \subset \mathbf{R}^n$, then pick a

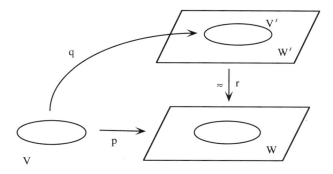

FIGURE II.8.4. Making an algebraic embedding into an algebraic subset

smooth function with compact support $\alpha \colon W \to \mathbf{R}^n$ so that $\alpha(p(x)) = x$ for all $x \in V$. Let $V' = \{\, (x, y) \in V \times W \mid y = p(x) \,\}$ and let $q \colon V \to V'$ be $q(x) = (x, p(x))$. Let $W'' = \{\, (\alpha(y), y) \in \mathbf{R}^n \times W \,\}$. Then W'' is a diffeomorphic copy of W containing the nonsingular algebraic set V'.

Assertion 2.8.7.1 *There is a smooth function $f \colon \mathbf{R}^n \times W \to \mathbf{R}^n$ transverse to $0 \in \mathbf{R}^n$ so that $f^{-1}(0) = W''$, and f restricts to an entire rational function $u \in \mathfrak{I}^r_{\mathbf{R}^n \times W}(V')$ outside of a compact set.*

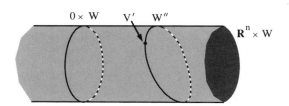

FIGURE II.8.5. A smooth approximation

Proof: Pick a proper polynomial $\rho \colon \mathbf{R}^n \times W \to \mathbf{R}$ with $\rho^{-1}(0) = V'$, and pick a smooth function $\beta \colon \mathbf{R}^n \times W \to [0, 1]$ with compact support such that $\beta(x, y) = 1$ when either $|x| < |\alpha(y)|$ or when (x, y) is in a neighborhood of $W'' - 0 \times W$. Then we let

$$f(x, y) = \beta(x, y)(x - \alpha(y)) + (1 - \beta(x, y))\rho^2(x, y)x.$$

It is clear that $W'' \subset f^{-1}(0)$. If $(x, y) \in f^{-1}(0) - W''$ then $x \neq \alpha(y)$. But then if $\beta(x, y) = 1$ we would have $0 = f(x, y) = x - \alpha(y) \neq 0$, hence $\beta(x, y) \neq 1$. We claim that $\beta(x, y) \neq 0$ also, otherwise $\rho^2(x, y)x = 0$ hence $x = 0$ so $|\alpha(y)| > 0 = |x|$ which implies $\beta(x, y) = 1$ contradiction. Hence $0 < \beta(x, y) < 1$, and by solving $f(x, y) = 0$ for $\alpha(y)$ we get $\alpha(y) = x(1 + \rho^2/\beta - \rho^2)$. Therefore $|\alpha(y)| > |x|$ implying $\beta(x, y) = 1$, a contradiction.

So we have shown that $W'' = f^{-1}(0)$. We leave the verification $f \pitchfork 0$ as an exercise. Now f equals $\rho^2(x,y)x$ outside of a compact set, so set $u = \rho^2(x,y)x$.

□

By Lemma 2.8.1, $f(x,y) - u(x,y)$ can be approximated by an entire rational function $\delta\colon (\mathbf{R}^n \times W, V') \to (\mathbf{R}^n, 0)$. Then $\gamma(x,y) = \delta(x,y) + u(x,y)$ is an entire rational function approximating $f(x,y)$ on $\mathbf{R}^n \times W$.

Let $W' = \gamma^{-1}(0)$, then $V' \subset W'$ and W' is ϵ-isotopic to W'' by transversality. Hence, $W' = \{ (\eta(y), y) \in \mathbf{R}^n \times W \}$ for some smooth $\eta\colon W \to \mathbf{R}^n$ with $\eta \approx \alpha$. We may now let $r\colon W' \to W$ be the map induced by projection $\mathbf{R}^n \times W \to W$.

□

One can ask when a smooth map f between nonsingular algebraic sets can be approximated by an entire rational function. The following theorem says that if f is homotopic to an entire rational function then f can be approximated by an entire rational function after relaxing the source by a rational diffeomorphism.

Proposition 2.8.8 ([AK8]) *Let V and W be nonsingular real Zopen sets with V compact. Let $f\colon V \to W$ be a smooth map which is homotopic to an entire rational function. Then there exist a nonsingular algebraic set V', an entire rational function $\varphi\colon V' \to W$, and a rational diffeomorphism $h\colon V' \to V$ such that $f \circ h$ is arbitrarily close to φ.*

Proof: By Lemma 2.1.4, V is birationally isomorphic to a real algebraic set, so we may as well assume V is a real algebraic set. Let $G\colon V \times [0,1] \to W$ be a homotopy with $G(x,0) = f(x)$ and $G(x,1) = \beta(x)$ where $\beta(x)$ is an entire rational function. By doubling G we get a smooth function $H\colon V \times S^1 \to W$ such that $H(x,a) = f(x)$ and $H(x,b) = \beta(x)$ for some $a,b \in S^1$. By Proposition 2.8.3 we can find an algebraic set $Z \subset (V \times S^1) \times \mathbf{R}^k$ for some k, an entire rational function $F\colon Z \to W$ and a nonsingular component Z_0 of Z such that the restriction of projection $\pi|_{Z_0}\colon Z_0 \to V \times S^1$ is a diffeomorphism, $H \circ \pi|_{Z_0}$ is close to $F|_{Z_0}$ and so $V \times b \subset Z_0$. Let $M = (\pi|_{Z_0})^{-1}(V \times a) \subset Z_0$. Then M is a diffeomorphic copy of V in Z_0 which is homologous to $V \times b \subset Z_0$. Hence by Theorem 2.8.2, M is ϵ-isotopic to a nonsingular algebraic subset V' of Z. Let $\varphi = F|_{V'}$ and $h = p \circ \pi|_{V'}\colon V' \to V$ where $p\colon V \times S^1 \to V$ is a projection. Then the entire rational function h is a diffeomorphism since $\pi(V')$ is a nearby copy of $V \times a \subset V \times S^1$. Since $f \circ p|_{V \times a} = H|_{V \times a}$ we have

$$f \circ h = f \circ p \circ \pi|_{V'} \sim H \circ \pi|_{V'} \sim F|_{V'} = \varphi$$

where $\sim$ means approximation.   □

The following is a relative version of Theorem 2.8.4 and Corollary 2.8.5. We delay its proof until after Lemma 2.8.12.

Theorem 2.8.9 ([AK3]) *Let M be a closed smooth submanifold of a nonsingular real Zopen set V with totally algebraic homology. Let M_i, $i = 1, \ldots, k$ be close smooth submanifolds of M in general position. Then there is an ϵ-isotopy of $V \times \mathbf{R}^n$ for some n, which simultaneously takes $M \times 0$ and all $M_i \times 0$ to nonsingular algebraic subsets.*

Corollary 2.8.10 *Let M be a closed smooth manifold, and M_i, $i = 1, \ldots, k$ be closed smooth submanifolds of M in general position. Then there exists a nonsingular algebraic set N and nonsingular algebraic subsets N_i, $i = 1, \ldots, k$ of N, and a diffeomorphism $\varphi \colon M \to N$ with $\varphi(M_i) = N_i$, $i = 1, \ldots, k$.*

Corollary 2.8.11 *If a closed smooth manifold is full, then it is diffeomorphic to a nonsingular algebraic set which has totally algebraic homology.*

(We define full in the next section, it just means that the homology is generated by embedded subsets.)

We first prove Theorem 2.8.9 up to cobordism, then deduce the proof from this weaker version.

Lemma 2.8.12 *Let M be a closed smooth manifold, and let M_i, $i = 1, \ldots, k$ be closed smooth submanifolds of M in general position. Let V be a nonsingular real Zopen set with totally algebraic homology, and $\varphi \colon M \to V$ be a map. Then there is a compact smooth manifold T with proper smooth submanifolds T_i, $i = 1, \ldots, k$ in general position and a map $\psi \colon T \to V$ such that*

a) $\partial T = U \cup M$ *where U is a disjoint union of nonsingular Zclosed subsets U^α of $V \times \mathbf{R}^n$ and α runs over all subsets of $\{1, \ldots, k\}$ such that $\bigcap_{i \in \alpha} M_i \neq \emptyset$.*

b) $\partial T \cap T_i = \partial T_i$, $M \cap T_i = M_i$ *and* $U^\alpha \cap T_i = U_i^\alpha$ *where U_i^α, $i = 1, \ldots, k$ are nonsingular algebraic subsets of U^α transverse to each other with $U_i^\alpha = \emptyset$ for $i \notin \alpha$.*

c) $\psi|_M = \varphi$, *and $\psi|_{U_\alpha}$ is induced by projection $V \times \mathbf{R}^n \to V$.*

d) *For each such α and $i \in \alpha$ there is an entire rational function $\lambda_i^\alpha \colon U_i^\alpha \to G(m, c_i)$ such that*

$$U^\alpha = \{ (x,y) \in U_i^\alpha \times \mathbf{R}^{m+1} \mid (\lambda_i^\alpha(x), y) \in E^*(m, c_i) \}$$

In particular λ_i^α classifies the normal bundle of U_i^α in U^α and U^α is a sphere bundle over U_i^α. Also c_i is the codimension of M_i in M.

Proof: For $\alpha \subset \{1, \ldots, k\}$ we denote $M_\alpha = \bigcap_{i \in \alpha} M_i$ and $M_\emptyset = M$. We prove the lemma by induction on the number of subsets α with the property $M_\alpha \neq \emptyset$. The case where all $M_\alpha = \emptyset$ is vacuous. Now pick $\alpha \subset \{1, \ldots, k\}$ so that $M_\alpha \neq \emptyset$ but $M_\beta = \emptyset$ for all β with $\alpha \subsetneq \beta$. The normal bundle maps $f_i \colon M_i \to G(m, c_i)$ for some large m define a map $f \colon M_\alpha \to \prod_{i \in \alpha} G(m, c_i)$. Let $g \colon M_\alpha \to V \times$

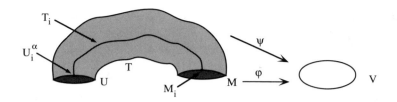

FIGURE II.8.6. The cobordism to the algebraic situation

$\prod_{i\in\alpha} G(m,c_i)$ be $g = (\varphi, f)$. By Lemma 2.7.1 there is a bordism $\gamma\colon K \to$ $V \times \prod_{i\in\alpha} G(m,c_i)$ such that ∂K is disjoint union of M_α and a nonsingular algebraic set $U \subset V \times \prod_{i\in\alpha} G(m,c_i) \times \mathbf{R}^a$ for some a with $\gamma|_{M_\alpha} = g$ and $\gamma|_U$ is the projection. Let $\mu^\alpha\colon K \to \prod_{i\in\alpha} G(m,c_i)$ be γ followed by projection, and define

$$S^\alpha = \{ (x,y) \in K \times (\mathbf{R}^{m+1})^{k'} \mid (\mu^\alpha(x),y) \in \prod_{i\in\alpha} E^*(m,c_i) \}$$

$$S_i^\alpha = \{ (x,y) \in S^\alpha \mid y_i = 0 \}$$

where k' is the number of elements of α. So S^α is the pullback

$$
\begin{array}{ccc}
S^\alpha & \longrightarrow & \prod_{i\in\alpha} E^*(m,c_i) \\
\downarrow & & \downarrow \\
K & \xrightarrow{\mu^\alpha} & \prod_{i\in\alpha} G(m,c_i)
\end{array}
$$

Notice S_i^α are in general position and $\bigcap_{i\in\alpha} S_i^\alpha = K$. Also note that if $\lambda_i^\alpha\colon S_i^\alpha \to$ $G(m,c_i)$ is the projection $S_i^\alpha \to K$ followed by the i-th coordinate map of μ^α, then S^α is the pull back of $E^*(m,c_i)$ by λ_i^α, i.e.,

$$S^\alpha = \{ (x,y) \in S_i^\alpha \times \mathbf{R}^{m+1} \mid (\lambda_i^\alpha(x),y) \in E^*(m,c_i) \}$$

S^α is a cobordism from a nonsingular algebraic set U^α to a smooth manifold $M^\alpha \supset M_\alpha$ such that $\partial S_i^\alpha \subset \partial S^\alpha = U^\alpha \sqcup M^\alpha$. Furthermore $U_i^\alpha = U^\alpha \cap \partial S_i^\alpha$ are nonsingular algebraic sets, with $\lambda_i^\alpha|_{U_i^\alpha}$ entire rational.

Since $\mu^\alpha|_{M_\alpha} = f$ there is a diffeomorphism $h\colon A \to B$ from a closed tubular neighborhood A of M_α in M^α to a closed tubular neighborhood B of M_α in M, such that $h(A \cap S_i^\alpha) = B \cap M_i$. In other words near M_α, (M, M_i) and $(M^\alpha, \partial S_i^\alpha \cap M^\alpha)$ are diffeomorphic.

Let $S = S^\alpha \cup M \times [0,1]$, where we identify $A \subset \partial S^\alpha$ with $B \times 1 \subset M \times [0,1]$ via h (corners smoothed), and let $S_i = S_i^\alpha \cup M_i \times [0,1] \subset S$.

Note that the S_i are in general position in S. Let $\theta\colon S \to V$ be the map obtained by first retracting S to $S^\alpha \cup M_\alpha \times [0,1] \cup M \times 0$, then mapping S^α by projection to K, mapping by γ to $V \times \prod_{i\in\alpha} G(m,c_i)$ and by projection to V, and mapping $M_\alpha \times [0,1] \cup M \times 0$ by projection to M and then by φ to V.

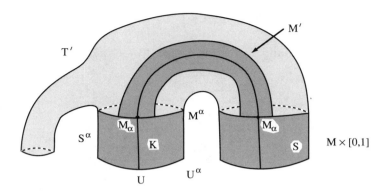

FIGURE II.8.7. The inductive step, reducing the number of nonempty M_α's

Notice $\theta|_{M \times 0} = \varphi$, $U^\alpha \subset U \times \mathbf{R}^{(m+1)k} \subset V \times \mathbf{R}^n$ for some large n, and $\theta|_{U^\alpha}$ is the projection. Define

$$M' = (M^\alpha - \text{Int}(A)) \bigcup_h (M - \text{Int}(B))$$

$$M'_i = M' \cap S_i$$

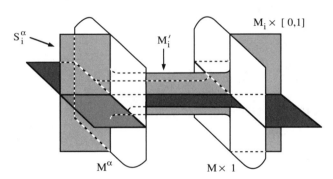

FIGURE II.8.8. Detail of the added handle

Notice $M'_\alpha = \emptyset$, and $M'_\beta = \emptyset$ when $M_\beta = \emptyset$. Let $\varphi' = \theta|_{M'}$, then by induction there are (T', T'_i) and a map $\psi': T' \to V$ satisfying the conclusions of the lemma. Let $T = S \cup T'$, $T_i = S_i \cup T'_i$, and $\psi = \psi' \cup \theta$. We are done.   □

We now give the delayed proof of Theorem 2.8.9.

Proof: (of Theorem 2.8.9) Let $\varphi \colon M \hookrightarrow V$ be the inclusion, and let $\psi \colon T \to V$ be the bordism obtained from Lemma 2.8.12, where $\psi|_M = \varphi$ and $\psi|_U$ is induced by projection $U \subset V \times \mathbf{R}^n \to V$. In the notation of the lemma, $U = \bigsqcup_\alpha U^\alpha$, for simplicity call $U_i = \bigsqcup_{i \in \alpha} U^\alpha_i$. By choosing n large enough ($n > 2 \dim T - \dim V$ suffices) we may pick a smooth function $\kappa \colon (T, U) \to (\mathbf{R}^n, 0)$ so that $(\psi, \kappa) \colon T \to$

$V \times \mathbf{R}^n$ is an embedding and $(\psi, \kappa)|_U$ is inclusion $U \subset V \times \mathbf{R}^n$. Let ψ' denote (ψ, κ).

By doubling T (and smoothing corners) we obtain smooth manifolds

$$
\begin{aligned}
X &= \partial(\psi'(T) \times [-1,1]) \subset (V \times \mathbf{R}^n) \times \mathbf{R} \\
X_i &= \partial(\psi'(T_i) \times [-1,1]) \subset X
\end{aligned}
$$

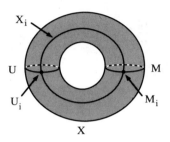

FIGURE II.8.9. X, the double of T

Identify U with $U \times 0$ and M with $M \times 0 \times 0$. The manifolds M, U and X_i are in general position in X. Furthermore M is homologous to U in X. Also notice that the germ of X at U is a germ of the nonsingular algebraic set $U \times \mathbf{R}$ at U. By Theorem 2.8.4 (with $L = P := U$, $M := X$, $S := V \times \mathbf{R}^n \times \mathbf{R}$ and $W :=$ a point) we may assume X is a nonsingular Zclosed subset of $V \times \mathbf{R}^n \times \mathbf{R}$.

Now let $\beta_i \colon X_i \to G(m, c_i)$ be the normal bundle map of X_i in X. By Lemma 2.8.12 we can assume that $\beta_i|_{U_i^\alpha} = \lambda_i^\alpha$ where λ_i^α are the entire rational functions defined by the lemma. Then Theorem 2.8.4 (with $L = P := U$, $M := X_i$, $S := V \times \mathbf{R}^n \times \mathbf{R}$, $W := G(m, c_i)$ and $f := \beta_i$) gives

$$
\begin{array}{ccc}
V \times \mathbf{R}^{n+1} \times \mathbf{R}^a \supset & Y_i' & \\
& \downarrow \rho & \searrow^{\gamma_i} \\
V \times \mathbf{R}^{n+1} \supset & X_i & \xrightarrow{\beta_i} \quad G(m, c_i)
\end{array}
$$

where Y_i' is a nonsingular algebraic set containing U_i and ρ is a diffeomorphism fixing U_i and γ_i is an entire rational function approximating $\beta_i \circ \rho$ with $\gamma_i|_{U_i} = \beta_i|$. We define

$$
Y_i'' = \{ (x, y) \in Y_i' \times \mathbf{R}^{m+1} \mid (\gamma_i(x), y) \in E^*(m, c_i) \}
$$

Since γ_i, and $\beta_i \circ \rho$ are homotopic, the normal bundle of Y_i' in Y_i'' is equivalent to the normal bundle of X_i in X.

Notice that $\{ (x, y) \in Y_i'' \mid x \in U_i \}$ is a union of components of U. Hence the germ of X at $U \cup X_i$ is diffeomorphic (in fact isotopic) to the germ of the nonsingular algebraic set Y_i'' at $U \cup Y_i'$. So X is ϵ-isotopic fixing U to a submanifold $Y_i \subset V \times \mathbf{R}^c$ for some c so that:

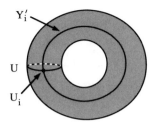

FIGURE II.8.10. Y_i''

1) $U \cup Y_i' \subset Y_i$.
2) The germ of Y_i at $U \cup Y_i'$ is the germ of the nonsingular algebraic set Y_i''.

So by Theorem 2.8.4 (with $L := U \cup Y_i'$, $P = Y_i''$, $M := Y_i$, $S := V \times \mathbf{R}^c$ and $W := $ a point) we may as well assume that Y_i is a nonsingular Zopen set. Note we have a diffeomorphism $\tau_i \colon (X, U \cup X_i) \to (Y_i, U \cup Y_i')$ such that: $\tau_i|_U = $ identity and τ_i is close to the identity.

Finally we apply Theorem 2.8.3 to the map $\tau \colon X \to Y_1 \times Y_2 \times \ldots \times Y_k$, where $\tau = (\tau_1, \ldots, \tau_k)$, and get an algebraic set $Z \subset X \times \mathbf{R}^b$ and an entire rational function $u \colon Z \to Y_1 \times Y_2 \times \ldots \times Y_k$ and a nonsingular component Z_0 of Z, such that if π is the projection $X \times \mathbf{R}^b \to X$ then

    i) $\pi|_{Z_0} \colon Z_0 \to X$ is a diffeomorphism.
    ii) $U \subset Z_0$ and $u|_U = (id, \ldots, id)$.
    iii) u approximates $\tau \circ \pi$ on Z_0.

Since U and $\pi^{-1}(M)$ (a copy of M in Z_0) are homologous, by Theorem 2.8.2 we know that $\pi^{-1}(M)$ is ϵ-isotopic in Z_0 to a nonsingular algebraic set N. If $u_i \colon Z_0 \to Y_i$ is the map u followed by the projection to the i-th factor, then $u_i^{-1}(Y_i')$ is a nearby copy of X_i. Let $N_i = N \cap u_i^{-1}(Y_i')$, then N_i are nonsingular algebraic sets in N and $(N, N_1, \ldots, N_k)$ is an isotopic copy of $(M, M_1, \ldots, M_k)$

$\square$

A useful application of Corollary 2.8.10 is the following theorem which classifies real algebraic sets with isolated singularities.

Proposition 2.8.13 ([AK1]) *A set X is homeomorphic to a real algebraic set V with isolated singularities if and only if X is obtained by taking a smooth compact manifold W with boundary $\partial W = \cup_{i=1}^r \Sigma_i$, where each Σ_i bounds a compact smooth manifold, then crushing some Σ_i's to points and deleting the remaining Σ_i's.*

Proof: Let V be a real algebraic set with isolated singularities. By Lemma 2.6.2 the one point compactification V^* is a real algebraic set with isolated singularities. Then Hironaka's resolution theorem (Theorem 2.5.11) applied to

V^* implies that V must be in the form of X as described in the statement of the theorem.

Conversely assume that we are given such $X = W \cup \bigcup_{i=1}^{r} \mathfrak{c}(\Sigma_i)$, with $\Sigma_i = \partial W_i$ for some compact smooth W_i's. We claim that after modifying the interiors of W_i's we can assume each W_i has a spine L_i consisting of a union of transversally intersecting closed smooth submanifolds, i.e., $W_i / L_i \approx \mathfrak{c}(\Sigma_i)$. Assuming this claim we can construct a smooth manifold $\tilde{X} = W \cup \bigcup_{i=1}^{r} W_i$. By Corollary 2.8.10 we can assume that $\tilde{X}$ is a nonsingular algebraic set so that each L_i is an algebraic subset of $\tilde{X}$. Note that X is obtained from $\tilde{X}$ by crushing some L_i's to points and deleting the remaining L_i's. But by Proposition 2.6.1 and Lemma 2.1.4 we can do these operations algebraically.

FIGURE II.8.11. Characterizing Zopen sets with isolated singularities

So it remains to prove the above claim about the W_i's. Let W be one of these W_i's and $\partial W = \Sigma^m$. We can pick smooth balls D_i, $i = 1, 2, ..., r$ in the interior of W such that:

   i) $\bigcup_i D_i$ is a spine of W.
   ii) The spheres $S_i = \partial D_i$ intersect transversally with each other.
   iii) $\bigcup D_i - \bigcup \partial D_i$ is a union of open balls $\bigcup_{j=1}^{s} B_j$.

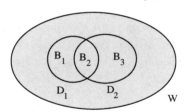

FIGURE II.8.12. Balls D_i making a spine of W

Pick smaller balls $B_j' \subset B_j$. Then $W_0 = W - \bigcup_{j=1}^{s} \text{Int}(B_j')$ is a manifold with spine $\bigcup S_i$, and

$$\partial W_0 = \Sigma \cup \bigcup_{j=1}^{s} \partial B_j'$$

where of course each $\partial B_j'$ is a sphere.

We can reindex the balls B_j' so that there is an arc from Σ to $\partial B_1'$ intersecting exactly one S_i. Furthermore this intersection is transverse and consists of a

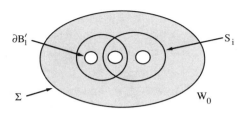

FIGURE II.8.13. $\bigcup S_i$ a spine of W_0

single point. Then we attach a 1-handle to ∂W_0 connecting Σ to $\partial B'_1$ and get $W_1 = W_0 \cup (1 - \text{handle})$ as in Figure II.8.14.

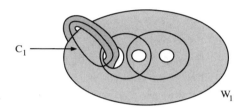

FIGURE II.8.14. Adding a handle to reduce the number of spheres in ∂W_i

Then $\partial W_1 = \Sigma \cup \bigcup_{j=2}^{s} \partial B'_j$ and $\bigcup S_i \cup C_1$ is a spine of W_1, where C_1 is the circle defined by the core of the 1-handle union of the arc. We have improved matters since there are fewer spheres $\partial B'_j$ in ∂W_1. By continuing in this way we get a manifold W_s with $\partial W_s = \Sigma$ and the spine of W_s is transversally intersecting codimension one spheres and circles $\bigcup S_i \cup \bigcup_{j=1}^{s} C_j$. Then W_s is the required manifold. (the first step of the proof of this claim came from a suggestion of L. Jones). $\square$

There is also an ambient version of this Proposition which says that if a closed smooth submanifold Σ of S^n is the boundary of a codimension at least one submanifold of S^n with trivial normal bundle, then it can be isotoped to the link of a real algebraic subset of $\mathbf{R}^{n+1}$ with an isolated singularity at the origin, [**AK18**].

9. Homology of Blowups

In this section we discuss how homology groups change under blowing up operations, and prove a resolution theorem for homology cycles of algebraic sets, which originally appeared in [**AK8**]. All homology and cohomology groups in this section will have $\mathbf{Z}/2\mathbf{Z}$ coefficients.

Definitions: For a smooth manifold M we define $H_*^{imb}(M)$ to be the subgroup of $H_*(M)$ generated by embedded closed smooth submanifolds. That is, every element of $H_i^{imb}(M)$ is a finite sum of classes represented by i-dimensional smooth submanifolds. We say M is *full* if $H_*^{imb}(M) = H_*(M)$.

We also want to define an algebraic analogue of fullness.

Definitions: Let V be a nonsingular Zopen set, we define the following subgroups of $H_*(V)$. Let $RH_*(V)$ be the subgroup generated by $\varphi_*[Z]$, where $\varphi\colon Z \to V$ is an entire rational function from a compact nonsingular algebraic set Z and φ is an embedding. Let $R_0H_*(V)$ be the same as above, except Z is allowed to be a compact nonsingular component of an algebraic set. In other words, $R_0H_*(V)$ is the subgroup generated by $\varphi_*[Z]$, where $\varphi\colon Z \to V$ is an entire rational function from a compact nonsingular component of an algebraic set Z and φ is an embedding. We say V is *algebraically full* if $RH_*(V) = H_*(V)$.

Clearly we have $RH_*(V) \subset H_*^A(V)$. Theorem 2.8.3 implies that $R_0H_*(V) = H_*^{imb}(M)$. Repeated applications of Theorem 2.8.7 along with Lemma 2.7.2 imply that, after changing V by a rational diffeomorphism $\widetilde{V} \xrightarrow{\approx} V$ we can identify $RH_*(\widetilde{V})$ as the subgroup generated by compact nonsingular algebraic subsets of $\widetilde{V}$. Another nice feature of $RH_*(V)$ is that its elements can be isotoped to make them transverse to any given smooth subcomplex of V. This last property follows from Theorem 2.8.8, because first we can isotop a given representative $\varphi\colon Z \hookrightarrow V$ of $RH_*(V)$ to a smooth $\varphi_0\colon Z \hookrightarrow V$ such that φ_0 is transverse to a given subcomplex then by Theorem 2.8.8 we can ϵ-isotop φ_0 to an entire rational function $\varphi_1\colon Z_1 \hookrightarrow V$ from a nonsingular algebraic set Z_1.

Before looking at the homology of blowups we need the following elementary lemmas.

Lemma 2.9.1 *Let $f\colon V \to W$ be a degree one map (in $\mathbf{Z}/2\mathbf{Z}$ coefficients) between closed smooth manifolds of the same dimension, then*

a) $H_*(W) = f_*H_*(V)$
b) $H_*^{imb}(W) \subset f_*H_*^{imb}(V)$
c) *If V and W are nonsingular Zopen sets and f is entire rational, then $RH_*(W) \subset f_*RH_*(V)$.*

Proof: Conclusion a) follows since f has degree one. To see b) we make f transverse to the generators of $H_*^{imb}(W)$. In particular, pick $[A] \in H_*^{imb}(W)$ where $A \subset W$ is a smooth submanifold. After isotoping A, we may as well assume f is transverse to A. Let $B = f^{-1}(A)$. Let α and β be the Poincare duals of $[A]$ and $[B]$ respectively. Clearly $f^*(\alpha) = \beta$, and since f has degree one:

$$f_*[B] = f_*(\beta \cap [V]) = f_*(f^*(\alpha) \cap [V]) = \alpha \cap f_*[V] = \alpha \cap [W] = [A].$$

To see c), pick $a \in RH_*(W)$ represented by a rational embedding $\varphi\colon D \to W$. Homotop f to a map $f_0\colon V \to W$ which is transverse to φ. By Theorem 2.8.8

there is a nonsingular algebraic set V' and a rational diffeomorphism $h\colon V' \overset{\approx}{\to} V$ and an entire rational function $f'\colon V' \to W$ such that $f_0 \circ h$ is close to f', in particular f' is transverse to φ. Now let C be the fiber product

$$C = \{\, (x,y) \in V' \times D \mid f'(x) = \varphi(y) \,\}.$$

Note that C is nonsingular by Lemma 2.2.14. Also projection $V' \times D \to V'$ restricts to an embedding $\pi\colon C \to V'$. Let $b = h_*\pi_*([C]) \in RH_*(V)$. Note that f^* of the Poincare dual of a is the Poincare dual of b. So as above we get $f_*(b) = a$. $\qquad\square$

The assumption that L be closed is not really necessary, but makes things easier and is all we need.

Lemma 2.9.2 *Let $p\colon E \to L$ be a k-dimensional vector bundle over a smooth closed manifold L and let $\pi\colon P(E) \to L$ be its projectivization, so $P(E)$ is the space of lines through the origin in fibres of p. Let $\eta\colon K \to P(E)$ be the canonical line bundle. Then there is a $H^*(L)$-module isomorphism*

$$\varphi\colon H^*(P(E)) \to H^*(L) \otimes H^*(\mathbf{RP}^{k-1})$$

so that if $\omega_1 \in H^1(P(E))$ is the first Stiefel-Whitney class of η, then

    a) *$\varphi\pi^*(\alpha) = \alpha \otimes 1$ for all $\alpha \in H^*(L)$*
    b) *$\varphi(\omega_1) = 1 \otimes \xi$ where ξ generates $H^1(\mathbf{RP}^{n-1})$.*
    c) *If L is full then $P(E)$ is full.*
    d) *Suppose $\pi\colon P(E) \to L$ is algebraic. That is, suppose we have $P(E) = \pi(E,L)^{-1}(L)$ where $\pi(E,L)\colon \mathfrak{B}(E,L) \to E$ is the blowup of nonsingular Zopen set along a nonsingular Zclosed subset L and π is the restriction of the blowing up projection $\pi(E,L)$. Then $P(E)$ is algebraically full if L is algebraically full.*

Proof: The existence of φ satisfying a) and b) is an immediate consequence of Theorem 5.7.9 of [**Sp**] since $\theta\colon H^*(\mathbf{RP}^{k-1}) \to H^*(P(E))$ is a cohomology extension of the fiber where $\theta(\xi^i) = \omega_1^i$.

Now suppose that L is full. Let $D\colon H^*(P(E)) \to H_*(P(E))$ be the Poincare duality isomorphism. To show $P(E)$ is full it suffices to show that

$$D\varphi^{-1}(\alpha \otimes \xi^m) \in H_*^{imb}(P(E))$$

for every m and $\alpha \in H^*(L)$. But we have

$$\varphi^{-1}(\alpha \otimes \xi^m) = \pi^*(\alpha) \cup \omega_1^m.$$

Now $D(\omega_1)$ is represented by an embedded submanifold $S = P(E) \cap P(E)'$ where $P(E)'$ is a section of $\eta\colon K \to P(E)$ transverse to the zero section $P(E)$. Let α be Poincare dual to $\sum[A_i]$ for embedded closed submanifolds $A_i \subset L$. Let

$S_1, \ldots, S_m \subset P(E)$ be isotopic copies of S in general position with themselves and with each $\pi^{-1}(A_i)$. Then if $S' = \cap S_i$ we know that

$$D(\pi^*(\alpha) \cup \omega_1^m) = \sum [\pi^{-1}(A_i) \cap S'] \in H_*^{imb}(P(E))$$

The proof of d) is very similar to the proof of c), the only difference is that we have embeddings of nonsingular Zopen sets as opposed to submanifolds. As pointed out in the introduction they obey transversality. Now the only place to modify the proof is that instead of $A_i \subset L$ we will have entire rational embeddings $\varphi_i \colon A_i \hookrightarrow L$, so in this case we replace $\pi^{-1}(A_i)$ by the fiber product B_i

$$
\begin{array}{ccc}
B_i & \longrightarrow & P(E) \\
\downarrow & & \downarrow \\
A_i & \xrightarrow{\varphi_i} & L
\end{array}
$$

$\square$

Lemma 2.9.3 *Let $\pi \colon B \to M$ be a blowup with center L, i.e., $B = \mathfrak{B}(M, L)$ and $\pi = \pi(M, L)$. Assume M is a closed smooth manifold and $\dim L < \dim M$, then the following hold:*

a) *If L is full then $\ker(\pi_*) \subset H_*^{imb}(B)$*

b) *If π is a blowup of nonsingular Zopen sets, and L is algebraically full, then $\ker(\pi_*) \subset RH_*(B)$*

c) *If M and L are full, then B is full. In the algebraic case if M and L are algebraically full, then B is algebraically full.*

d) *$\ker(\pi_*) \subset j_* H_*(\pi^{-1}(L))$ where $j \colon \pi^{-1}(L) \to B$ is inclusion. In fact $\ker(\pi_*) = j_* \left(\pi|_*^{-1}(\ker(H_*(L) \to H_*(M))) \right)$.*

Proof: Let P denote $\pi^{-1}(L)$. We compare the following exact sequences where the vertical maps are induced by π:

$$
\begin{array}{ccccccc}
\cdots \longrightarrow & H_i(P) & \xrightarrow{j_*} & H_i(B) & \xrightarrow{k} & H_i(B, P) & \longrightarrow \cdots \\
& \downarrow & & \downarrow \pi_* & & \downarrow \approx & \\
\cdots \longrightarrow & H_i(L) & \longrightarrow & H_i(M) & \longrightarrow & H_i(M, L) & \longrightarrow \cdots
\end{array}
$$

The third vertical map is an isomorphism by excision. Hence if $\alpha \in \ker(\pi_*)$ we must have $k(\alpha) = 0$, so $\alpha \in \text{Image}(j_*)$, so d) is true. Then Lemma 2.9.2 implies a) and b). To see c), take any $\alpha \in H_i(B)$. By Lemma 2.9.1b or c, there is a $\beta \in H_i^{imb}(B)$ or $RH_i(B)$ so that $\pi_*(\alpha) = \pi_*(\beta)$. But then $\alpha - \beta \in \ker \pi_*$ so a) and b) imply c). $\square$

Note the following more general definition of uzunblowup supersedes that in [**AK4**].

Definitions: Let V be a nonsingular Zopen set. We call a map $\pi\colon \widetilde{V} \to V$ an *uzunblowup* if π is the composition

$$\widetilde{V} = V_n \xrightarrow{\pi_n} V_{n-1} \xrightarrow{\pi_{n-1}} \cdots \longrightarrow V_1 \xrightarrow{\pi_1} V_0 = V$$

where each π_{i+1} is either a rational diffeomorphism or a blowup of V_i along a some nonsingular center $L_i \subset V_i$. We call $\{L_i\}$ the centers of $\pi\colon \widetilde{V} \to V$. A *multiblowup* is an uzunblowup where each π_i is a blowup. We call an uzunblowup *full* or *algebraically full* if all the centers are full or algebraically full. (For the curious reader: *uzun* means long in Turkish.) We call a multiblowup or uzunblowup *nondegenerate* if all centers L_i have $\dim L_i < \dim V$.

Now we are ready to prove an algebraic version of the Steenrod representability theorem for homology classes.

Lemma 2.9.4 *If V is a compact nonsingular Zopen set, then there is a nondegenerate multiblowup $\pi\colon \widetilde{V} \to V$ with:*

a) $\pi_* H_*^{imb}(\widetilde{V}) = H_*(V)$.
b) $\pi_* RH_*(\widetilde{V}) = H_*^A(V)$.

Proof: By Lemma 2.9.1, π_* is onto and each element of $H_*^{imb}(V)$ lifts to an element of $H_*^{imb}(\widetilde{V})$. Hence a) follows by repeated applications of the following assertion.

Assertion 2.9.4.1 *Let $0 \neq \theta \in H_k(V)$. Then there exists a multiblowup $\pi\colon \widetilde{V} \to V$, a k-dimensional nonsingular Zclosed subset Z of $\widetilde{V}$ and a component Z_0 of Z such that $\pi_*[Z_0] = \theta$.*

Proof: By the Steenrod representability theorem [**Th**], we can find a map $f\colon M \to V$ such that M is a closed smooth manifold, and $f_*[M] = \theta$. By transversality we can assume that f is one to one almost everywhere. By Theorem 2.8.3 we can find an algebraic set Q, an entire rational function $\varphi\colon Q \to V$, a component of Q_0 of Q and a diffeomorphism $h\colon M \to Q_0$ so that $\varphi \circ h \sim f$, in particular $\varphi_*[Q_0] = \theta$ and φ is one to one almost everywhere on Q_0.

Now, $\varphi(Q_0)$ is a semialgebraic set of dimension k, hence if Y is the Zariski closure of $\varphi(Q_0)$ in V then $\dim Y = k$. Now by Theorem 2.5.11 there is a multiblowup $\pi\colon \widetilde{V} \to V$ such that the strict preimage Z of Y is a k-dimensional nonsingular Zclosed subset of $\widetilde{V}$ and the centers lie over $\operatorname{Sing} Y$. By chapter 0, section 5 of [**H**] (c.f. Proposition 6.2.7) there is a multiblowup $p\colon \widetilde{Q} \to Q$ and a rational function $\widetilde{\varphi}\colon \widetilde{Q} \to Z$ so that $\pi\widetilde{\varphi} = \varphi p$, i.e., the following commutes:

$$\widetilde{Q} \xrightarrow{\;\widetilde{\varphi}\;} Z$$

$$p \downarrow \qquad\qquad \downarrow \pi$$

$$Q \xrightarrow{\;\varphi\;} Y$$

(Alternatively, one could avoid using chapter 0 section 5 of **[H]** by using Lemma 2.5.9 and Proposition 2.8.8 and settling for an uzunblowup instead of a multiblowup.)

For simplicity assume that Q_0 is connected. Hence $p^{-1}(Q_0) = \widetilde{Q}_0$ is connected. Let $Z_0 = \widetilde{\varphi}(\widetilde{Q}_0)$, then Z_0 is connected. Notice $\widetilde{\varphi}$ is one to one almost everywhere on $\widetilde{Q}_0$ since φ is one to one almost everywhere on Q_0. Therefore $[Z_0] = \widetilde{\varphi}_*[\widetilde{Q}_0]$. Hence

$$\pi_*[Z_0] = \pi_*\widetilde{\varphi}_*[\widetilde{Q}_0] = \varphi_* p_*[\widetilde{Q}_0] = \varphi_*[Q_0] = \theta$$

Since $\dim Z = k$ and Z_0 is a connected subset representing a non-zero k-dimensional homology class, Z_0 must be the whole component of Z.   □

The proof of b) is similar, but simpler since in this case we can take $Q = Q_0$ and $Z = Z_0$.   □

The following is a homology version of the resolution Theorem 2.5.11. It says in particular that we can make all $\mathbf{Z}/2\mathbf{Z}$-cycles of smooth manifolds embedded after blowing up operations.

Theorem 2.9.5 *Every compact nonsingular real Zopen set V admits a nondegenerate full uzunblowup $\pi\colon \widetilde{V} \to V$ such that:*

a) $\widetilde{V}$ *is full.*
b) $\pi_* RH_*\big(\widetilde{V}\big) = H_*^A(V)$.

Proof: Consider the following assertions:

Assertion $\mathcal{A}(v)$: *Every compact nonsingular real Zopen set V with $\dim V = v$ admits a nondegenerate full uzunblowup $\pi\colon \widetilde{V} \to V$ such that $\widetilde{V}$ is full and so that $\pi_* RH_*\big(\widetilde{V}\big) = H_*^A(V)$.*

Assertion $\mathcal{C}(v)$: *Suppose that V is a compact nonsingular real Zopen set with $\dim V = v$ and $p\colon \overline{V} \to V$ is a nondegenerate multiblowup. Then there exists a nondegenerate full uzunblowup $\pi\colon \widetilde{V} \to V$ and an entire rational function $g\colon \widetilde{V} \to \overline{V}$ such that the following diagram commutes up to $\mathbf{Z}/2\mathbf{Z}$ homology*

$$\overline{V} \xleftarrow{\;g\;} \widetilde{V}$$

$$p \downarrow \qquad \nearrow \pi \qquad\qquad (*)$$

$$V$$

It suffices to prove that $\mathcal{C}(v)$ implies $\mathcal{A}(v)$ and to prove that $\mathcal{A}(i)$ for all $i < v$ implies $\mathcal{C}(v)$.

Proof: (of $\mathcal{C}(v) \implies \mathcal{A}(v)$) Suppose dim $V = v$. From Lemma 2.9.4 we get a nondegenerate multiblowup $p\colon \overline{V} \to V$ such that:

1) $p_* H_*^{imb}\left(\overline{V}\right) = H_*(V)$.
2) $p_* RH_*\left(\overline{V}\right) = H_*^A(V)$.

By $\mathcal{C}(v)$ there exists a nondegenerate full uzunblowup $\pi\colon \widetilde{V} \to V$ and an entire rational function $g\colon \widetilde{V} \to \overline{V}$ making the diagram $(*)$ commute up to $\mathbf{Z}/2\mathbf{Z}$ homology. By Lemma 2.9.1, $H_*^{imb}\left(\overline{V}\right) \subset g_* H_*^{imb}\left(\widetilde{V}\right)$ and $RH_*\left(\overline{V}\right) \subset g_* RH_*\left(\widetilde{V}\right)$, hence

$$\pi_* H_*^{imb}\left(\widetilde{V}\right) = p_* g_* H_*^{imb}\left(\widetilde{V}\right) \supset p_* H_*^{imb}\left(\overline{V}\right) = H_*(V)$$

implying $\pi_* H_*^{imb}\left(\widetilde{V}\right) = H_*(V)$. Similarly we get $\pi_* RH_*\left(\widetilde{V}\right) = H_*^A(V)$.

Furthermore the surjectivity of π_* implies that $H_*\left(\widetilde{V}\right) = \ker(\pi_*) \oplus G$ where $\pi_*\colon G \to H_*(V)$ is an isomorphism and $G \subset H_*^{imb}\left(\widetilde{V}\right)$. By Lemma 2.9.3, $\ker(\pi_*) \subset H_*^{imb}\left(\widetilde{V}\right)$. Therefore $H_*\left(\widetilde{V}\right) = H_*^{imb}\left(\widetilde{V}\right)$, i.e., $\widetilde{V}$ is full. □

Proof: (of $\mathcal{A}(i)$ for all $i < v \implies \mathcal{C}(v)$) Pick V with dim $V = v$ and $p\colon \overline{V} \to V$, which is a composition:

$$\overline{V} = V_n \xrightarrow{p_n} V_{n-1} \xrightarrow{p_{n-1}} \cdots \longrightarrow V_1 \xrightarrow{p_1} V_0 = V$$

such that each $V_{i+1} \xrightarrow{p_{i+1}} V_i$ is a blowup along a nonsingular center $L_i \subset V_i$ with dim $L_i < $ dim $V_i = $ dim V.

Assume we have constructed a nondegenerate full uzunblowup $\pi(i)\colon Z_i \to V$ and an entire rational function $g_i\colon Z_i \to V_i$, such that the following commutes up to $\mathbf{Z}/2\mathbf{Z}$ homology

$$V_i \xleftarrow{\;g_i\;} Z_i$$

$$p(i) \downarrow \quad \swarrow \quad \pi(i)$$

$$V$$

where $p(i) = p_1 \circ p_2 \circ \ldots \circ p_i$, $Z_0 = V$ and $g_0 = \pi(0) = id$. If we call this statement $\mathcal{B}(i)$, by induction it suffices to prove $\mathcal{B}(i) \implies \mathcal{B}(i+1)$.

By Theorem 2.8.8 we can find a nonsingular algebraic set Z_i' and a rational diffeomorphism $\alpha\colon Z_i' \xrightarrow{\approx} Z_i$ and an entire rational function $\varphi\colon Z_i' \to V_i$ which is transverse to L_i, and $\varphi_* = (g_i)_* \circ \alpha_*$ in homology. Then $N_i' = \varphi^{-1}(L_i)$ is a nonsingular algebraic set of dimension less than v in Z_i'.

By $\mathcal{A}(\dim N_i')$ there is a nondegenerate full uzunblowup $N_i'' \to N_i'$ with $H_*(N_i'') = H_*^{imb}(N_i'')$. Let $\psi\colon Z_i'' \to Z_i'$ be the induced full uzunblowup with the same centers. Consider the blowups $\beta\colon Q = \mathcal{B}(Z_i', N_i') \longrightarrow Z_i'$ and

$\gamma\colon Z_{i+1} = \mathfrak{B}\left(Z_i'', N_i''\right) \longrightarrow Z_i''$. By Lemma 2.5.9 and Lemma 2.5.10 we have entire rational functions δ_1 and δ_2 making the following diagram commute in homology.

$$
\begin{array}{ccccc}
 & & Q & \overset{\delta_2}{\longleftarrow} & Z_{i+1} \\
 & \overset{\delta_1}{\diagup} & \downarrow & & \downarrow \gamma \\
V_{i+1} & & Z_i' & \overset{\psi}{\longleftarrow} & Z_i'' \\
\downarrow & \overset{\varphi}{\diagup} & \approx \downarrow \alpha & & \\
V_i & \longleftarrow & Z_i & & \\
\downarrow & \diagup & & & \\
V & & & &
\end{array}
$$

Then $\pi\,(i+1)\colon Z_{i+1} \to V$ is a full-uzunblowup, where $\pi\,(i+1) = \pi\,(i)\circ\alpha\circ\psi\circ\gamma$. If we let $g_{i+1} = \delta_1 \circ \delta_2$ we get the homology commutative diagram:

$$
\begin{array}{ccc}
V_{i+1} & \overset{g_{i+1}}{\longleftarrow} & Z_{i+1} \\
p_{i+1} \downarrow & \diagup & \pi(i+1) \\
V & &
\end{array}
$$

$\square$

In the above theorem we can not hope to make $H_*\big(\widetilde{V}\big) = RH_*\big(\widetilde{V}\big)$, since this would imply $H_*\,(V) = H_*^A\,(V)$ contradicting Theorem 2.7.3. However we can get :

Theorem 2.9.6 *Every algebraic set with totally algebraic homology admits a nondegenerate algebraically full uzunblowup* $\pi\colon \widetilde{V} \to V$, *such that* $H_*\big(\widetilde{V}\big) = RH_*\big(\widetilde{V}\big)$.

Proof: The proof is the same as the proof of Theorem 2.9.5 except in the statement of the assertions we take $\pi\colon \widetilde{V} \to V$ to be an algebraically full uzunblowup.

$\square$

The recent simple proof of resolution of singularities in [**BM**] allows a simplification in the proof of Theorem 2.9.5 and probably allows a strengthening of the conclusion, namely that π can be an algebraic multiblowup rather than just an uzunblowup. The reason the proof is simpler is that in [**BM**], a complexity is assigned to each point in an algebraic set and then one does resolution of singularities by blowing up with center the points of maximum complexity. But if one blows up a center contained in the set of maximum complexity, the resulting complexity does not increase and in fact, with a little bit of care, it decreases everywhere except on the strict transform of the set of maximal complexity. Consequently, it is not necessary to use homology commutative triangles, but instead give a simpler direct proof of Theorem 2.9.5.

10. Isotoping Submanifolds to Algebraic Subsets

When is a closed smooth submanifold M^m of $\mathbf{R}^n$ isotopic to an algebraic subset of $\mathbf{R}^n$? Seifert [S] showed that if M has a trivial normal bundle then it can be isotoped to a nonsingular component Z_0 of an algebraic subset Z. Furthermore, he showed that one can take $Z = Z_0$ if either $n - m = 1$, or $n - m = 2$ and M is orientable (c.f., Theorem 2.8.2). His method in fact gives Z to be a complete intersection in $\mathbf{R}^n$. This conclusion makes his result the best possible; because it turns out that there are homotopy theoretical obstructions to isotoping submanifolds with trivial normal bundle to complete intersections [AK7].

In [N] Nash showed that any closed smooth submanifold M can be ϵ-isotoped to a smooth sheet of an algebraic subset of $\mathbf{R}^n$ (the sheets might intersect each other). Then by a normalization process he was able to separate the sheets in $\mathbf{R}^n \times \mathbf{R}^k$ for some k, thereby isotoping M to a nonsingular component of an algebraic set in $\mathbf{R}^n \times \mathbf{R}^k$. Then by generic projections he was able to obtain the result that every $M^m \subset \mathbf{R}^n$ is isotopic to a nonsingular component of an algebraic subset provided $n \geq 2m + 1$. He conjectured that the dimension restriction is not necessary.

This conjecture has a long history. Tognoli [To1] strengthen the Nash's result by showing that when $n \geq 2m + 1$ the extra components in the conclusion of the Nash's theorem can be removed (Corollary 2.8.6). In literature sometimes [To1] is incorrectly referred to as the solution of the Nash Conjecture. Furthermore in [I] and [To1] it was shown that Nash's theorem can be improved to $n \geq 3m/2$. There have been also two published incorrect proofs of the conjecture ([W], [To2]). Finally the conjecture was proven in [AK12] and generalized to immersions $M \looparrowright \mathbf{R}^n$. Moreover [AK13] shows that the extra components of the algebraic set can be eliminated if and only if this immersion has an algebraic representative. In this section we will present a proof of Nash's conjecture by summarizing the results of [AK12]. One can ask for a generalization of these results when $\mathbf{R}^n$ is replaced by any nonsingular algebraic set. We will discuss some of this and refrain from the general case by referring the reader to the above papers, since we don't need these results for the rest of the book.

We refer to some algebraic geometric notions such as normalization and the finiteness of rational functions which we have not yet introduced. See [Sh] for their definitions.

Definitions: If $X \subset \mathbf{C}^n$ is a complex algebraic set we let $X_{\mathbf{R}}$ denote $X \cap \mathbf{R}^n$. We let $\overline{X}$ denote the complex conjugate of X, i.e., $\overline{X} = \{\, z \in \mathbf{C}^n \mid \overline{z} \in X \,\}$.

We say that X is defined over $\mathbf{R}$ if $X = \overline{X}$. If X and Y are complex algebraic sets defined over $\mathbf{R}$, we say a rational function $\varphi\colon X \to Y$ is defined over $\mathbf{R}$ if $\varphi(\overline{z}) = \overline{\varphi(z)}$ for all $z \in X$.

We need to prove a complex version of Lemma 2.8.1 which allows us to approximate a smooth function with a complex polynomial in some situations. Proposition 2.10.2 provides an answer, but first we have an elementary result.

Lemma 2.10.1 *Let $F \subset \mathbf{C}^m$ be a finite set of points, and $y \in \mathbf{C}^m$ such that $y \notin F$ and $\overline{y} \notin F$. Then there is a polynomial $q\colon \mathbf{C}^m \to \mathbf{C}$ defined over $\mathbf{R}$, such that q is C^∞ close to 0 near F and C^∞ close to 1 near y and $\overline{y}$.*

Proof: Pick any integer k. By picking a linear projection $\pi\colon \mathbf{C}^m \to \mathbf{C}$ defined over $\mathbf{R}$ so that $\pi(y) \notin \pi(F)$ and $\pi(\overline{y}) \notin \pi(F)$ and replacing q by $\pi \circ q$ it suffices to consider the case $n = 1$. If $F = \{\, \alpha_i + \beta_i \sqrt{-1} \mid i = 1, \dots, b \,\}$, let $\sigma_i(z) = (z - \alpha_i)^2 + \beta_i^2$ and $\sigma(z) = \Pi(\sigma_i(z))^{k+1}$. Then σ is C^k close to 0 near F and $\sigma(y) \neq 0$. Suppose $\sigma(y) = \alpha + \beta\sqrt{-1}$, then

$$q(z) = 1 - \left(\frac{(\sigma(z) - \alpha)^2 + \beta^2}{\alpha^2 + \beta^2} \right)^{k+1}$$

has the required properties and its first k derivatives are close. $\square$

Proposition 2.10.2 *Let $\theta\colon \mathbf{C}^m \to \mathbf{C}^n$ be a polynomial map defined over $\mathbf{R}$. Let T be a compact subset of $\theta^{-1}(\mathbf{R}^n)$ with $\overline{T} = T$, and let $f\colon T \to \mathbf{C}$ be a continuous function with $f(\overline{z}) = \overline{f(z)}$ for all $z \in T$. Suppose $\theta|_T$ is finite to one. Then there is a polynomial $h\colon \mathbf{C}^m \to \mathbf{C}$ defined over $\mathbf{R}$ so that $h|_T$ is a C^0 approximation of f. Furthermore if for some $S \subset T$ with $\overline{S} = S$, $f|_S$ can be locally C^k approximated by polynomials defined over $\mathbf{R}$, then we can also conclude that $h|_S$ is a C^k approximation to $f|_S$.*

Proof: Suppose for each $x \in \mathbf{R}^n$ there is a neighborhood U_x of x in $\mathbf{C}^n$ and a polynomial $h_x\colon \mathbf{C}^m \to \mathbf{C}$ such that h_x is a C^k approximation of f on $\theta^{-1}(U_x) \cap T$. Then by compactness we can cover T by a finite number of $\theta^{-1}(U_{x_i})$'s, $i = 1, 2 \dots, b$. Let $\psi_i\colon \mathbf{R}^m \to [0,1]$ be a partition of unity with $supp(\psi_i) \subset U_{x_i}$. Then approximate ψ_i by real polynomials p_i on T. Think of these polynomials as complex polynomials defined over $\mathbf{R}$. We let

$$h(z) = \sum_{i=1}^{b} p_i(\theta(z)) h_{x_i}(z)$$

So it suffices to find the polynomials h_x. Let $y \in \theta^{-1}(x) \cap T$. Then if $y \in S$ let g_y be a polynomial defined over $\mathbf{R}$ which is C^k close to f on some neighborhood V_y of y, otherwise let g_y be the constant $f(y)$. We may choose $g_y = g_{\overline{y}}$, since for $z \in \overline{V_y} \cap T$, $g_y(z) = \overline{g_y(\overline{z})}$ which is close to $\overline{f(\overline{z})} = f(z)$. Now by Lemma

2.10.1 we can choose polynomials $q_y \colon \mathbf{C}^m \to \mathbf{C}$ defined over $\mathbf{R}$, such that q_y approximates to 0 on $\theta^{-1}(x) \cap T - \{y, \overline{y}\}$ and 1 on $\{y, \overline{y}\}$. Then we let

$$h_x = \sum_{y \in \theta^{-1}(x) \cap T} g_y q_y$$

$\square$

We will be talking about isotoping immersed submanifolds to algebraic subsets. To be able to to do that we need to make the following definition for algebraic sets. It gives the correct notion of nonsingularity for algebraic sets that are images of immersions of smooth manifolds:

Definition: We say x is an *almost nonsingular point* of an algebraic set X of dimension d, if a neighborhood of x in X is a union of analytic manifolds of dimension d, and furthermore the complexification of these analytic manifolds form a neighborhood of x in the complexification $X_{\mathbf{C}}$ of X. An algebraic set consisting entirely of almost nonsingular points of dimension d is called an *almost nonsingular algebraic set.*

Hence all nonsingular points are almost nonsingular, and the converse is true if X is normal. Recall that the image of a real algebraic set under a polynomial map may not be real algebraic. The following Proposition gives a sufficient conditions which imply this.

Proposition 2.10.3 *Let $Z \subset \mathbf{R}^m$ be a real algebraic set and $Z_{\mathbf{C}} \subset \mathbf{C}^m$ be its complexification. Let $\psi \colon Z_{\mathbf{C}} \to \mathbf{C}^n$ be a proper polynomial map defined over $\mathbf{R}$. Let $Y = Cl_{\mathbf{R}}(\psi(Z))$. Suppose:*

a) *Nonsing Z is closed.*
b) *ψ immerses Nonsing Z and restricts to an embedding on an open dense subset of Nonsing Z.*
c) *$\psi^{-1}\psi(\text{Nonsing }Z) = \text{Nonsing }Z$.*

Then $\psi(\text{Nonsing }Z)$ is the set of almost nonsingular points of Y. If in addition we know that ψ is an embedding on Nonsing Z then we get $\psi(\text{Nonsing }Z) = \text{Nonsing }Y$.

Proof: Let $k = \dim Z$ and $Z_0 = \text{Nonsing }Z$. By applying the proposition to each k-dimensional irreducible component of Z, we can assume Z is irreducible.

Since ψ is proper, $X = \psi(Z_{\mathbf{C}})$ is a complex algebraic set. Since $\psi(Z) \subset \psi(Z_{\mathbf{C}}) \cap \mathbf{R}^n = X_{\mathbf{R}}$, we know that $Y \subset X_{\mathbf{R}} \subset X$. Hence $Y_{\mathbf{C}} \subset X$. Also since

$$Z \subset \psi^{-1}\psi(Z) \subset \psi^{-1}(Y) \subset \psi^{-1}(Y_{\mathbf{C}})$$

we have $Z_{\mathbf{C}} \subset \psi^{-1}(Y_{\mathbf{C}})$. Therefore

$$X = \psi(Z_{\mathbf{C}}) \subset \psi\psi^{-1}(Y_{\mathbf{C}}) \subset Y_{\mathbf{C}} \subset X.$$

So $X = Y_{\mathbf{C}}$ and $Y = X \cap \mathbf{R}^n$.

Let $z \in Z_0$. Then by b) and c), $\psi^{-1}\psi(z)$ is a finite set of points of Z_0. Let $\psi^{-1}\psi(z) = \{z_1 \ldots z_b\}$. By b), $d\psi_{z_i}$ has rank k, so there are small open neighborhoods U_i of z_i in Z_0 such that $\psi|_{U_i}$ is a complex embedding. By properness $\psi(Z_{\mathbf{C}} - \cup_{i=1}^b U_i)$ is closed hence $\psi(\cup_{i=1}^b U_i)$ is a neighborhood of $\psi(z)$ in X. Since each $\psi(U_i)$ is a complexification of the real analytic manifold $\psi(U_i \cap Z)$, $\psi(z)$ is an almost nonsingular point of Y.

If $\psi|_{Z_0}$ is an embedding then $b = 1$, hence $\psi(z) \in \text{Nonsing}\, Y$. So $\psi(Z_0) \subset \text{Nonsing}\, Y$. We claim $\text{Nonsing}\, Y \subset \psi(Z_0)$. If this were not true then we could pick $x \in \text{Nonsing}\, Y - \psi(Z_0)$. Hence x has an open k-dimensional neighborhood U in Y. Since $\psi(Z_0)$ is closed (ψ is proper) we may assume $U \subset \text{Nonsing}\, Y - \psi(Z_0)$. In particular $\dim(Y - \psi(Z_0)) = k$. Also since $\psi \colon Z \to Y$ embeds an open dense subset Z_0, it has degree 1. By Corollary 2.3.3 $\dim(Y - \psi(Z)) < \dim(Z) = k$. Also $\dim(\psi(\text{Sing}\, Z) \le \dim(\text{Sing}\, Z) < k$. But $Y - \psi(Z_0) = (Y - \psi(Z)) \cup \psi(\text{Sing}\, Z)$, hence $\dim(Y - \psi(Z_0)) < k$, a contradiction. Therefore $\psi(Z_0) = \text{Nonsing}\, Y$. $\qquad\square$

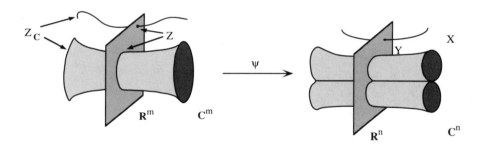

FIGURE II.10.1. An almost nonsingular immersion

Next we need to prove a certain generalization of the Theorem 2.8.4.

Theorem 2.10.4 *Let $f \colon M \leftrightarrowtail V$ be a smooth immersion from a closed smooth manifold M to a nonsingular algebraic set V. Assume that the bordism class of f is algebraic. Then there are a real algebraic set Z and a diffeomorphism $h \colon M \to \text{Nonsing}\, Z$ and a polynomial $\rho \colon Z \to V$ such that:*

   a) *$\rho \circ h \colon M \to V$ approximates f.*

   b) *The complexification $\rho_{\mathbf{C}} \colon Z_{\mathbf{C}} \to V_{\mathbf{C}}$ of ρ is a finite regular map to its image.*

   c) *$\text{Nonsing}\, Z$ is a union of connected components of $\rho_{\mathbf{C}}^{-1}(V)$.*

Furthermore in case $V = \mathbf{R}^n$ (and hence the cobordism condition is always satisfied) we may take $Z_{\mathbf{C}}$ to be nonsingular.

Proof: We first show that conclusion c) is a consequence of a) and b). Nonsing Z is closed in $\rho_{\mathbf{C}}^{-1}(V)$ since it is the image of a compact set M. Hence it suffices to show Nonsing Z is open in $\rho_{\mathbf{C}}^{-1}(V)$. Let $m = \dim M$. Pick $x \in$ Nonsing Z, note that $d\rho_{\mathbf{C}}$ has rank m at x. Hence, after an analytic coordinate change defined over $\mathbf{R}$, $\rho_{\mathbf{C}}$ is locally an injective linear map defined over $\mathbf{R}$. So locally $\rho_{\mathbf{C}}$ takes nonreal points to nonreal points. Hence Nonsing Z is open in $\rho_{\mathbf{C}}^{-1}(V)$.

To see a) and b) we apply Theorem 2.8.4 and get a nonsingular algebraic set X and a diffeomorphism $g\colon M \to X$ and an entire rational function $\psi\colon X \to V$ so that $\psi \circ g$ approximates f. The only problem is $\psi_{\mathbf{C}}$ may not be finite. Let us show how to make $\psi_{\mathbf{C}}$ finite when $V = \mathbf{R}^n$. By Proposition 2.2.15 we may assume $X_{\mathbf{C}}$ is nonsingular. Suppose $X_{\mathbf{C}} \subset \mathbf{C}^k$. Let $Z = \{\, (x,y) \in X \times \mathbf{R}^n \mid y = \psi(x) \,\}$, then $Z_{\mathbf{C}} = \{\, (x,y) \in X_{\mathbf{C}} \times \mathbf{C}^n \mid y = \psi_{\mathbf{C}}(x) \,\} \subset \mathbf{C}^k \times \mathbf{C}^n$. By the proof of Theorem 10 in Chapter 1 and Section 5 of [**Sh**] the restriction to $Z_{\mathbf{C}}$ of a generic linear retraction $\mathbf{C}^k \times \mathbf{C}^n \to \mathbf{C}^n$ is finite. So choose a linear retraction $\pi\colon \mathbf{C}^k \times \mathbf{C}^n \to \mathbf{C}^n$, defined over $\mathbf{R}$ and close to the standard projection, so that $\pi|\colon Z_{\mathbf{C}} \to \pi(Z_{\mathbf{C}})$ is finite. Then by letting $h(x) = (g(x), \psi g(x))$ and $\rho = \pi|_Z$ we are finished.

The proof for the general V (which we don't need for the proof of the Nash Conjecture) is a more sophisticated version of the above procedure. We refer the interested reader to [**AK12**] □

We are now ready to prove a generalized version of the Conjecture of Nash.

Theorem 2.10.5 *Let* $f\colon M \looparrowright \mathbf{R}^n$ *be a smooth immersion of a closed smooth manifold. Then* f *is* ϵ-*regularly homotopic to an immersion* $f'\colon M \looparrowright \mathbf{R}^n$ *such that* $f'(M)$ *is the set of almost nonsingular points of an algebraic subset of* $\mathbf{R}^n$. *In particular* $f'(M)$ *is a union of components of a real algebraic subset of* $\mathbf{R}^n$.

Proof: We apply Theorem 2.10.4 to the immersion f and obtain a nonsingular real algebraic set Z, a diffeomorphism $h\colon M \xrightarrow{\approx} Z$, and a polynomial $\rho\colon Z \to \mathbf{R}^n$ such that if $Y = \rho_{\mathbf{C}}(Z_{\mathbf{C}})$ then:

1) $\rho \circ h\colon M \to \mathbf{R}^n$ approximates f
2) $\rho_{\mathbf{C}}\colon Z_{\mathbf{C}} \to Y$ is a finite regular map
3) Z is a union of connected components of $\rho_{\mathbf{C}}^{-1}(\mathbf{R}^n)$

Take a generic projection $\pi\colon \mathbf{C}^n \to L$ onto a codimension one linear subspace L defined over $\mathbf{R}$ such that $\pi|_Y$ is finite [**Sh**]. Let v be a unit vector perpendicular to L in $\mathbf{R}^n$, then

$$\rho_{\mathbf{C}}(z) = \pi(\rho_{\mathbf{C}}(z)) + \langle \rho_{\mathbf{C}}(z), v \rangle v$$

Let K be a compact neighborhood of $\pi\rho(Z)$ in $L \cap \mathbf{R}^n$ and let $T = \rho_{\mathbf{C}}^{-1}\pi^{-1}(K)$. Since $\pi \circ \rho_{\mathbf{C}}$ is finite it is proper, hence T is compact. We know that Z has a neighborhood U in $Z_{\mathbf{C}}$ such that $\rho_{\mathbf{C}}|_U$ is a complex analytic immersion and hence $\rho_{\mathbf{C}}(U - Z)$ has no real points.

Pick a smooth function $\alpha\colon T \to [0,1]$ such that $\alpha = 0$ on a neighborhood U' of Z in $T \cap U$, $\alpha = 1$ on $T - U$ and $\alpha(z) = \alpha(\bar{z})$. By Proposition 2.10.2 we can find a polynomial $\alpha'\colon Z_{\mathbf{C}} \to \mathbf{C}$ defined over $\mathbf{R}$ so that $\alpha'|_T$ approximates α and this approximation is C^1 on U'. Let b be the maximum of $|\langle \rho_{\mathbf{C}}(z), v \rangle|$ for $z \in T$, then we define $\psi\colon Z_{\mathbf{C}} \to \mathbf{C}^n$ by:

$$
\begin{aligned}
\psi(z) &= \rho_{\mathbf{C}}(z) + 2(b+1)\alpha'(z)v \\
&= \pi\rho_{\mathbf{C}}(z) + [\langle \rho_{\mathbf{C}}(z), v \rangle + 2(b+1)\alpha'(z)]v
\end{aligned}
$$

Since $\pi \circ \psi = \pi \circ \rho_{\mathbf{C}}$, ψ is proper. Also ψ is defined over $\mathbf{R}$. Since ψ is C^1 close to $\rho_{\mathbf{C}}$ on U', ψ is an immersion on U'. Hence Proposition 2.10.3 would imply the theorem once we can show $\psi^{-1}\psi(z) \subset Z$ for all $z \in Z$.

Suppose $\psi(z) = \psi(u)$ for $z \in Z$ and $u \in Z_{\mathbf{C}} - Z$. Since $\pi\rho_{\mathbf{C}}(u) = \pi\psi(u) = \pi\psi(z) = \pi\rho_{\mathbf{C}}(z) = \pi\rho(z) \in K$ we have $u \in T$. Since ψ is defined over $\mathbf{R}$ and an immersion on U', after an analytic coordinate change over $\mathbf{R}$ it becomes locally an injective linear map defined over $\mathbf{R}$. Hence ψ takes nonreal points of U' to nonreal points. So $\psi(U' - Z)$ has no real points. Therefore $u \in T - U'$. If $u \in U \cap T - U'$ then:

$$
\psi(u) \sim \rho_{\mathbf{C}}(u) + 2(b+1)\alpha(u)v
$$

where $\sim$ means close to. Notice v and $\alpha(u)$ are real and $\rho_{\mathbf{C}}(U - Z)$ has no real points. Therefore $\psi(u)$ is not real, a contradiction. So $u \in T - U$. Then

$$
\psi(u) - \pi\rho_{\mathbf{C}}(u) \sim [\langle \rho_{\mathbf{C}}(u), v \rangle + 2b + 2]v
$$

On the other hand

$$
\psi(u) - \pi\rho_{\mathbf{C}}(u) = \psi(z) - \pi\rho_{\mathbf{C}}(z) \sim \langle \rho_{\mathbf{C}}(z), v \rangle v
$$

This gives a contradiction, because $|\langle \rho_{\mathbf{C}}(z), v \rangle| \leq b$. But

$$
|\langle \rho_{\mathbf{C}}(u), v \rangle + 2b + 2| \geq 2(b+1) - |\langle \rho_{\mathbf{C}}(z), v \rangle| \geq b + 1.
$$

Therefore $\psi^{-1}\psi(z) \subset Z$ for all $z \in Z$. $\square$

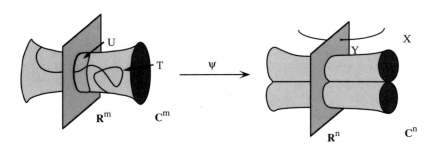

FIGURE II.10.2. Making an immersion algebraic

As a special case of this theorem we obtain:

Corollary 2.10.6 *Any closed smooth submanifold $M \subset \mathbf{R}^n$ is ϵ-isotopic to the nonsingular points of an algebraic subset of $\mathbf{R}^n$. In particular, M is isotopic to a union of components of a real algebraic subset of $\mathbf{R}^n$.*

By applying the proof of Lemma 2.1.4 we get:

Corollary 2.10.7 *Any closed smooth submanifold $M \subset \mathbf{R}^n$ is isotopic to a nonsingular real algebraic subset of $\mathbf{R}^{n+1}$.*

We can try to generalize Theorem 2.10.5 by replacing $\mathbf{R}^n$ by any nonsingular algebraic set V. Of course in this case we must have the bordism class of f algebraic. Also we need the notion of projection V onto a hyperplane. To be able to do this we replace V by $V \times \mathbf{R}$.

Theorem 2.10.8 *Let $f: M \hookrightarrow V$ be an immersion of a closed smooth manifold into a nonsingular algebraic set V such that the bordism class of f is algebraic. Then f is ϵ-regularly homotopic to an immersion $f': M \hookrightarrow V \times \mathbf{R}$ such that $f'(M)$ is the set of almost nonsingular points of an algebraic set. Hence $f'(M)$ is a union of components of an algebraic subset of $V \times \mathbf{R}$.*

Proof: As before we apply Theorem 2.10.4 to the immersion f and get: an algebraic set Z, a diffeomorphism $h: M \xrightarrow{\approx} \text{Nonsing}\, Z$, and a polynomial $\rho: Z \to \mathbf{R}^n$ satisfying the conclusions a), b), c) of Theorem 2.10.4.

Let $K \subset V$ be a compact set containing a neighborhood of $\rho(\text{Nonsing}\, Z)$. Then since $\rho_{\mathbf{C}}$ is proper the set $T = \rho_{\mathbf{C}}^{-1}(K)$ is compact. Define $g: T \to \mathbf{C}$ by $g(z) = 0$ for all $z \in \text{Nonsing}\, Z$, and $g(z) = 2$ for all $z \in T - \text{Nonsing}\, Z$. By Proposition 2.10.2 there is a polynomial $\gamma: Z_{\mathbf{C}} \to \mathbf{C}$ defined over $\mathbf{R}$ such that $\gamma|_T$ approximates g. Let $\psi = (\rho_{\mathbf{C}}, \gamma): Z_{\mathbf{C}} \to Y_{\mathbf{C}} \times \mathbf{C}$ where $Y_{\mathbf{C}} = \rho_{\mathbf{C}}(Z_{\mathbf{C}})$. Then ψ is proper, hence $\psi(Z_{\mathbf{C}})$ is a complex algebraic subset of $Y_{\mathbf{C}} \times \mathbf{C}$. Since $\psi \circ h$ is close to $(f, 0)$, they are ϵ-regularly homotopic to $\psi(\text{Nonsing}\, Z)$. To conclude the proof we apply Proposition 2.10.3. So we only have to check c) of Proposition 2.10.3.

Let $z \in \text{Nonsing}\, Z$. If $\psi(z) = \psi(w)$ then $\rho_{\mathbf{C}}(z) = \rho_{\mathbf{C}}(w)$, hence $w \in T$. But if $w \in T - \text{Nonsing}\, Z$, then $\gamma(w)$ is near 2, but $\gamma(z)$ is near 0, hence we can not have $\gamma(z) = \gamma(w)$. Therefore $w \in \text{Nonsing}\, Z$. Hence $\varphi^{-1}\varphi(z) \subset \text{Nonsing}\, Z$. We are done. $\square$

Again by applying the proof of Lemma 2.1.4 we get:

Corollary 2.10.9 *If $M \subset V$ is a smooth closed submanifold of a nonsingular algebraic set V and the bordism class of the inclusion $f: M \to V$ is algebraic, then M is ϵ-isotopic to a nonsingular algebraic subset in $V \times \mathbf{R}^2$.*

Finally we state without proof the following theorem which says that the extra components of the algebraic set in the conclusion of Theorem 2.10.5 can be removed if they can be removed up to immersed cobordism:

Theorem 2.10.10 ([AK13])   *Let $f\colon M \to \mathbf{R}^n$ be a smooth immersion of a smooth closed manifold M. Then f is ϵ-regularly homotopic to an immersion onto an almost nonsingular real algebraic subset of $\mathbf{R}^n$ if and only if f is immersion cobordant to a degree one immersion onto an almost nonsingular real algebraic set in $\mathbf{R}^n$.*

CHAPTER III

TICOS

In this chapter we study ticos which are the basic topological objects underlying our work. Before defining them we will give a few results on smooth functions.

1. Some Results about Smooth Functions

Lemma 3.1.1 *Let* $f : (\mathbf{R}^n, \mathbf{R}_1^n) \to (\mathbf{R}, 0)$ *be a smooth function. Then there is a smooth map* $g \colon \mathbf{R}^n \to \mathbf{R}$ *so that* $f(x) = x_1 g(x)$ *for all* $x \in \mathbf{R}$. *Furthermore, the map* $f \mapsto f/x_1$ *from* $\mathfrak{I}^\infty(\mathbf{R}_1^n)$ *to* $C^\infty(\mathbf{R}^n)$ *is continuous in the* C^∞ *topology.*

Proof: The existence of g follows easily from Lemma 2.5.4. Suppose now that $f_i \in \mathfrak{I}^\infty(\mathbf{R}_1^n)$ is a sequence of smooth functions and $f_i \to 0$ in the C^∞ topology. We have not explained which C^∞ topology we are using, but the Lemma is true for all the ones we know of. We will give a proof for the weak topology which is sufficient for our needs. The proof is similar for the strong topology where $|Df(x)| < \epsilon(x)$ for all D of order less than n, $\epsilon \colon \mathbf{R}^n \to (0, \infty)$ gives neighborhoods of 0. Let

$$g_i(x) = f_i(x)/x_1 = \int_0^1 \partial f_i/\partial x_1 (tx_1, x_2, \ldots, x_n) \ dt.$$

We need to show that $g_i \to 0$. Pick a compactum K and an $\epsilon > 0$ and an n. We need to find an N so that $|Dg_i(x)| < \epsilon$ for all $x \in K$, for all $i > N$ and all partial derivatives D of order less than or equal to n. Pick a ball B with center at the origin so that $K \subset B$. Pick an N so that $|Df_i(x)| < \epsilon$ for all $i > N$, all $x \in B$ and all partial derivatives D of order $\leq n + 1$. Then if $i > N$ and D is a partial derivative of order $\leq n$ and $x \in B$ we have

$$
\begin{aligned}
|Dg_i(x)| &= \left| \int_0^1 t^\alpha \frac{\partial}{\partial x_1} Df_i(tx_1, x_2, \ldots, x_n) \ dt \right| \\
&\leq \int_0^1 \left| \frac{\partial}{\partial x_1} Df_i(tx_1, \ldots, x_n) \right| \ dt \leq \int_0^1 \epsilon \ dt = \epsilon
\end{aligned}
$$

where α is the number of $\partial/\partial x_1$'s in D.   □

The next lemma shows that for example, there is a smooth function $f\colon \mathbf{R}^3 \to$ $\mathbf{R}$ so that $f(x,y,0) = \sin(x+y^2)$, $f(x,0,z) = \sin(xe^z)$ and $f(0,y,z) = yz^2 + e^{-z}\sin(y^2)$. It is a special case of a more general result, c.f., [**L1**].

Lemma 3.1.2 *Let $f_i\colon \mathbf{R}_i^n \to \mathbf{R}$, $i = 1,\dots,k$ be smooth functions which agree on overlaps, i.e., so that $f_i|_{\mathbf{R}_i^n \cap \mathbf{R}_j^n} = f_j|_{\mathbf{R}_i^n \cap \mathbf{R}_j^n}$ for all $i,j = 1,\dots,k$. Then there is a smooth $f\colon \mathbf{R}^n \to \mathbf{R}$ so that $f|_{\mathbf{R}_i^n} = f_i$ for all $i = 1,\dots,k$.*

Proof: By induction on k there is an $f'\colon \mathbf{R}^n \to \mathbf{R}$ so that $f'| = f_i$ for $i = 1,\dots,k-1$. Let $g\colon \mathbf{R}_k^n \to \mathbf{R}$ be $f_k - f'|$. Then $g|_{\mathbf{R}_i^n \cap \mathbf{R}_k^n} = 0$ for $i = 1,\dots,k-1$. So by Lemma 3.1.1 we know that

$$g(x) = \Big(\prod_{i=1}^{k-1} x_i\Big)h(x)$$

for some smooth $h\colon \mathbf{R}_k^n \to \mathbf{R}$. Let $p\colon \mathbf{R}^n \to \mathbf{R}_k^n$ be orthogonal projection. Now let

$$f(x) = f'(x) + \Big(\prod_{i=1}^{k-1} x_i\Big)hp(x).$$

Then $f|_{\mathbf{R}_i^n} = f'| = f_i$ for $i < k$ and $f|_{\mathbf{R}_k^n} = f'| + \prod_{i=1}^{k-1} x_i \cdot h = f'| + g = f_k$. So we are done.   □

The next lemma shows that for example, if $f\colon \mathbf{R}^2 \to \mathbf{R}$ and $g\colon \mathbf{R}^2 \to \mathbf{R}$ are somewhere zero smooth functions and $fg = xy$, then either f or g is x times a nowhere zero function. We presume that this is a special case of a more general result where the x_i are replaced by arbitrary irreducible analytic germs, but we do not know a reference. In any case, this Lemma has an elementary proof.

Lemma 3.1.3 *Suppose $f_i\colon \mathbf{R}^n \to \mathbf{R}$, $i = 1,\dots,m$ are smooth functions so that $\prod_{i=1}^m f_i(x) = \prod_{j=1}^n x_j^{a_j}$ for all x, where a_j are non-negative integers. Then there are non-negative integers b_{ij} and nowhere zero smooth functions $g_i\colon \mathbf{R}^n \to \mathbf{R}-0$ so that $f_i(x) = \prod_{j=1}^n x_j^{b_{ij}} g_i(x)$ for all x and i.*

Proof: It suffices to prove the case $m = 2$. Pick maximal exponents b_{ij} so that $f_i(x) = \prod_{j=1}^n x_j^{b_{ij}} g_i(x)$ for some smooth functions $g_i\colon \mathbf{R}^n \to \mathbf{R}$, $i = 1,2$. Suppose $\mathbf{R}_j^n \subset g_i^{-1}(0)$ for some i,j. Then by Lemma 3.1.1, $g_i(x) = x_j g_i'(x)$ for some smooth g_i' which violates the maximality of the b_{ij}'s. So we may assume that $\mathbf{R}_j^n \not\subset g_i^{-1}(0)$ for all i,j. Let $c_j = a_j - b_{1j} - b_{2j}$. Then $g_1(x) g_2(x) = \prod_{j=1}^n x_j^{c_j}$. If $c_j = 0$ for all j then we are done since $g_1(x) g_2(x) = 1$ so $g_i(\mathbf{R}^n) \subset \mathbf{R} - 0$. We must also have $c_j \geq 0$ for all j, otherwise g_1 or g_2 would approach ∞ near $\mathbf{R}_j^n$. So we may assume that $c_k > 0$ for some k. We will arrive at a contradiction and thus prove the lemma.

Let $b = (b_1, \dots, b_n)$ be an n-tuple of non-negative integers. Then let $D_b g$ denote the partial derivative $\partial^{\Sigma b_i} g / \partial x_1^{b_1} \partial x_2^{b_2} \cdots \partial x_n^{b_n}$. We claim that if $z \in \mathbf{R}_k^n - g_2^{-1}(0)$ and $b_k < c_k$ then $D_b g_1(z) = 0$. To see this, note that $D_b \left(\prod_{j=1}^n x_j^{c_j} \right)$ is 0 at z, so

$$0 = D_b(g_1 g_2)(z) = \sum_d C_{db} D_d g_1(z) D_{b-d} g_2(z)$$

for some non-zero constants $C_{db} = \prod \binom{b_i}{d_i}$ where d runs over all n-tuples with $0 \le d_j \le b_j$ for all j. By induction we may assume that $D_d g_1(z) = 0$ for all such d except possibly $d = b$. Then we have $0 = C_{bb} D_b g_1(z) g_2(z)$. So $D_b g_1(z) = 0$ as we wish. Similarly, $D_b g_2(z) = 0$ if $z \in \mathbf{R}_k^n - g_1^{-1}(0)$ and $b_k < c_k$.

Suppose now that $c_k > 0$ and z is in the frontier of $\mathbf{R}_k^n - g_2^{-1}(0)$, i.e., $z \in \mathrm{Cl}\left(\mathbf{R}_k^n - g_2^{-1}(0) \right) \cap g_2^{-1}(0)$. By continuity we know that $D_b g_1(z) = 0$ for all b with $b_k < c_k$. Hence

$$c_k! \prod_{\substack{j=1 \\ j \ne k}}^n z_j^{c_j} = \partial^{c_k} g_1 g_2 / \partial x_k^{c_k}(z) = g_2(z) \partial^{c_k} g_1 / \partial x_k^{c_k}(z) = 0.$$

So $z_j = 0$ for some $j \ne k$ with $c_j > 0$. So the frontier of $\mathbf{R}_k^n - g_2^{-1}(0)$ is contained in a union of coordinate hyperplanes, hence $\mathrm{Cl}\left(\mathbf{R}_k^n - g_2^{-1}(0) \right)$ is a union of orthants. Since we are assuming that $\mathbf{R}_k^n - g_2^{-1}(0)$ is non-empty, this implies that $0 \in \mathrm{Cl}\left(\mathbf{R}_k^n - g_2^{-1}(0) \right)$. Similarly, $0 \in \mathrm{Cl}\left(\mathbf{R}_k^n - g_1^{-1}(0) \right)$. Consequently $D_b g_i(0) = 0$ for all b with $b_k < c_k$ for some k and all $i = 1, 2$. So

$$\prod_{j=1}^n c_j! = D_c g_1 g_2(0) = \sum_d C_{dc} D_d g_1(0) D_{c-d} g_2(0) = 0,$$

a contradiction. So $c_j = 0$ for all j and we are finished.   □

2. Ticos

The fundamental entity we will use is called a tico (an acronym for transversally intersecting codimension one, also similar to the greek $\tau o \tilde{\iota} \chi o s$ meaning wall). This section develops the fundamental properties of ticos and tico maps.

Definitions: A *tico* $\mathcal{A}$ in a smooth manifold M is a finite collection of proper immersed codimension one smooth submanifolds in general position. Let $|\mathcal{A}| = \bigcup_{A \in \mathcal{A}} A$ denote the union of all the immersed submanifolds. We call $|\mathcal{A}|$ the *realization* of $\mathcal{A}$. Notice that $|\mathcal{A}|$ is an immersed submanifold and $\{|\mathcal{A}|\}$ is a tico also. Let $\bigcap \mathcal{A}$ denote $\bigcap_{A \in \mathcal{A}} A$. The elements of $\mathcal{A}$ are called *sheets* of the tico. For convenience we allow sheets to be the empty submanifold.

Let us explain our terms *proper* and *general position* more carefully. They are equivalent to the following. Take any $z \in M$. Then if $z \notin \partial M$ we ask that there be a smooth coordinate chart $\psi \colon (\mathbf{R}^n, 0) \to (M, z)$ so that $\psi^{-1}(|\mathcal{A}|) =$

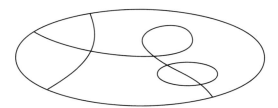

FIGURE III.2.1. A tico in the disc

$\bigcup_{i=1}^{a} \mathbf{R}_i^n$ for some $a > 0$. If $z \in \partial M$ we ask that there be a smooth chart
$\psi \colon (\mathbf{R}^{n-1} \times [0,1), 0) \to (M, z)$ so that $\psi^{-1}(|\mathcal{A}|) = \bigcup_{i=1}^{a} \mathbf{R}_i^{n-1} \times [0,1)$. (We
allow $a = 0$ in case $z \notin |\mathcal{A}|$.) We call such a chart a *tico chart.*

To simplify proofs, we will consistently ignore boundary points. If we were
to take them into account, it would only mean that whenever we refer to a tico
chart we would also have to consider the case of the half space tico charts. Doing
so would only clutter up the argument and help to obscure the ideas in the proof.
We only use boundaries in Chapter VII where we must understand bordism.

If $\mathcal{A}$ is a tico in M we will sometimes find it convenient to abuse notation by
saying $(M, \mathcal{A})$ is a tico.

We have a naturally defined stratification of M, called the *tico stratification,*
where a codimension d stratum is connected and $|\mathcal{A}|$ is a union of strata. In par-
ticular, let I_d be the set of d-fold self intersections of $|\mathcal{A}|$, i.e., the set of $q \in M$
so that there is a tico chart $\psi \colon (\mathbf{R}^m, 0) \to (M, q)$ (or $\psi \colon (\mathbf{R}^{n-1} \times [0,1), 0) \to$
$(M, q))$ with $\psi^{-1}(|\mathcal{A}|) = \bigcup_{i=1}^{d} \mathbf{R}_i^m$ (or $\psi^{-1}(|\mathcal{A}|) = \bigcup_{i=1}^{d} \mathbf{R}_i^{n-1} \times [0,1)$). Then
the codimension d strata are the connected components of I_d. We should empha-
size that all strata are connected. We will sometimes use the phrase "connected
stratum" to remind the reader of this fact. Note that in the bounded case, the
strata are manifolds with boundary.

Definition: We say that a tico $\mathcal{A}$ is *finite* if this stratification has only a finite
number of strata. In other words, each I_d has a finite number of connected
components $d = 0, \dots, m$. Of course if M is compact, any tico in M is finite.

Definition: If S is a sheet of a tico $\mathcal{A}$ in M then the *immersion associated to*
S is the proper immersion $h \colon S' \to M$ so that $h(S') = S$ and the restriction
$h| \colon h^{-1}(S - Q) \to S - Q$ is one to one where as above, $Q = \bigcup_{d=2}^{\infty} I_d$ is the set of
self intersections of $|\mathcal{A}|$. Any other immersion with image S is just a covering of
S'. Note that if S is embedded, $S' = S$ and h is inclusion.

Definitions: A tico $\mathcal{A}$ is *regular* if each sheet of $\mathcal{A}$ is an embedded submanifold.
A tico $\mathcal{A}$ in M is *algebraic* if M is a nonsingular real Zopen set, $\mathcal{A}$ is finite and
each sheet $A \in \mathcal{A}$ is a nonsingular real Zclosed subset of M.

Allowing nonregular ticos is merely a technical convenience (since local con-
structions are then automatically global). One should really think of a tico as

being regular. Thus, if assuming regularity will make a definition or result easier and we do not need the nonregular form, we will not hesitate to assume regularity.

Note that any algebraic tico is regular by definition and also has empty boundary. By Theorem 2.8.9, any regular tico in a closed manifold is isomorphic to an algebraic tico in a nonsingular Zopen set. In fact, one can show that any regular tico $(M, \mathcal{A})$ in a compact manifold with boundary can be made algebraic in the sense that there is an algebraic tico $\mathcal{B}$ in a nonsingular real algebraic set V isomorphic the interior of $(M, \mathcal{A})$. Hence for sufficiently large balls B we have $(V \cap B, \mathcal{B} \cap B)$ isomorphic to $(M, \mathcal{A})$.

Definition: If $(M, \mathcal{A})$ and $(N, \mathcal{B})$ are ticos then a smooth map $f \colon M \to N$ is called a *tico map* if it satisfies the following property. Pick any $q \in M$. Then we ask that there be tico charts $\psi \colon (\mathbf{R}^m, 0) \to (M, q)$ and $\theta \colon (\mathbf{R}^n, 0) \to (N, f(q))$ so that:

1) $\psi^{-1}(|\mathcal{A}|) = \bigcup_{i=1}^{a} \mathbf{R}_i^m$, $\theta^{-1}(|\mathcal{B}|) = \bigcup_{i=1}^{b} \mathbf{R}_i^n$ and $f\psi(\mathbf{R}^m) \subset \theta(\mathbf{R}^n)$.
2) Let $\kappa_i(x)$ be the i-th coordinate of $\theta^{-1}f\psi(x)$, then there are nonnegative integers α_{ij}, $1 \le i \le b$, $1 \le j \le a$ and smooth functions $\varphi_i \colon \mathbf{R}^m \to \mathbf{R}$ so that $\varphi_i(0) \ne 0$ and $\kappa_i(x) = \prod_{j=1}^{a} x_j^{\alpha_{ij}} \varphi_i(x)$ for all x near 0 and $i = 1, \dots, b$.

For example, let $M = \mathbf{R}^3$, $N = \mathbf{R}^2$, $\mathcal{A} = \{\mathbf{R}_1^3, \mathbf{R}_2^3\}$ and $\mathcal{B} = \{\mathbf{R}_1^2\}$. Then $f(x, y, z) = (x^3 y, e^x \sin(y + z))$ is a tico map. The only relevant fact to check is that the first coordinate is $x^3 y$, a monomial. Other examples of tico maps are $(x^3, 0)$, (xye^z, z) and $(\ln(1 + x^2 y^2), \cos z)$.

Lemma 3.2.1 *Tico maps have the following properties:*

a) *Up to permutations of $j = 1, \dots, a$ and $i = 1, \dots, b$ the exponents α_{ij} above depend only on the point q, not on ψ and θ. Furthermore, for any tico charts satisfying 1) above, κ_i will satisfy 2).*
b) *$|\mathcal{A}| \supset f^{-1}(|\mathcal{B}|)$. In fact, if $\mathcal{A}$ is regular then $f^{-1}(|\mathcal{B}|)$ is a union of components of sheets of $\mathcal{A}$.*
c) *Given any tico charts ψ and φ there is a unique smooth function φ_i so that $\kappa_i(x) = \prod_{j=1}^{a} x_j^{\alpha_{ij}} \varphi_i(x)$ for all $x \in \mathbf{R}^m$ and so that $\varphi_i^{-1}(0)$ is empty.*

Proof: First let us prove a). Suppose we had other tico charts $\psi' \colon (\mathbf{R}^m, 0) \to (M, q)$ and $\theta' \colon (\mathbf{R}^n, 0) \to (N, f(q))$ so that $\psi'^{-1}(|\mathcal{A}|) = \bigcup_{i=1}^{a'} \mathbf{R}_i^m$, $\theta'^{-1}(|\mathcal{B}|) = \bigcup_{i=1}^{b'} \mathbf{R}_i^n$ and $f\psi'(\mathbf{R}^m) \subset \theta'(\mathbf{R}^n)$. Clearly $a = a'$ and $b = b'$. Also, after permuting the coordinates we may as well assume that $\theta^{-1}\theta'(\mathbf{R}_i^n)$ and $\psi^{-1}\psi'(\mathbf{R}_j^m)$ are open subsets of $\mathbf{R}_i^n$ and $\mathbf{R}_j^m$ respectively $i = 1, \dots, b$ and $j = 1, \dots, a$. Let $\lambda_i(x)$ be the i-th coordinate of $\theta'^{-1}\theta(x)$ and let $\mu_j(x)$ be the j-th coordinate of $\psi^{-1}\psi'(x)$. Then since $\lambda_i(\mathbf{R}_i^n) = 0$ for $i = 1, \dots, b$ and $\mu_j(\mathbf{R}_j^m) = 0$ for $j = 1, \dots, a$ we know by Lemma 3.1.1 that $\lambda_i(x) = x_i \lambda_i'(x)$ for $0 \le i \le b$ and

$\mu_j(x) = x_j \mu'_j(x)$ for $0 \le j \le a$ for some smooth functions λ'_i and μ'_j. Then the i-th coordinate of ${\theta'}^{-1} f\psi'(x)$ is

$$
\begin{aligned}
\lambda_i \left(\theta^{-1} f\psi'(x) \right) &= \kappa_i \left(\psi^{-1}\psi'(x) \right) \cdot \lambda'_i \theta^{-1} f\psi'(x) \\
&= \prod_{j=1}^{a} \mu_j(x)^{\alpha_{ij}} \cdot \varphi_i \psi^{-1}\psi'(x) \cdot \lambda'_i \theta^{-1} f\psi'(x) \\
&= \prod_{j=1}^{a} x_j^{\alpha_{ij}} \prod_{j=1}^{a} \mu'_j(x)^{\alpha_{ij}} \varphi_i \psi^{-1}\psi'(x) \cdot \lambda'_i \theta^{-1} f\psi'(x).
\end{aligned}
$$

So we need to show that

$$
\prod_{j=1}^{a} \mu'_j(0)^{\alpha_{ij}} \cdot \varphi_i \psi^{-1}\psi'(0) \cdot \lambda'_i \theta^{-1} f\psi'(0) \ne 0.
$$

But $\varphi_i \psi^{-1}\psi'(0) = \varphi_i(0) \ne 0$ and $\lambda'_i \theta^{-1} f\psi'(0) = \lambda'_i \theta^{-1} f(q) = \lambda'_i(0)$. So we need only show that $\lambda'_i(0) \ne 0$ and $\mu'_j(0) \ne 0$ for all i and j. But the Jacobian matrix of $\psi^{-1}\psi'$ at 0 is of the form $\begin{pmatrix} D & 0 \\ * & * \end{pmatrix}$ where D is an $a \times a$ diagonal matrix with j-th diagonal entry $\mu'_j(0)$. Since this matrix is nonsingular, we must have $\mu'_j(0) \ne 0$ for $j = 1, \ldots, a$. Likewise by looking at the Jacobian matrix of $\theta^{-1}\theta'$ we see that $\lambda'_i(0) \ne 0$ for $i = 1, \ldots, b$. So the exponents α_{ij} are independent of ψ and θ.

Now we will prove part b). In fact we will show that if $i \le b$ and $A = \psi^{-1} f^{-1} \theta(\mathbf{R}_i^n)$ then $A = \bigcup_{j \in J} \mathbf{R}_j^m$ where $J = \{\, j \mid 1 \le j \le a \text{ and } \alpha_{ij} > 0 \,\}$. This will certainly imply part b).

Pick $z \in A$ and let $B_z = \{\, j \in \mathbf{Z} \mid 1 \le j \le a \text{ and } z_j = 0 \,\}$. Define a chart $\psi_z \colon (\mathbf{R}^m, 0) \to (\mathbf{R}^m, z)$ where the j-th coordinate of $\psi_z(x)$ is x_j if $j \in B_z$, $x_j + z_j$ if $j > a$ and $z_j \exp x_j$ otherwise. Then $(\psi\psi_z)^{-1}(|\mathcal{A}|) = \bigcup_{i \in B_z} \mathbf{R}_i^n$ so $\psi\psi_z$ is a tico chart centered at $\psi(z)$. By a) and the definition of tico map there are non-negative integers α_{ij}^z, $j \in B_z$ and smooth functions $\varphi_i^z \colon \mathbf{R}^m \to \mathbf{R}$ so that $\varphi_i^z(0) \ne 0$ and the i-th coordinate of $\theta^{-1} f\psi\psi_z(x)$ is $\prod_{j \in B_z} x_j^{\alpha_{ij}^z} \varphi_i^z(x)$ for x near 0 and $i \le b$. In particular, the functions $z \mapsto \alpha_{ij}^z$ are continuous on $A \cap \mathbf{R}_j^m$, since if we take z' near z in A then $B_{z'} \subset B_z$ and if $z' \in A \cap \mathbf{R}_j^m$ then $\alpha_{ij}^{z'} = \alpha_{ij}^z$ and

$$
\varphi_i^{z'}(x) = \prod_{j \in B_z - B_{z'}} (z'_j)^{\alpha_{ij}^z} \exp\left(x_j \alpha_{ij}^z \right) \varphi_i^z \left(\psi_z^{-1}\psi_{z'}(x) \right).
$$

Notice also that for some neighborhood U of z,

$$
U \cap A = \{\, x \in U \mid x_j = 0 \text{ for some } j \in B_z \text{ with } \alpha_{ij}^z > 0 \,\}.
$$

Thus $A \subset \bigcup_{j=1}^{a} \mathbf{R}_j^m$. Also if we pick any $j = 1, \ldots, a$ then $\mathbf{R}_j^m \cap A - \bigcup_{k \ne j} \mathbf{R}_k^m$ is an open subset of $\mathbf{R}_j^m - \bigcup_{k \ne j} \mathbf{R}_k^m$. Hence it is a union of connected components of $\mathbf{R}_j^m - \bigcup_{k \ne j} \mathbf{R}_k^m$ since A is closed. But every connected component of $\mathbf{R}_j^m -$

$\bigcup_{k\neq j} \mathbf{R}_k^m$ contains a point near 0. If $\alpha_{ij} > 0$ then every point of $\mathbf{R}_j^m$ near 0 is in A, hence $\mathbf{R}_j^m - \bigcup_{k\neq j} \mathbf{R}_k^m \subset A$ so $\mathbf{R}_j^m \subset A$. If $\alpha_{ij} = 0$ then no point of $\mathbf{R}_j^m - \bigcup_{k\neq j} \mathbf{R}_k^m$ near 0 is in A so $A \subset \bigcup_{k\neq j} \mathbf{R}_k^m$. Consequently $A = \bigcup_{j\in J} \mathbf{R}_j^m$ and b) is proven.

Now let us prove c). Since each $A \cap \mathbf{R}_j^m$ is connected and each α_{ij}^z is integer-valued and continuous in z, α_{ij}^z is constant. So $\alpha_{ij}^z = \alpha_{ij}$ for all $z \in A \cap \mathbf{R}_j^m$.

Consequently, for every $z \in A$ the function $\prod_{j=1}^a x_j^{-\alpha_{ij}} \kappa_i (x)$ is well-defined, smooth and non-zero on a neighborhood of z since it is just

$$\prod_{j\notin B_z} x_j^{-\alpha_{ij}} \varphi_i^z \left(\psi_z^{-1}(x)\right).$$

But for $x \notin A$, we have $\prod_{j=1}^a x_j^{\alpha_{ij}} \neq 0$ so $\prod_{j=1}^a x_j^{-\alpha_{ij}} \kappa_i(x)$ is also smooth and non-zero. Hence we may take φ_i to be the unique function so that $\varphi_i(x) = \prod_{j=1}^a x_j^{-\alpha_{ij}} \kappa_i(x)$ for $x \notin A$. Then φ_i is well-defined, smooth and nowhere zero. $\square$

Definition: If $\mathcal{A}$ and $\mathcal{B}$ are regular ticos and $f\colon (M,\mathcal{A}) \to (N,\mathcal{B})$ is a tico map then to each $S \in \mathcal{A}$ and $T \in \mathcal{B}$ we may assign a function $\alpha_{ST}\colon S \to \mathbf{Z}$ as follows. Pick $q \in S$. If $f(q) \notin T$ set $\alpha_{ST}(q) = 0$. If $f(q) \in T$ pick tico charts $\psi\colon (\mathbf{R}^m, 0) \to (M, q)$ and $\theta\colon (\mathbf{R}^n, 0) \to (N, f(q))$ so that $\psi^{-1}(|\mathcal{A}|) = \bigcup_{i=1}^a \mathbf{R}_i^m$, $\psi^{-1}(S) = \mathbf{R}_1^m$ and $\theta^{-1}(T) = \mathbf{R}_1^n$. Then we know that the first coordinate of $\theta^{-1} f \psi(x)$ is $\prod_{i=1}^a x_i^{\alpha_i} \varphi(x)$ where $\varphi(0) \neq 0$. We then define $\alpha_{ST}(q) = \alpha_1$. We call α_{ST} the *local exponent map* for f.

Note that if $\mathcal{A}$ is not regular (but $\mathcal{B}$ is), we still have a local exponent map $\alpha_{ST}\colon S' \to \mathbf{Z}$ where $h\colon S' \to M$ is the immersion associated to S. We will avoid using it however.

The following Lemma is an easy consequence of Lemma 3.2.1.

Lemma 3.2.2 *Let* $f\colon (M,\mathcal{A}) \to (N,\mathcal{B})$ *be a tico map where* $\mathcal{A}$ *and* $\mathcal{B}$ *are regular. For sheets* $S \in \mathcal{A}$ *and* $T \in \mathcal{B}$ *let* $\alpha_{ST}\colon S \to \mathbf{Z}$ *be the local exponent map for* f. *Then* α_{ST} *is continuous, hence it is constant on any connected component* S' *of* S. *Also,* $f^{-1}(T) = \bigcup_{S\in\mathcal{A}} \alpha_{ST}^{-1}(\mathbf{Z} - 0)$.

In the algebraic case we get a stronger result below.

Lemma 3.2.3 *Let* $f\colon (V,\mathcal{A}) \to (W,\mathcal{B})$ *be an entire rational tico map between algebraic ticos (i.e.,* f *is both an entire rational function and a tico map). For sheets* $S \in \mathcal{A}$ *and* $T \in \mathcal{B}$ *let* $\alpha_{ST}\colon S \to \mathbf{Z}$ *be the local exponent map for* f. *Then* α_{ST} *is constant on any irreducible component* S' *of* S.

Proof: First suppose that $f(S') \not\subset T$. Then $\dim\left(S' \cap f^{-1}(T)\right) < \dim S'$ by Lemma 2.2.9. In particular, $S' - f^{-1}(T)$ is dense in S'. But $\alpha_{ST}(x) = 0$ for $x \in S - f^{-1}(T)$, so $\alpha_{ST}(x) = 0$ for all $x \in S'$ by continuity.

Now suppose that $f(S') \subset T$. Pick any $x \in S' - |\mathcal{A} - \{S\}|$, i.e., $x \in S'$ is not a double point of $|\mathcal{A}|$. Pick polynomials $p\colon (V, S) \to (\mathbf{R}, 0)$ and $q\colon (W, T) \to (\mathbf{R}, 0)$ so that x is a regular point of p and $f(x)$ is a regular point of q. We may pick tico charts $\psi\colon (\mathbf{R}^n, 0) \to (V, x)$ and $\theta\colon (\mathbf{R}^m, 0) \to (W, f(x))$ so that $p\psi(z) = z_1$ for all $z \in \mathbf{R}^n$ near 0 and $q\theta(z) = z_1$ for all $z \in \mathbf{R}^m$ near 0. Since f is a tico map we have $qf(y) = p(y)^{\alpha'} r'(y)$ for y near x where $\alpha' = \alpha_{ST}(x)$ and r' is some smooth function defined on a neighborhood of x with $r'(x) \neq 0$. By repeated applications of Lemma 2.2.11 there is a Zopen neighborhood Z of x in V and an entire rational function $r\colon Z \to \mathbf{R}$ so that

$$qf(y) = p(y)^\alpha r(y)$$

for all $y \in Z$, and so that r does not vanish identically on $S' \cap Z$. By Lemma 2.2.9, $S' \cap r^{-1}(0) \cap Z$ is nowhere dense in S'. But for y near x,

$$r(y) = p(y)^{\alpha' - \alpha} r'(y)$$

so $\alpha = \alpha' = \alpha_{ST}(x)$ and $r(x) \neq 0$.

Let

$$Y = S' \cap Z - \left(r^{-1}(0) \cup \Sigma p \cup f^{-1}(\Sigma q) \cup |\mathcal{A} - \{S\}| \right)$$

where Σp and Σq are the sets of critical points of p and q respectively. Then Y is a Zopen neighborhood of x in S' and $\dim(S' - Y) < \dim S'$ by Lemma 2.2.9. But $\alpha_{ST}(y) = \alpha$ for all $y \in Y$ since we may take local charts at y and $f(y)$ so that one of the coordinates in V is p and one of the coordinates in W is q. Hence by continuity we know that $\alpha_{ST}(y) = \alpha$ for all $y \in S'$. $\square$

Definitions: If the functions α_{ST} are constant on all of S for all $S \in \mathcal{A}$ and $T \in \mathcal{B}$ we say that the map f has *constant exponents*. In this case we say that the constant value of α_{ST} is the *exponent* of S in T for the map f.

One consequence of a map having constant exponents is that $f(S) \subset T$ if and only if the exponent of S in T is positive. This is because $f^{-1}(T) = \bigcup_{S \in \mathcal{A}} \alpha_{ST}^{-1}(\mathbf{Z} - 0) = |\mathcal{A}_T|$ where $\mathcal{A}_T = \{S \in \mathcal{A} \mid \alpha_{ST} > 0\}$.

Exercise: Show that if $\mathcal{A}$ and $\mathcal{B}$ are regular ticos then $f\colon (M, \mathcal{A}) \to (N, \mathcal{B})$ is a tico map with constant exponents α_{ST} if and only if

$$f^*(\mathfrak{I}_N^\infty(T)) = \prod_{S \in \mathcal{A}} (\mathfrak{I}_M^\infty(S))^{\alpha_{ST}}$$

for all $T \in \mathcal{B}$. $\diamond$

We now show that the composition of tico maps is a tico map, and the exponents are given by matrix multiplication.

Lemma 3.2.4 *Let* $f\colon (M,\mathcal{A}) \to (N,\mathcal{B})$ *and* $g\colon (N,\mathcal{B}) \to (P,\mathcal{C})$ *be tico maps.*

  a) *Then* $g \circ f\colon (M,\mathcal{A}) \to (P,\mathcal{C})$ *is a tico map.*

  b) *Suppose* $\mathcal{A}$, $\mathcal{B}$ *and* $\mathcal{C}$ *are regular ticos. For* $A \in \mathcal{A}$, $B \in \mathcal{B}$ *and* $C \in \mathcal{C}$ *let* $\alpha_{AB}\colon A \to \mathbf{Z}$, $\alpha_{BC}\colon B \to \mathbf{Z}$ *and* $\alpha_{AC}\colon A \to \mathbf{Z}$ *be the exponent maps for* f, g *and* $g \circ f$. *Extend each* α_{BC} *to any perhaps discontinuous function on* N. *Then for all* $x \in A$,

$$\alpha_{AC}(x) = \sum_{B \in \mathcal{B}} \alpha_{AB}(x) \cdot \alpha_{BC}(f(x)).$$

In particular, if f *and* g *have constant exponents then* $g \circ f$ *has constant exponents also.*

Proof: Let us first prove part a). Pick any $x \in M$. As usual, pick tico charts $\psi\colon (\mathbf{R}^m,0) \to (M,x)$, $\theta\colon (\mathbf{R}^n,0) \to (N,f(x))$ and $\varphi\colon (\mathbf{R}^p,0) \to (P,gf(x))$ so that $f\psi(\mathbf{R}^m) \subset \theta(\mathbf{R}^n)$, $g\theta(\mathbf{R}^n) \subset \varphi(\mathbf{R}^p)$, $\psi^{-1}(|\mathcal{A}|) = \bigcup_{i=1}^a \mathbf{R}_i^m$, $\theta^{-1}(|\mathcal{B}|) = \bigcup_{i=1}^b \mathbf{R}_i^n$ and $\psi^{-1}(|\mathcal{C}|) = \bigcup_{i=1}^c \mathbf{R}_i^p$. Let $f_i(y)$ be the i-th coordinate of $\theta^{-1}f\psi(y)$, let $g_i(y)$ be the i-th coordinate of $\varphi^{-1}g\theta(y)$ and let $h_i(y)$ be the i-th coordinate of $\varphi^{-1}gf\psi(y)$. Then we know $f_i(y) = \prod_{j=1}^a y_j^{\beta_{ij}} \eta_i(y)$ and $\eta_i(0) \neq 0$ for $i \leq b$. Also $g_i(y) = \prod_{j=1}^b y_j^{\gamma_{ij}} \lambda_i(y)$ and $\lambda_i(0) \neq 0$ for $i \leq c$. Hence $h_i(y) = \prod_{j=1}^a y_j^{\epsilon_{ij}} \mu_j(y)$ for $i \leq c$ where the exponents ϵ_{ij} are given by $\epsilon_{ij} = \sum_{k=1}^b \gamma_{ik}\beta_{kj}$ and where $\mu_i(y) = \lambda_i(\theta^{-1}f\psi(y)) \prod_{j=1}^b \eta_j(y)^{\gamma_{ij}}$ so $\mu_i(0) \neq 0$. So we have shown that gf is a tico map.

Now let us prove part b). Pick $x \in A$ and let ψ, φ and θ be as above. Let $A_i \in \mathcal{A}$ for $i = 1, \ldots, a$, $B_i \in \mathcal{B}$ for $i = 1, \ldots, b$ and $C_i \in \mathcal{C}$ for $i = 1, \ldots, c$ be such that $\psi^{-1}(A_i) = \mathbf{R}_i^m$, $\theta^{-1}(B_i) = \mathbf{R}_i^n$ and $\varphi^{-1}(C_i) = \mathbf{R}_i^p$. Note A is some A_j, say A_1. Then $\beta_{k1} = \alpha_{AB_k}(x)$, $\gamma_{ik} = \alpha_{B_kC_i}(f(x))$ and $\epsilon_{i1} = \alpha_{AC_i}(x)$. Also $\alpha_{AB}(x) = 0$ if $B \in \mathcal{B} - \{B_1, \ldots, B_b\}$ and $\alpha_{AC}(x) = 0 = \alpha_{B_iC}(f(x))$ if $C \in \mathcal{C} - \{C_1, \ldots, C_c\}$ by Lemma 3.2.2. So

$$\alpha_{AC_i}(x) = \epsilon_{i1} = \sum_{k=1}^b \gamma_{ik}\beta_{k1} = \sum_{k=1}^b \alpha_{B_kC_i}(f(x))\alpha_{AB_k}(x)$$
$$= \sum_{B \in \mathcal{B}} \alpha_{AB}(x)\alpha_{BC_i}(f(x))$$

as we desire. If $C \in \mathcal{C} - \{C_1, \ldots, C_c\}$ then $\sum_{B \in \mathcal{B}} \alpha_{AB}(x)\alpha_{BC}(f(x)) = 0 = \alpha_{AC}(x)$ since if $B \in \mathcal{B} - \{B_1, \ldots, B_b\}$ then $\alpha_{AB}(x) = 0$ and if $B = B_i$ then $\alpha_{B_iC}(f(x)) = 0$. So in either case $\alpha_{AC}(x) = \sum_{B \in \mathcal{B}} \alpha_{AB}(x) \cdot \alpha_{BC}(f(x))$. This formula then implies that if f and g have constant exponents then gf has constant exponents. $\qquad\square$

Except for the notion of a tico map having constant exponents (which we need in Chapter V) and a submanifold being unskewed (defined later) it does not really matter how we divide up $|\mathcal{A}|$ into sheets. For all other purposes, if $|\mathcal{A}| = |\mathcal{B}|$ the ticos $\mathcal{A}$ and $\mathcal{B}$ are interchangeable. For example one consequence of Lemma 3.2.5 below is that the question of whether a map is a tico map or not is independent of the sheet decomposition.

Lemma 3.2.5 *Let $f\colon (M, \mathcal{A}) \to (N, \mathcal{B})$ be a tico map. Let $\mathcal{D}$ be a tico in N with $|\mathcal{D}| \subset |\mathcal{B}|$. Then there is a tico $\mathcal{C}$ in M with $|\mathcal{A}| \supset |\mathcal{C}| = f^{-1}(|\mathcal{D}|)$. Furthermore, if $\mathcal{E}$ is any tico in M with $|\mathcal{E}| \supset f^{-1}(|\mathcal{D}|)$ we know that $f\colon (M, \mathcal{E}) \to (N, \mathcal{D})$ is a tico map.*

Proof: Pick any $x \in M$. We may choose tico charts $\psi\colon (\mathbf{R}^m, 0) \to (M, x)$ and $\theta\colon (\mathbf{R}^n, 0) \to (N, f(x))$ so that $f\psi(\mathbf{R}^m) \subset \theta(\mathbf{R}^n)$, $\psi^{-1}(|\mathcal{A}|) = \bigcup_{i=1}^{a} \mathbf{R}_i^m$, $\theta^{-1}(|\mathcal{B}|) = \bigcup_{i=1}^{b} \mathbf{R}_i^n$ and $\theta^{-1}(|\mathcal{D}|) = \bigcup_{i=1}^{d} \mathbf{R}_i^n$. Let $f_i(z)$ be the i-th coordinate of $\theta^{-1}f\psi(z)$. Then $f_i(z) = \prod_{j=1}^{a} z_j^{\alpha_{ij}} \varphi_i(z)$ for some smooth $\varphi_i\colon \mathbf{R}^m \to \mathbf{R} - 0$, $i = 1, \ldots, b$. Notice then that $\psi^{-1}\left(f^{-1}(|\mathcal{D}|)\right)$ is $\bigcup_{j \in J} \mathbf{R}_j^m$ where $J = \{\, j \mid j \leq a$ and there is an $i \leq d$ so that $\alpha_{ij} > 0\,\}$. So $\{f^{-1}(|\mathcal{D}|)\} = \mathcal{C}$ is a tico (with a single sheet).

After reordering we may assume that $J = \{1, \ldots, c\}$ for some $c \leq a$. Then we may take another chart $\eta\colon (\mathbf{R}^m, 0) \to (M, x)$ so that $\eta^{-1}(|\mathcal{E}|) = \bigcup_{i=1}^{e} \mathbf{R}_i^m$ and the j-th coordinate of $\psi^{-1}\eta(z)$ is z_j for all $j \leq c$. Let $g_i(z)$ be the i-th coordinate of $\theta^{-1}f\eta(z)$. Then for $i \leq d$, $f_i(z) = \prod_{j=1}^{c} z_j^{\alpha_{ij}} \varphi_i(z)$ so $g_i(z) = \prod_{j=1}^{c} z_j^{\alpha_{ij}} \varphi_i \psi^{-1} \eta(z)$. So $f\colon (M, \mathcal{E}) \to (N, \mathcal{D})$ is a tico map. $\square$

Lemma 3.2.6 *Let $f\colon (M, \mathcal{A}) \to (N, \mathcal{B})$ be a tico map. Let S be any stratum of the tico stratification of M. Then for every stratum T of the tico stratification of N either $f(S) \subset T$ or $f(S) \cap T$ is empty. In other words, $f(S) \cap T$ is empty for all but one stratum T of N.*

Proof: Let T' be a stratum of N so that $f(S) \cap T'$ is nonempty and furthermore T' has smallest dimension among strata intersecting $f(S)$. Then $f(S) \cap T' = f(S) \cap \mathrm{Cl}(T')$ so $S \cap f^{-1}(T') = S \cap f^{-1}(\mathrm{Cl}(T'))$ is a closed subset of S. We will show that $S \cap f^{-1}(T')$ is also open in S, hence by connectivity of S we know that $S \cap f^{-1}(T') = S$ so our result follows.

Pick any $q \in S \cap f^{-1}(T')$. Pick tico charts $\psi\colon (\mathbf{R}^m, 0) \to (M, q)$ and $\theta\colon (\mathbf{R}^n, 0) \to (N, f(q))$ so that $f\psi(\mathbf{R}^m) \subset \theta(\mathbf{R}^n)$, $\psi^{-1}(|\mathcal{A}|) = \bigcup_{i=1}^{a} \mathbf{R}_i^m$ and $\theta^{-1}(|\mathcal{B}|) = \bigcup_{i=1}^{b} \mathbf{R}_i^n$. Then we know that $\psi^{-1}(S) = \bigcap_{i=1}^{a} \mathbf{R}_i^m$ and $\theta^{-1}(T') = \bigcap_{i=1}^{b} \mathbf{R}_i^n$. Let $f_i(x)$ be the i-th coordinate of $\theta^{-1}f\psi(x)$. Then for some smooth $\varphi_i\colon \mathbf{R}^m \to \mathbf{R} - 0$ we have $f_i(x) = \prod_{j=1}^{a} x_j^{\alpha_{ij}} \varphi_i(x)$ for all $i \leq b$. Since $\theta^{-1}f\psi(0) = \theta^{-1}f(q) = 0$ we know that $f_i(0) = 0$ for all i, hence for each $i \leq b$ there is a $j \leq a$ so that $\alpha_{ij} > 0$. But then $f_i(x) = 0$ for all $x \in \bigcap_{j=1}^{a} \mathbf{R}_j^m = \psi^{-1}(S)$. So $\psi(\bigcap_{i=1}^{a} \mathbf{R}_i^m) \subset f^{-1}(T')$. So our result follows since $\psi(\bigcap_{i=1}^{a} \mathbf{R}_i^m)$ is a neighborhood of q in S. $\square$

The following Lemma 3.2.7 is a bit surprising. Basically, it says the lifting of a tico map must be a tico map. The similar statement, where f and h are tico maps and you wish to conclude that g is a tico map, is false.

Lemma 3.2.7 *Let* $(M, \mathcal{A})$, $(N, \mathcal{B})$ *and* $(P, \mathcal{C})$ *be ticos. Let* $f \colon (M, \mathcal{A}) \to (P, \mathcal{C})$ *and* $g \colon (N, \mathcal{B}) \to (P, \mathcal{C})$ *be tico maps and let* $h \colon M \to N$ *be a smooth map so that* $g \circ h = f$. *If* $g^{-1}(|\mathcal{C}|) = |\mathcal{B}|$ *then* $h \colon (M, \mathcal{A}) \to (N, \mathcal{B})$ *is a tico map.*

$$(N, \mathcal{B})$$

$$h \nearrow \qquad \downarrow g$$

$$(M, \mathcal{A}) \xrightarrow{\ f\ } (P, \mathcal{C})$$

Proof: Pick any $q \in M$ and pick charts $\psi \colon (\mathbf{R}^m, 0) \to (M, q)$ and $\theta \colon (\mathbf{R}^n, 0) \to (N, h(q))$ so that $h\psi(\mathbf{R}^m) \subset \theta(\mathbf{R}^n)$, $\psi^{-1}(|\mathcal{A}|) = \bigcup_{i=1}^{a} \mathbf{R}_i^m$ and $\theta^{-1}(|\mathcal{B}|) = \bigcup_{i=1}^{b} \mathbf{R}_i^n$. After perhaps shrinking ψ and θ we may also assume there is a chart $\eta \colon (\mathbf{R}^k, 0) \to (P, f(q))$ so that $g\theta(\mathbf{R}^n) \subset \eta(\mathbf{R}^k)$ and $\eta^{-1}(|\mathcal{C}|) = \bigcup_{i=1}^{c} \mathbf{R}_i^k$. Let $f_i(x)$, $g_i(x)$ and $h_i(x)$ be the i-th coordinates of $\eta^{-1}f\psi(x)$, $\eta^{-1}g\theta(x)$ and $\theta^{-1}h\psi(x)$ respectively. Then we know that $f_i(x) = \prod_{j=1}^{a} x_j^{\alpha_{ij}} \lambda_i(x)$ and $g_i(x) = \prod_{k=1}^{b} x_k^{\beta_{ik}} \mu_i(x)$ for some smooth functions $\lambda_i \colon \mathbf{R}^m \to \mathbf{R} - 0$ and $\mu_i \colon \mathbf{R}^n \to \mathbf{R} - 0$, for $i \leq c$.

Since $f = gh$ we know that

$$\prod_{j=1}^{a} x_j^{\alpha_{ij}} \lambda_i(x) = \prod_{k=1}^{b} h_k(x)^{\beta_{ik}} \mu_i \theta^{-1} h\psi(x)$$

so

$$\prod_{j=1}^{a} x_j^{\alpha_{ij}} = \prod_{k=1}^{b} h_k(x)^{\beta_{ik}} \mu_i \theta^{-1} h\psi(x) / \lambda_i(x).$$

So by Lemma 3.1.3 we know that if $\beta_{ik} > 0$ for some i then

$$h_k(x) = \prod_{j=1}^{a} x_j^{\gamma_{kj}} \varphi_k(x)$$

for all $x \in \mathbf{R}^m$ and $k = 1, \ldots, b$ where $\varphi_k \colon \mathbf{R}^m \to \mathbf{R} - 0$ is some smooth function. But if $\beta_{ik} = 0$ for some $k \leq b$ and all $i \leq c$ then $\mathbf{R}_k^n \not\subset \bigcup_{i=1}^{c} g_i^{-1}(0) = \theta^{-1}g^{-1}(|\mathcal{C}|) = \theta^{-1}(|\mathcal{B}|) = \bigcup_{i=1}^{b} \mathbf{R}_i^n$, a contradiction. So h is a tico map. □

Definition: If $\mathcal{A}$ and $\mathcal{B}$ are ticos in M and N respectively we may define the *product tico* $\mathcal{A} \times N \cup M \times \mathcal{B}$ in $M \times N$ where $\mathcal{A} \times N = \{\, A \times N \mid A \in \mathcal{A} \,\}$ and $M \times \mathcal{B} = \{\, M \times B \mid B \in \mathcal{B} \,\}$.

Definition: A *tico homotopy* $f_t \colon (M, \mathcal{A}) \to (N, \mathcal{B})$, $t \in I$, is a homotopy $f_t \colon M \to N$ so that $F \colon (M \times I, \mathcal{A} \times I) \to (N, \mathcal{B})$ is a tico map where $F(x, t) = f_t(x)$.

Lemma 3.2.8 *Let $f \colon (M, \mathcal{A}) \to (N, \mathcal{B})$ be a tico map so that all local exponents are either 0 or 1. Then any ϵ-homotopy of f rel $|\mathcal{A}|$ is a tico homotopy. (An ϵ-homotopy just means a homotopy which is C^∞ small.)*

Proof: Let $f_t \colon M \to N$, $t \in [0, 1]$ be an ϵ-homotopy rel $|\mathcal{A}|$ such that $f_0 = f$. Pick any $v \in M$ and pick tico charts $\theta \colon (\mathbf{R}^n, 0) \to (M, v)$ and $\psi \colon (\mathbf{R}^n, 0) \to (N, f(v))$ so that $\theta^{-1}(|\mathcal{A}|) = \bigcup_{i=1}^{a} \mathbf{R}_i^n$ and $\psi^{-1}(|\mathcal{B}|) = \bigcup_{i=1}^{b} \mathbf{R}_i^m$. Let $\kappa_j(x, t)$ be the j-th coordinate of $\psi^{-1} f_t \theta(x)$. Then since f is a tico map we know that for $j \leq b$, $\kappa_j(x, 0) = \prod_{i \in \mathcal{C}_j} x_i \varphi_j(x)$ where $\mathcal{C}_j \subset \{1, 2, \ldots, a\}$ and $\varphi_j(x)$ is never 0. Now $f_t(x) = f(x)$ if $x \in |\mathcal{A}|$ so $\kappa_j(x, t) = 0$ if $x_i = 0$ for some $i \in \mathcal{C}_j$. Consequently, $\kappa_j(x, t) = \prod_{i \in \mathcal{C}_j} x_i \lambda_j(x, t)$ for some smooth $\lambda_j(x, t)$ by Lemma 3.1.1. But since each f_t is close to f, Lemma 3.1.1 tells us that $\lambda_j(x, t)$ is close to $\lambda_j(x, 0) = \varphi_j(x) \neq 0$. Hence $\lambda_j(x, t) \neq 0$ so f_t is a tico homotopy. $\square$

3. Tico Blowups

If $\mathcal{A}$ is a tico in a manifold M it will be useful to have a good notion of when some blowup of M is compatible with the tico. This is what we will discuss in this section. The reader acquainted with Hironaka's resolution of singularities [H] will recognize that a tico blowup is just one where $|\mathcal{A}|$ has only normal crossings with the center.

The important results of this section are Proposition 3.3.8 and 3.3.9. Proposition 3.3.8 says you can blow up a tico to make it regular. Proposition 3.3.9 says that tico blowups can be pulled back by tico maps. Because of Proposition 3.3.8, when we state a theorem we need not hesitate to add the hypothesis that a tico be regular if it will make the proof easier.

Definitions: Let $\mathcal{A}$ be a tico in M and let $f \colon N \to M$ be a map. We say that f *intersects* $\mathcal{A}$ *cleanly* if for any $z \in N$ there is a neighborhood U of z in N and a tico chart $\psi \colon (\mathbf{R}^m, 0) \to (M, f(z))$ so that:

1) $\psi^{-1}(|\mathcal{A}|) = \bigcup_{i=1}^{a} \mathbf{R}_i^m$.
2) $f(U) \subset \psi(\bigcap_{i=b+1}^{a} \mathbf{R}_i^m)$.
3) f is transverse to $\psi(\bigcap_{i=1}^{b} \mathbf{R}_i^m)$ at z.

We say that f is *transverse* to $\mathcal{A}$ if f is transverse to $|\mathcal{A}|$, i.e., f intersects $\mathcal{A}$ cleanly and we may take $b = a$ above.

Thus if $\mathcal{A}$ is regular and N is connected, then f intersects $\mathcal{A}$ cleanly if there is a $\mathcal{B} \subset \mathcal{A}$ so that $f(N) \subset \bigcap \mathcal{B}$ and f is transverse to $\mathcal{A} - \mathcal{B}$.

Definition: Let $\mathcal{A}$ be a tico in M and let $f\colon N \to M$ be a map intersecting $\mathcal{A}$ cleanly. Then we have an *induced tico* $f^*(\mathcal{A})$ in N as follows. Let $S \in \mathcal{A}$ be any sheet. We will define a sheet $f^*(S) \subset N$ and then let $f^*(\mathcal{A}) = \{ f^*(S) \mid S \in \mathcal{A} \}$. If S is embedded then $f^*(S)$ is easy to describe. Let N' be a connected component of N. Then $f^*(S) \cap N' = f^{-1}(S) \cap N'$ if $f|_{N'}$ is transverse to S and $f^*(S) \cap N' = \emptyset$ if $f(N') \subset S$. (These are the only two possibilities.) If S is not embedded we just do the same thing, only locally. Let $g\colon S' \to M$ be the immersion associated to S. Then $f^*(S)$ is the set of $x \in N$ so that for some open $V \subset S'$, $g|_V$ is an embedding, f is transverse to $g(V)$ at x and $f(x) \in g(V)$. In terms of our local chart ψ above we thus have $U \cap |f^*(\mathcal{A})| = U \cap f^{-1}\psi\left(\bigcup_{i=1}^{b} \mathbf{R}_i^m\right)$.

Lemma 3.3.1 *Let $\mathcal{A}$ be a tico in M and suppose $f\colon N \to M$ intersects $\mathcal{A}$ cleanly. Then $f^*(\mathcal{A})$ is a tico in N and $f^*(\mathcal{A})$ is regular if $\mathcal{A}$ is regular. If $\mathcal{C}$ is any tico in M with $|\mathcal{C}| \subset |\mathcal{A}|$ then f intersects $\mathcal{C}$ cleanly. If f is transverse to $\mathcal{A}$ then $f\colon (N, f^*(\mathcal{A})) \to (M, \mathcal{A})$ is a tico map. If f is transverse to $\mathcal{A}$ and $\mathcal{A}$ is regular then f has constant exponents, $\alpha_{ST} = 1$ if $S = f^*(T)$ and $\alpha_{ST} = 0$ if $S \neq f^*(T)$.*

Proof: Pick any $z \in N$. Pick a neighborhood U of z in N and a tico chart $\psi\colon (\mathbf{R}^m, 0) \to (M, f(z))$ so that $\psi^{-1}(|\mathcal{A}|) = \bigcup_{i=1}^{a} \mathbf{R}_i^m$, $f(U) \subset \psi\left(\bigcap_{i>b} \mathbf{R}_i^m\right)$ and f is transverse to $\psi\left(\bigcap_{i=1}^{b} \mathbf{R}_i^m\right)$ at z. Since $\psi^{-1}f$ is transverse to $\bigcap_{i=1}^{b} \mathbf{R}_i^m$, we may pick a chart $\eta\colon (\mathbf{R}^n, 0) \to (U, z)$ so that the i-th coordinate of $\psi^{-1}f\eta(z)$ is z_i for all $i \leq b$ and all z near 0. Hence $\eta^{-1}(|f^*(\mathcal{A})|) = \bigcup_{i=1}^{b} \mathbf{R}_i^n$ near 0 so $f^*(\mathcal{A})$ is a tico.

If $\mathcal{A}$ is regular then $\eta^{-1}(f^*(A)) = \mathbf{R}_i^n$ if $A \in \mathcal{A}$ and $\psi^{-1}(A) = \mathbf{R}_i^m$ for some $i \leq b$. Otherwise $\eta^{-1}(f^*(A)) = \emptyset$. So $f^*(A)$ is an embedded submanifold so $f^*(\mathcal{A})$ is regular.

If $\mathcal{C}$ is a tico with $|\mathcal{C}| \subset |\mathcal{A}|$, let $J \subset \{1, \cdots, a\}$ be the index subset so that $\psi^{-1}(|\mathcal{C}|) = \bigcup_{i \in J} \mathbf{R}_i^m$. Then $f(U) \subset \psi\left(\bigcap_{i \in J'} \mathbf{R}_i^m\right)$ and f is transverse to $\psi\left(\bigcap_{i \in J''} \mathbf{R}_i^m\right)$ where $J' = \{i \in J \mid i > b\}$ and $J'' = J - J'$. Hence f intersects $\mathcal{C}$ cleanly.

If f is transverse to $\mathcal{A}$ then $b = a$ so since the i-th coordinate of $\psi^{-1}f\eta(a)$ is z_i for all $i \leq a$ we know that f is a tico map. The exponent $\alpha_{ij} = 0$ for $i \neq j$ and $\alpha_{ii} = 1$ so if $\mathcal{A}$ is regular then the required formulae for α_{ST} hold.   □

Lemma 3.3.2 *Let $\mathcal{A}$ and $\mathcal{C}$ be ticos in M with $\mathcal{A} \subset \mathcal{C}$. Pick any $S \in \mathcal{C} - \mathcal{A}$ and let $g\colon S' \to M$ be the immersion associated to S. Then g is transverse to $\mathcal{A}$. (Hence $g\colon (S', g^*(\mathcal{A})) \to (M, \mathcal{A})$ is a tico map by Lemma 3.3.1.)*

Proof: Pick any $x \in S'$. Pick a tico chart $\psi\colon (\mathbf{R}^m, 0) \to (M, g(x))$ so that $\psi^{-1}(|\mathcal{A}|) = \bigcup_{i=1}^{a} \mathbf{R}_i^m$ and $\psi^{-1}(|\mathcal{C}|) = \bigcup_{i=1}^{c} \mathbf{R}_i^m$. Let d be such that $\psi^{-1}g(U) = \mathbf{R}_d^m$ for some neighborhood U of x in S'. Since $S \notin \mathcal{A}$ we know that $d > a$ so $\psi^{-1}g|_U$ is transverse to $\bigcap_{i=1}^{a} \mathbf{R}_i^m$. Hence g is transverse to $\mathcal{A}$.   □

Definitions: If $\mathcal{A}$ is a tico in M and $L \subset M$ is a proper smooth submanifold, we say that L intersects $\mathcal{A}$ *cleanly* if the embedding map $L \hookrightarrow M$ hits $\mathcal{A}$ cleanly, i.e., for each $z \in L$ there is a chart $\psi \colon (\mathbf{R}^m, 0) \to (M, z)$ so that $\psi^{-1}(|\mathcal{A}|) = \bigcup_{i=1}^{a} \mathbf{R}_i^m$ and $\psi^{-1}(L) = \bigcap_{i=b}^{c} \mathbf{R}_i^m$. We say that L is *unobtrusive* if L hits $\mathcal{A}$ cleanly and $L \subset |\mathcal{A}|$, i.e., we can always take $b \leq a$ in the above chart. We say that L is *unskewed* if for every connected component L' of L and every $S \in \mathcal{A}$ then $L' \subset S$ implies $L \subset S$. In other words, either $L \subset S$ or $\dim(L \cap S) < \dim L$. We say that L is *fat* if it is unskewed and it is a union of connected components of strata, (so, we can always take $c \leq a$ in the above). For example M is fat but not unobtrusive.

If $\mathcal{A}$ is regular this means that L hits $\mathcal{A}$ cleanly if for each component L' of L there is a $\mathcal{B} \subset \mathcal{A}$ so that L' is transverse to $|\mathcal{A} - \mathcal{B}|$ and $L' \subset \cap \mathcal{B}$. If L is unskewed then $\mathcal{B}$ is independent of L', i.e., L is transverse to $|\mathcal{A} - \mathcal{B}|$ and $L \subset \cap \mathcal{B}$. If L is fat then L is a union of connected components of $\bigcap \mathcal{B}$.

Lemma 3.3.3. *Let $(X, \mathcal{A})$ be an algebraic tico and suppose $Y \subset X$ is an irreducible nonsingular Zclosed subset of X which intersects $\mathcal{A}$ cleanly. Then Y is unskewed.*

Proof: Suppose K is some connected component of Y and S is some sheet of $\mathcal{A}$. If $K \subset S$ then $\dim Y \cap S = \dim Y$. Hence $Y \cap S = Y$ by irreducibility of Y and Lemma 2.2.9. So $Y \subset S$, thus Y is unskewed. $\square$

Definition: If L is any submanifold of M such that the inclusion map $i \colon L \to M$ hits $\mathcal{A}$ cleanly we let $\mathcal{A} \cap L$ or $L \cap \mathcal{A}$ denote $i^*(\mathcal{A})$. It is worth emphasizing the case where L is an open subset of M. Then the inclusion $i \colon L \to M$ always hits $\mathcal{A}$ cleanly (in fact it is transverse to $\mathcal{A}$) and $L \cap \mathcal{A} = \{ L \cap A \mid A \in \mathcal{A} \}$. Also, if $\mathcal{A}$ is regular and $S \in \mathcal{A}$ then $S \cap \mathcal{A}$ is the tico $\{ S \cap T \mid T \in \mathcal{A}, \ T \neq S \}$.

Definitions: We say that $\pi \colon (M', \mathcal{A}') \to (M, \mathcal{A})$ is a *tico blowup* with center L if L is a proper smooth submanifold of M, L intersects $\mathcal{A}$ cleanly, $M' = \mathfrak{B}(M, L)$ is the blowup of M with center L, $\pi = \pi(M, L) \colon M' \to M$ is the blowup map and $\mathcal{A}'$ is the tico

$$\mathcal{A}' = \{ \pi^{-1}(L) \} \cup \{ \mathrm{Cl}\left(\pi^{-1}(A - L) \right) \mid A \in \mathcal{A} \}.$$

Lemma 3.3.4 below will show that $\mathcal{A}'$ is a tico and π is a tico map. We call $\mathcal{A}'$ the *total transform* of $\mathcal{A}$ and we call $\{ \mathrm{Cl}\left(\pi^{-1}(A - L) \right) \mid A \in \mathcal{A} \}$ the *strict transform* of $\mathcal{A}$. We sometimes use the notation $\mathcal{A}' = \mathfrak{B}(\mathcal{A}, L)$. We say $\pi \colon (M', \mathcal{A}') \to (M, \mathcal{A})$ is a *tico multiblowup* if there is a sequence of tico blowups $\pi_i \colon (M_i, \mathcal{A}_i) \to (M_{i-1}, \mathcal{A}_{i-1})$ with centers $L_i \subset M_{i-1}$ for $i = 1, \ldots, n$ so that $(M_0, \mathcal{A}_0) = (M, \mathcal{A})$, $(M_n, \mathcal{A}_n) = (M', \mathcal{A}')$ and $\pi = \pi_1 \circ \cdots \circ \pi_n$. We say the centers of π *lie over* a subset $K \subset M$ if $\pi_1 \pi_2 \cdots \pi_{i-1}(L_i) \subset K$ for all $i = 1, \ldots, n$. We say π is a *fat multiblowup*, an *unobtrusive multiblowup* or an *unskewed multiblowup* if each center L_i is respectively fat, unobtrusive or unskewed. We say that $\mathcal{A}'$ is the *total transform* of $\mathcal{A}$. If A is a sheet of $\mathcal{A}$ then the *strict preimage* or *strict transform* of A is a sheet $A_n \in \mathcal{A}'$ where $A_0 = A$ and A_i is the strict preimage of A_{i-1} for $i = 1, \ldots, n$. The *strict transform* of $\mathcal{A}$ is the set of strict preimages of sheets $A \in \mathcal{A}$. By Lemmas 3.2.4 and 3.3.4, a tico multiblowup is a tico map.

Definition: If $(X, \mathcal{A})$ is an algebraic tico then an *algebraic tico multiblowup* is a tico multiblowup $\pi\colon (X', \mathcal{A}') \to (X, \mathcal{A})$ where each center is a nonsingular Zclosed subset. (In particular, $(X', \mathcal{A}')$ is an algebraic tico also.)

Lemma 3.3.4 *Let* $\pi\colon (M', \mathcal{A}') \to (M, \mathcal{A})$ *be a tico blowup with center L. Then $\mathcal{A}'$ is a tico, π is a tico map and $|\mathcal{A}'| = \pi^{-1}(|\mathcal{A}| \cup L)$. Furthermore, if $\mathcal{A}$ is regular then $\mathcal{A}'$ is regular and the exponent map $\alpha_{ST}\colon S \to \mathbf{Z}$ for $S \in \mathcal{A}'$, $T \in \mathcal{A}$ is as follows:*

$$
\alpha_{ST}(z) = \begin{cases}
1 & \text{if } S = \mathcal{B}(T, T \cap L) \\
0 & \text{if } S = \mathcal{B}(T', T' \cap L),\ T \neq T' \\
1 & \text{if } S = \pi^{-1}(L) \quad \text{and} \quad L_z \subset T \text{ where } L_z \\
  & \text{is the component of } L \text{ containing } \pi(z) \\
0 & \text{if } S = \pi^{-1}(L) \quad \text{and} \quad L_z \not\subset T
\end{cases}
$$

In particular, if L is unskewed then π has constant exponents.

Proof: Let $E = \pi^{-1}(L)$. We know that $\pi|\colon M' - E \to M - L$ is a diffeomorphism and $\mathcal{A}' \cap (M' - E) = (\pi|)^*(\mathcal{A}) \cup \{\emptyset\}$. Hence the restriction map

$$
\pi|\colon (M' - E,\ \mathcal{A}' \cap (M' - E)) \to (M - L,\ \mathcal{A} \cap (M - L))
$$

is a tico map by Lemmas 3.3.1 and 3.2.5. If $\mathcal{A}$ is regular, $S = \mathcal{B}(T', T' \cap L)$ and $z \in S - E$ we may apply Lemma 3.3.1 to $\pi|$ and obtain:

$$
\alpha_{ST}(z) = \begin{cases}
1 & \text{if } T = T' \\
0 & \text{if } T \neq T'
\end{cases}
$$

Since $S - E$ is dense in S, the above formulae hold for all $z \in S$. Consequently, we only need to show that π is a tico map at points of E and the α_{ST} formulae for $S = E$ hold if $\mathcal{A}$ is regular.

So pick any $z \in E$. Pick a chart $\theta\colon (\mathbf{R}^m, 0) \to (M, \pi(z))$ so that $\theta^{-1}(|\mathcal{A}|) = \bigcup_{i=1}^{a} \mathbf{R}_i^m$ and $\theta^{-1}(L) = \bigcap_{i=b}^{c} \mathbf{R}_i^m$. By Lemma 2.5.1 we have charts $\psi_j\colon \mathbf{R}^m \to M'$, $j = b, \ldots, c$ covering $\pi^{-1}\theta(\mathbf{R}^m)$ so that if $\lambda_{ji}(y)$ is the i-th coordinate of $\theta^{-1}\pi\psi_j(y)$ then $\lambda_{ji}(y) = y_i$ for $i < b$ or $i > c$ or $i = j$ and $\lambda_{ji}(y) = y_i y_j$ for $b \leq i \leq c$, $i \neq j$.

Choose some j so that $z \in \psi_j(\mathbf{R}^m)$. Notice $\psi_j^{-1}(E) = \mathbf{R}_j^m$. For any $A \in \mathcal{A}$ let $J_A = \{i \in \mathbf{Z} \mid 1 \leq i \leq a \text{ and } \theta(\mathbf{R}_i^m) \subset A\}$. (In the regular case, of course

J_A has at most one element). Then

$$
\begin{aligned}
\psi_j^{-1}\left(\mathfrak{B}\left(A, A \cap L\right)\right) &= \psi_j^{-1}\left(\mathrm{Cl}\left(\pi^{-1}\left(A - L\right)\right)\right) \\
&= \mathrm{Cl}\left\{y \in \mathbf{R}^m \mid \lambda_{ji}\left(y\right) = 0 \text{ for some } i \in J_A \right. \\
&\quad\quad \left. \text{and } \lambda_{ji}\left(y\right) \neq 0 \text{ for some } i \in \{b, \dots, c\}\right\} \\
&= \mathrm{Cl}\left\{y \in \mathbf{R}^m - \mathbf{R}_j^m \mid y_i = 0 \text{ for some } i \in J_A \right\} \\
&= \bigcup_{i \in J_A - j} \mathbf{R}_i^m.
\end{aligned}
$$

So $\mathcal{A}'$ is a tico and $\psi_j^{-1}\left(|\mathcal{A}'|\right) = \bigcup_{i=1}^a \mathbf{R}_i^m \cup \mathbf{R}_j^m = \psi_j^{-1}\pi^{-1}\left(|\mathcal{A}| \cup L\right)$, so $|\mathcal{A}'| = \pi^{-1}\left(|\mathcal{A}| \cup L\right)$. Furthermore since the λ_{ji} maps are all monomials we know that π is a tico map.

Now suppose $\mathcal{A}$ is regular. Let $T_1, \dots, T_a \in \mathcal{A}$ be the sheets so that $\theta^{-1}\left(T_i\right) = \mathbf{R}_i^m$. Let $S_i = \mathfrak{B}\left(T_i, T_i \cap L\right)$. Then $\psi_j^{-1}\left(S_i\right) = \mathbf{R}_i^m$ if $i \neq j$, $\psi_j^{-1}\left(E\right) = \mathbf{R}_j^m$ and $\psi_j^{-1}\left(S\right)$ is empty for all $S \in \mathcal{A}'$ with $S \neq E$ and $S \neq S_i$, $1 \leq i \leq a$, $i \neq j$. Suppose $z \in E \cap \psi_j\left(\mathbf{R}^m\right)$. Then looking at the exponents of $\lambda_{ji}\left(y\right)$ we see that $\alpha_{ET}\left(z\right) = 1$ if $T = T_i$ for some $i \in \{b, \dots, a\}$ and $\alpha_{ET}\left(z\right) = 0$ otherwise. So our exponent map is just what we wish for $z \in E = \pi^{-1}\left(L\right)$. So we are done. $\square$

Lemma 3.3.5 *Let $\pi \colon (N', \mathcal{B}') \to (N, \mathcal{B})$ be a tico blowup with fat center L and let $f \colon (M, \mathcal{A}) \to (N, \mathcal{B})$ be a tico map. Suppose there is a map $g \colon M \to N'$ so that $\pi g = f$. Then $g \colon (M, \mathcal{A}) \to (N', \mathcal{B}')$ is a tico map.*

Furthermore, if $\mathcal{A}$ and $\mathcal{B}$ are regular, L is an intersection of sheets of $\mathcal{B}$ and f has constant exponents then g has constant exponents also. In particular, if $S \in \mathcal{A}$ and $T \in \mathcal{B}$ let α_{ST} be the exponent of S in T. Let σ_S be the minimum of α_{ST} over all $T \in \mathcal{B}$ with $L \subset T$. Then the exponents β_{SU} of g are as follows.

$$
\beta_{SU} = \begin{cases}
\alpha_{ST} & \text{if } U = \mathfrak{B}\left(T, T \cap L\right) \text{ and } L \not\subset T \\
\alpha_{ST} - \sigma_S & \text{if } U = \mathfrak{B}\left(T, L\right) \text{ and } L \subset T \\
\sigma_S & \text{if } U = \pi^{-1}\left(L\right)
\end{cases}
$$

Proof: Lemmas 3.2.7 and 3.3.4 imply g is a tico map since $|\mathcal{B}'| = \pi^{-1}\left(|\mathcal{B}|\right)$ by Lemma 3.3.4. (Note either $L \subset |\mathcal{B}|$ or else $\dim L = \dim N$ so $\pi^{-1}\left(L\right) = \emptyset$.)

Now suppose $\mathcal{A}$ and $\mathcal{B}$ are regular, $L = \cap \mathcal{L}$ for some $\mathcal{L} \subset \mathcal{B}$ and f has constant exponents. Let α_{ST}, β_{SU} and γ_{UT} be the exponent maps of f, g and π respectively. Notice α_{ST} is constant and γ_{ST} is constant by Lemma 3.3.4. But by Lemma 3.2.4 we know that $\alpha_{ST} = \sum_{U \in \mathcal{B}'} \beta_{SU}\left(x\right) \cdot \gamma_{UT}$ for any $x \in S$. By Lemma 3.3.4 we know that $\gamma_{UT} = 1$ if $U = \mathfrak{B}\left(T, T \cap L\right)$ or if $U = \pi^{-1}\left(L\right)$ and $T \in \mathcal{L}$. Otherwise $\gamma_{UT} = 0$. Hence $\alpha_{ST} = \beta_{ST'}\left(x\right)$ if $T \notin \mathcal{L}$ and $T' = \mathfrak{B}\left(T, T \cap L\right)$. Also $\alpha_{ST} = \beta_{ST'}\left(x\right) + \beta_{SE}\left(x\right)$ if $T \in \mathcal{L}$, $T' = \mathfrak{B}\left(T, L\right)$ and E denotes $\pi^{-1}\left(L\right)$. So we only need show $\beta_{SE}\left(x\right) = \sigma_S$ for all S and $x \in S$.

Suppose that $f\left(S'\right) \not\subset L$ for some component S' of a sheet $S \in \mathcal{A}$. Then by Lemma 3.2.2 we know that $\alpha_{ST''} = 0$ for some $T'' \in \mathcal{L}$, hence $\sigma_S = 0$ and

$f(S'') \not\subset L$ for all components S'' of S. But then we must also have $g(S'') \not\subset E$ for all components S'' of S. Hence by Lemma 3.2.2 we must have $\beta_{SE}(x) = 0 = \sigma_S$ for all $x \in S$.

Now suppose that $f(S') \subset L$ for some component S' of $S \in \mathcal{A}$. Then by Lemma 3.2.2 we know that $\alpha_{ST} > 0$ for all $T \in \mathcal{L}$ and hence $f(S) \subset L$ and $\sigma_S > 0$. Suppose that $\beta_{SE}(x) < \sigma_S$ for some $x \in S$. Then $\beta_{ST'}(x) > 0$ for all T' in the strict transform $\mathcal{L}'$ of $\mathcal{L}$ since $\beta_{ST'}(x) = \alpha_{ST} - \beta_{SE}(x)$. Hence $g(x) \in \bigcap_{T \in \mathcal{L}} \mathfrak{B}(T, L) = \mathfrak{B}(\cap \mathcal{L}, L) = \mathfrak{B}(L, L) = \emptyset$, a contradiction. So $\beta_{SE}(x) \geq \sigma_S$ for all $x \in S$. But if $\beta_{SE}(x) > \sigma_S$ then pick $T \in \mathcal{L}$ with $\sigma_S = \alpha_{ST}$ and let $T' = \mathfrak{B}(T, L)$. Then $\beta_{ST'}(x) = \alpha_{ST} - \beta_{SE}(x) = \sigma_S - \beta_{SE}(x) < 0$, contradicting the positivity of $\beta_{ST'}$. Hence $\beta_{SE}(x) = \sigma_S$ for all $x \in S$. □

If $\pi\colon (M', \mathcal{A}') \to (M, \mathcal{A})$ is a tico multiblowup, then π is a tico map by Lemma 3.3.4. Hence Lemma 3.2.6 implies that if T' is a stratum of M', then $\pi(T')$ is contained in a stratum of M. A fat tico multiblowup has a stronger property as the next lemma shows.

Lemma 3.3.6 Let $\pi\colon (M', \mathcal{A}') \to (M, \mathcal{A})$ be a fat tico multiblowup. Let T' be any stratum of the tico stratification of M' and let T be the stratum of M such that $\pi(T') \subset T$. Then $\pi|\colon T' \to T$ is a locally trivial fibration (in fact the fibre is a disjoint union of Euclidean spaces, so T' is a vector bundle over a finite cover of T). In particular $\dim T' \geq \dim T$.

Proof: It suffices to consider the case where π is a single blowup with center L. If $T \not\subset L$ then $T \cap L = \emptyset$. So $\pi|\colon T' \to T$ is a diffeomorphism, hence a fibration. On the other hand, if $T \subset L$ then $\pi|\colon \pi^{-1}(T) \to T$ is a fibration with projective space fiber.

Furthermore, the stratification on $\pi^{-1}(T)$ is obtained by taking the fiber $\mathbf{RP}^{k-1}$, taking k codimension one linear subspaces in $\mathbf{RP}^{k-1}$ in general position, taking the resulting tico stratification and then forming some bundle over T. Here k is the codimension of L and the linear subspaces are the tangent directions in the various sheets containing L. Hence $\pi|\colon T' \to T$ is a fibration with fiber diffeomorphic to a disjoint union of Euclidean spaces $\mathbf{R}^n$, since it is easy to see that each stratum of $\mathbf{RP}^{k-1}$ is diffeomorphic to some Euclidean space.

For an example of this lemma, suppose $\dim M = 3$ and L is a circle with trivial normal bundle. Suppose also that L is a stratum, i.e., there are no triple points of $|\mathcal{A}|$ on L. Then a neighborhood of L in $|\mathcal{A}|$ is an X bundle over L where X is taken literally, i.e., it is two lines with midpoints identified. Suppose this bundle is nontrivial and as you go around L once, X rotates $90°$. Then $\pi^{-1}(L)$ is $L \times \mathbf{RP}^1 = S^1 \times S^1$. Now $\pi^{-1}(L)$ intersected with the strict transform of $|\mathcal{A}|$ is a curve which runs twice around the torus $\pi^{-1}(L)$. This stratum of M' double covers L. The other stratum over L has fiber $\mathbf{RP}^1$ with two points deleted, i.e., two copies of $\mathbf{R}$. □

Lemma 3.3.7 *Let* $\pi\colon (M',\mathcal{A}') \to (M,\mathcal{A})$ *be a fat tico multiblowup with* $\mathcal{A}$ *regular and suppose* $\mathcal{B}$ *is a tico in* M *such that* $|\mathcal{A}| \subset |\mathcal{B}|$ *and each sheet of* $\mathcal{A}$ *is unskewed in* $\mathcal{B}$. *Then* $\pi\colon (M',\mathcal{B}') \to (M,\mathcal{B})$ *is a fat tico multiblowup, i.e., all the centers are fat with respect to* $\mathcal{B}$ *and the total transforms of* $\mathcal{B}$.

Proof: We will prove this by induction on the number of blowups in π. Suppose $\pi = \pi'' \circ \pi'$ where $\pi''\colon (M'',\mathcal{A}'') \to (M,\mathcal{A})$ is a tico blowup with fat center L. For each $z \in L$, pick a chart $\psi\colon (\mathbf{R}^m,0) \to (M,z)$ so that $\psi^{-1}(|\mathcal{A}|) = \bigcup_{i=1}^a \mathbf{R}_i^m$ and $\psi^{-1}(|\mathcal{B}|) = \bigcup_{i=1}^b \mathbf{R}_i^m$. Since $|\mathcal{B}| \supset |\mathcal{A}|$ we know that $b \geq a$. Since L is fat (with respect to $\mathcal{A}$), $\psi^{-1}(L) = \bigcap_{i \in C} \mathbf{R}_i^m$ for some $C \subset \{1,\dots,a\}$. Hence L is a union of strata (with respect to the $\mathcal{B}$ stratification).

So we must show that L is unskewed (with respect to $\mathcal{B}$). Take a connected component L' of L. Suppose $L' \subset B \in \mathcal{B}$. By the above local analysis there is a sheet $A \in \mathcal{A}$ so that $L' \subset A$. But then $L \subset A$ by unskewedness of L in $\mathcal{A}$. Let A' be the component of A containing L'. Then $A' \subset B$ so $A \subset B$ by unskewedness. So $L \subset A \subset B$ so L is unskewed in $\mathcal{B}$.

So if $\mathcal{B}''$ is the total transform of $\mathcal{B}$ we must show $|\mathcal{A}''| \subset |\mathcal{B}''|$ and each sheet of $\mathcal{A}''$ is unskewed in $\mathcal{B}''$. But $|\mathcal{B}''| = {\pi''}^{-1}(|\mathcal{B}|) \supset {\pi''}^{-1}(|\mathcal{A}|) = |\mathcal{A}''|$. Also

$$\mathcal{B}'' = \{\,\mathfrak{B}(B, B \cap L) \mid B \in \mathcal{B}\,\} \cup \{{\pi''}^{-1}(L)\} \ \text{ and}$$
$$\mathcal{A}'' = \{\,\mathfrak{B}(A, A \cap L) \mid A \in \mathcal{A}\,\} \cup \{{\pi''}^{-1}(L)\}.$$

Certainly ${\pi''}^{-1}(L)$ is unskewed in $\mathcal{B}''$. A component of $\mathfrak{B}(A, A \cap L)$ for $A \in \mathcal{A}$ is $\mathfrak{B}(A', A' \cap L)$ for some component A' of A (with $A' \not\subset L$). If $\mathfrak{B}(A', A' \cap L) \subset \mathfrak{B}(B, B \cap L)$ for $B \in \mathcal{B}$ then $A' \subset B$ so $A \subset B$ so $\mathfrak{B}(A, A \cap L)$ is unskewed.

By induction, $\pi'\colon (M',\mathcal{B}') \to (M'', \mathfrak{B}(\mathcal{B},L))$ is a fat tico multiblowup so we are done. $\square$

Proposition 3.3.8 *Let* $(M,\mathcal{A})$ *be a tico. Then there is a tico multiblowup* $\pi\colon (M',\mathcal{A}') \to (M,\mathcal{A})$ *with fat unobtrusive centers so that* $\mathcal{A}'$ *is regular.*

Proof: For example let $M = \mathbf{R}^2$ and $|\mathcal{A}| =$ a figure eight. After blowing up the double point of the figure eight we see M' is a Möbius strip, $\pi^{-1}(L)$ is the center circle and the strict transform of the figure eight is an embedded circle intersecting $\pi^{-1}(L)$ in two points.

We now prove the proposition. For any $S \in \mathcal{A}$ let $\beta(S)$ be the depth of S, i.e., the maximum number of self intersections of S. Let $d(\mathcal{A}) = \max_{S \in \mathcal{A}} \beta(S)$ and let $e(\mathcal{A})$ be the number of sheets $S \in \mathcal{A}$ with $\beta(S) = d(\mathcal{A})$. Our proof will be by induction on $(d(\mathcal{A}), e(\mathcal{A}))$ with lexicographical order. Let $d = d(\mathcal{A})$. Notice that if $d = 1$ we are done, since all sheets are then embedded. So assume $d > 1$. Pick $S_0 \in \mathcal{A}$ with $\beta(S_0) = d$ and let $T \subset S_0$ be the points where S_0 has d-fold self intersection. Then T is a fat unobtrusive proper submanifold of M (with codimension d). Let $N = \mathfrak{B}(M,T)$, $p = \pi(M,T)$ and let $\mathcal{B}$ be the total transform of $\mathcal{A}$. Then if $S \in \mathcal{A}$, $S \neq S_0$ and S' is the strict preimage of S we see that $\beta(S') = \beta(S)$. If S_0' is the strict preimage of S_0 then $\beta(S_0') = d - 1$. Also $\beta(p^{-1}(T)) = 1$ since $p^{-1}(T)$ is embedded. Hence if $e(\mathcal{A}) > 1$ we have $d(\mathcal{B}) = d(\mathcal{A})$ and $e(\mathcal{B}) = e(\mathcal{A}) - 1$. If $e(\mathcal{A}) = 1$ we have $d(\mathcal{B}) = d(\mathcal{A}) - 1$. So by induction we are done (c.f., Lemma 2.3 of [**AK8**]). $\square$

The next result, Proposition 3.3.9, is very important since it allows us to lift a tico map to a multiblowup. Actually one can weaken the fatness hypothesis, but it does make the proof easier.

Proposition 3.3.9 Let $f\colon (M, \mathcal{A}) \to (N, \mathcal{B})$ be a tico map where $\mathcal{A}$ is a finite tico and let $\pi\colon (N', \mathcal{B}') \to (N, \mathcal{B})$ be a fat tico multiblowup with centers lying over some $Z \subset N$. Then there is a fat tico multiblowup $p\colon (M', \mathcal{A}') \to (M, \mathcal{A})$ and a tico map $g\colon (M', \mathcal{A}') \to (N', \mathcal{B}')$ so that $\pi g = fp$ and the centers of p lie over $f^{-1}(Z)$.

If the multiblowup p is unobtrusive we may guarantee that the centers of p are unobtrusive. If $(M, \mathcal{A})$, $(N, \mathcal{B})$ and π are algebraic and f is an entire rational function, we may guarantee that p is an algebraic tico multiblowup and g is an entire rational function.

Proof: It suffices to consider the case where π consists of a single blowup with fat center L and $L = Z$. The general result then follows by repeated applications of the single blowup case.

First suppose that L is not unobtrusive. Then L is a union of components of N and $N' = N - L$. We just let $K = f^{-1}(L)$, then K is a union of components of M and we may let $M' = M - K = \mathfrak{B}(M, K)$.

So we may as well assume that L is unobtrusive. Recall that by Lemma 2.5.8 our goal is to make $p^* f^* \mathfrak{I}^{\infty}(L)$ locally principal.

Before doing the general proof, we will give an example. We let $M = \mathbf{R}^4$, $\mathcal{A} = \{\mathbf{R}_1^4, \mathbf{R}_2^4, \mathbf{R}_3^4, \mathbf{R}_4^4\}$, $N = \mathbf{R}^2$, $\mathcal{B} = \{\mathbf{R}_1^2, \mathbf{R}_2^2\}$, $L = (0,0)$ and $f(x) = (x_3^3 x_4, x_1 x_2^2 x_3 x_4^2)$. Let $A_i = \mathbf{R}_i^4$.

Now f does not lift to the blowup because

$$f^*(\mathfrak{I}(L)) = \langle x_3^3 x_4, x_1 x_2^2 x_3 x_4^2 \rangle = x_3 x_4 \langle x_3^2, x_1 x_2^2 x_4 \rangle$$

is not locally principal at 0. We describe the map f by the following data:

$$1234 \qquad \begin{smallmatrix} 0 & 0 & 3 & 1 \\ 1 & 2 & 1 & 2 \end{smallmatrix} \qquad 4 \qquad 1$$

The 1234 means that the center of this chart is the point intersection $A_1 \cap A_2 \cap A_3 \cap A_4$. The next entry gives the exponents of f on the chart. The next two entries give $\gamma(D)$ and $\lambda(D)$ for $D = \{1, 2, 3, 4\}$ where γ and λ are as in the proof below. To see where $\gamma(D)$ comes from, look at the ideal $\langle x_3^2, x_1 x_2^2 x_4 \rangle$ which is the reduced form of $f^*(\mathfrak{I}(L))$ and note that the highest exponent in the first coordinate is x_3^2 and the highest exponent in the second is x_2^2, then $\gamma(D) = 2 + 2$

is the sum of the exponents. The next entry $\lambda(D)$ is the number of pairs which give $\gamma(D)$, in this case only the one pair $(3, 2)$.

We now blow up $A_2 \cap A_3$ because the pair $(3, 2)$ gave the largest $\gamma(D)$. In the blowup, the tico transform of $\mathcal{A}$ has five sheets, the strict transforms of A_i for $i = 1, 2, 3, 4$ (which we just call A_i again) and the inverse image of the center which we call A_5. We describe the map f on the blowup by the following data:

$$1245 \qquad \begin{smallmatrix} 0 & 0 & 1 & 3 \\ 1 & 2 & 2 & 3 \end{smallmatrix} \qquad *$$

$$1345 \qquad \begin{smallmatrix} 0 & 3 & 1 & 3 \\ 1 & 1 & 2 & 3 \end{smallmatrix} \qquad 3 \qquad 2$$

The $*$ indicates that on the first chart the pullback of the ideal has become locally principal, it is just $\langle x_4 x_5^3 \rangle$. So we can ignore this chart and concentrate on the chart 1345, i.e., the chart with center $A_1 \cap A_3 \cap A_4 \cap A_5$.

Now either $(3, 1)$ or $(3, 4)$ gives the maximal γ, we choose one of them for our blowup, say $(3, 4)$. So we blow up $A_3 \cap A_4$. Then f can be described by:

$$1356 \qquad \begin{smallmatrix} 0 & 3 & 3 & 4 \\ 1 & 1 & 3 & 3 \end{smallmatrix} \qquad 3 \qquad 1$$

$$1456 \qquad \begin{smallmatrix} 0 & 1 & 3 & 4 \\ 1 & 2 & 3 & 3 \end{smallmatrix} \qquad 2 \qquad 2$$

We ignore the blowup of the chart 1245, since f will lift on that chart.

Next we blow up $A_1 \cap A_3$ since that pair gives the maximal γ. We obtain:

$$1567 \qquad \begin{smallmatrix} 0 & 3 & 4 & 3 \\ 1 & 3 & 3 & 2 \end{smallmatrix} \qquad 2 \qquad 2$$

$$3567 \qquad \begin{smallmatrix} 3 & 3 & 4 & 3 \\ 1 & 3 & 3 & 2 \end{smallmatrix} \qquad *$$

$$1456 \qquad \begin{smallmatrix} 0 & 1 & 3 & 4 \\ 1 & 2 & 3 & 3 \end{smallmatrix} \qquad 2 \qquad 2$$

Next we blow up $A_1 \cap A_6$ and obtain:

$$5678 \qquad \begin{smallmatrix} 3 & 4 & 3 & 4 \\ 3 & 3 & 2 & 4 \end{smallmatrix} \qquad *$$

$$1578 \qquad \begin{smallmatrix} 0 & 3 & 3 & 4 \\ 1 & 3 & 2 & 4 \end{smallmatrix} \qquad 2 \qquad 1$$

$$4568 \qquad \begin{smallmatrix} 1 & 3 & 4 & 4 \\ 2 & 3 & 3 & 4 \end{smallmatrix} \qquad 2 \qquad 1$$

$$1458 \qquad \begin{smallmatrix} 0 & 1 & 3 & 4 \\ 1 & 2 & 3 & 4 \end{smallmatrix} \qquad *$$

Next we blow up $A_1 \cap A_7$ to obtain:

$$5789 \qquad \begin{smallmatrix} 3 & 3 & 4 & 3 \\ 3 & 2 & 4 & 3 \end{smallmatrix} \qquad *$$

$$1589 \qquad \begin{smallmatrix} 0 & 3 & 4 & 3 \\ 1 & 3 & 4 & 3 \end{smallmatrix} \qquad *$$

$$4568 \qquad \begin{smallmatrix} 1 & 3 & 4 & 4 \\ 2 & 3 & 3 & 4 \end{smallmatrix} \qquad 2 \qquad 1$$

Finally we blow up $A_4 \cap A_6$ to obtain:

$$5680 \qquad \begin{smallmatrix} 3 & 4 & 4 & 5 \\ 3 & 3 & 4 & 5 \end{smallmatrix} \qquad *$$

$$4580 \qquad \begin{smallmatrix} 1 & 3 & 4 & 5 \\ 2 & 3 & 4 & 5 \end{smallmatrix} \qquad *$$

So at last we know f will lift everywhere on the blowup.

The preceeding example gives the main idea behind this proof, we must just exercise a certain amount of care to make sure everything works. We now give the proof.

Assertion 3.3.9.1 *After perhaps replacing $(M, \mathcal{A})$ with some tico multiblowup, we may as well assume that $f^{-1}(L) = |\mathcal{C}|$ for some regular tico $\mathcal{C}$. Furthermore, for each sheet $S \in \mathcal{C}$ there is a regular tico $\mathcal{A}_S$ in S so that $|\mathcal{A}_S| = |S \cap \mathcal{A}|$.*

Proof: If $\mathcal{A}$ is regular, this is easy since $f^{-1}(L)$ is a finite union of fat submanifolds of M. After blowing these up one by one $f^{-1}(L)$ becomes a finite union of codimension one fat submanifolds, i.e., the realization of a regular tico. The second part follows by taking $\mathcal{A}_S = S \cap \mathcal{A}$.

The nonregular case is similar, but now $f^{-1}(L)$ is a finite union of things which are like fat submanifolds, only immersed. So you turn them into embedded fat submanifolds by first blowing up their d-fold self intersections as we did in Proposition 3.3.8. Then you blow them up one by one as above and get $f^{-1}(L)$ to be the realization of a regular tico. Now for any sheet S, the tico $(S, S \cap \mathcal{A})$ may not be regular but by Proposition 3.3.8 there is a fat unobtrusive tico multiblowup of S which makes it regular. Using the same centers, you obtain a fat unobtrusive tico multiblowup of M. Do this for each sheet in $f^{-1}(L)$ and eventually the required condition is reached.

The reader dissatisfied with this proof outline could alternatively first blow up $(M, \mathcal{A})$ using Proposition 3.3.8 to make it regular, then proceed as in the above regular case. We lose the conclusion that the centers lie over $f^{-1}(L)$, but they still lie over $|\mathcal{A}| \cup f^{-1}(L)$ which is all we need in the applications in this book. □

Assume that the conclusions of Assertion 3.3.9.1 hold. Let $\mathcal{U}(f)$ denote the set of points at which the ideal $f^* \mathcal{I}^\infty(L)$ is locally principal.

Assertion 3.3.9.2 *$\mathcal{U}(f)$ is open and is a union of strata of M. If $\pi \colon M' \to M$ is any multiblowup then $\pi^{-1}(\mathcal{U}(f)) \subset \mathcal{U}(f\pi)$.*

Proof: Pick any $y \in \mathcal{U}(f)$. As usual, pick tico charts $\theta \colon (\mathbf{R}^n, 0) \to (N, f(y))$ and $\psi \colon (\mathbf{R}^m, 0) \to (M, y)$ so that $\theta^{-1}(L) = \bigcap_{i=1}^c \mathbf{R}_i^n$, $\theta^{-1}(|\mathcal{B}|) = \bigcup_{i=1}^b \mathbf{R}_i^n$, $c \le b$, $\psi^{-1}(|\mathcal{A}|) = \bigcup_{j=1}^a \mathbf{R}_j^m$ and so that $f\psi(\mathbf{R}^m) \subset \theta(\mathbf{R}^n)$. Let $f_i(x)$ be the i-th coordinate of $\theta^{-1}f\psi(x)$. Then $f_i(x) = \prod_{j=1}^a x_j^{\alpha_{ij}} \varphi_i(x)$ and $\varphi_i(0) \ne 0$ for $i \le b$. Notice that $\psi^* f^* \mathcal{I}^\infty(L)$ is the ideal with the c generators $\prod_{j=1}^a x_j^{\alpha_{ij}}$ for $i = 1, \dots, c$. Now $f^* \mathcal{I}^\infty(L)$ is locally principal at y if and only if $\psi^* f^* \mathcal{I}^\infty(L)$ is locally principal at 0, i.e., for some $i_0 \le c$, we have $a_{i_0 j} \le a_{ij}$ for all $i = 1, \dots, c$ and $j = 1, \dots, a$. But then $\psi^* f^* \mathcal{I}^\infty(L)$ is principal so $\psi(\mathbf{R}^m) \subset \mathcal{U}(f)$. We also see that any stratum containing a point of $\psi(\mathbf{R}^m)$ must be in $\mathcal{U}(f)$ since the exponents α_{ij} are locally constant.

Finally, if π is a multiblowup (or indeed any map) and $\pi(x) \in \mathcal{U}(f)$ then locally near x, we know $(f\pi)^* \mathcal{I}^\infty(L) = \langle g \circ \pi \rangle$ if $f^* \mathcal{I}^\infty(L) = \langle g \rangle$ near $\pi(x)$. So $\pi^{-1}(\mathcal{U}(f)) \subset \mathcal{U}(f\pi)$. □

If $\mathcal{B}$ were a regular tico then the following proof could be simplified a bit since we could talk about the exponent map for f. But we need to apply this in the situation where $\mathcal{B}$ is not regular, so we only have the exponents locally, which entails a more complicated proof. In particular the proof will proceed by taking a stratum T_1 of M, then finding a multiblowup $\pi_1 \colon M_1 \to M$ so that $\pi_1^{-1}(T_1) \subset \mathcal{U}(f\pi_1)$. We then pick another stratum T_2 and find a multiblowup $\pi_2 \colon M_2 \to M_1$ so that $\pi_2^{-1}\pi_1^{-1}(T_2) \subset \mathcal{U}(f\pi_1\pi_2)$. Since $\mathcal{A}$ is finite this process eventually terminates. By Lemma 3.3.6 we know that π_i maps each stratum onto its target stratum. So to show that $\pi_2^{-1}\pi_1^{-1}(T_2) \subset \mathcal{U}(f\pi_1\pi_2)$ for example, we only need to show that $\pi_2^{-1}\pi_1^{-1}(y) \subset \mathcal{U}(f\pi_1\pi_2)$ for some point $y \in T_2$. Thus we essentially reduce to a local question.

So we may suppose we are in the following situation. We have a fat unobtrusive tico multiblowup $\pi_1 \colon (M_1, \mathcal{A}_1) \to (M, \mathcal{A})$ so that all centers of π_1 are allowable. We will define what an allowable center is below. Pick a stratum T of M and a point $y \in T$. Then we want to find a fat unobtrusive tico multiblowup $\pi_2 \colon (M_2, \mathcal{A}_2) \to (M_1, \mathcal{A}_1)$ so that all centers of π_1 are allowable and so that for each stratum T' of M_2 with $\pi_1\pi_2(T') \subset T$ we have $T' \cap \pi_2^{-1}\pi_1^{-1}(y) \subset \mathcal{U}(f\pi_1\pi_2)$. If we can do this we have proven our Proposition. So let us try to find this multiblowup π_2.

First we say what we mean for a center L' to be allowable. Of course L' must be fat and unobtrusive and lie over $f^{-1}(L)$. In addition, let $\mathcal{C}_1$ be the tico transform of $\mathcal{C}$ and let S be a sheet of $\mathcal{C}_1$ which contains the center L'. We know that $|S \cap \mathcal{A}_1| = |\mathcal{A}_{S1}|$ for some regular tico $\mathcal{A}_{S1}$ in S. Then, unless we are in the algebraic case, we ask that L' be a connected component of a sheet of $\mathcal{A}_{S1}$. In the algebraic case we ask that L' be an irreducible component of a sheet of $\mathcal{A}_{S1}$. In particular L' has codimension 2 and is unskewed.

Now pick tico charts θ and ψ exactly as we did in the proof of Assertion 3.3.9.2. Let the exponents α_{ij} for $i \le c$ and $j \le a$ be as in the proof also. After reordering the coordinates we may assume that $\psi^{-1}(|\mathcal{C}|) = \bigcup_{j=d}^{a} \mathbf{R}_j^m$. So for each $j < d$ we must have $\alpha_{ij} = 0$ for some $i \le c$, otherwise we would have $\mathbf{R}_j^m \subset \psi^{-1}f^{-1}(L) = \psi^{-1}(|\mathcal{C}|)$. Likewise, for each $j \ge d$ we must have $\alpha_{ij} > 0$ for all $i \le c$.

Let $V = \psi(\mathbf{R}^m)$, $V_1 = \pi_1^{-1}(V)$ and $W = \theta(\mathbf{R}^n)$. Let $A_i = \psi(\mathbf{R}_i^m)$ for $i = 1, \ldots, a$ and $B_i = \theta(\mathbf{R}_i^n)$ for $i = 1, \ldots, c$. Then $\mathcal{A}' = \{A_i\}$ and $\mathcal{B}' = \{B_i\}$ are ticos in V and W respectively. Note that $|V \cap \mathcal{A}| = |\mathcal{A}'|$. Suppose that the multiblowup π_1 consists of s blowups. Then the total transform of $\mathcal{A}'$ is a tico $\mathcal{A}_1'$ in V_1 where $\mathcal{A}_1' = \{A_1', \ldots, A_{a+s}'\}$ and A_i' is the strict transform of A_i for $i \le a$ and A_{a+i}' is the strict transform of the inverse image of the i-th blowup for $i = 1, \ldots, s$.

The restriction $f| \colon (V, \mathcal{A}') \to (W, \mathcal{B}')$ is a tico map with constant exponents. Hence by Lemmas 3.2.4 and 3.3.4 we know that $f\pi_1| \colon (V_1, \mathcal{A}_1') \to (W, \mathcal{B}')$ has constant exponents. Let α_{ji} be the exponent of A_i' in B_j. Lemmas 3.2.4 and 3.3.4 imply that this agrees with our earlier notation when $i \le a$.

Let

$$\mathcal{D}(\pi_1) \;=\; \{D \subset \{1,\dots,a+s\} \mid \bigcap_{i \in D} A_i' \neq \emptyset \text{ and } D \not\subset \{1,\dots,d-1\}\}$$

$$\mathcal{D}'(\pi_1) \;=\; \{D \in \mathcal{D}(\pi_1) \mid \bigcap_{i \in D} A_i' \cap \mathcal{U}(f\pi_1) = \emptyset\}$$

Thus $\mathcal{D}(\pi_1)$ parameterizes the fat submanifolds of V_1 which are contained in $\pi_1^{-1}f^{-1}(L)$. The following result shows that in fact, $\mathcal{D}(\pi_1)$ also parameterizes components of fat submanifolds in $\pi_1^{-1}f^{-1}(L)$ which hit V_1.

Assertion 3.3.9.3 Suppose $D \in \mathcal{D}(\pi_1)$. Then there is a unique fat submanifold $K \subset M_1$ so that $K \cap V_1 = \bigcap_{i \in D} A_i'$ and so that K is connected (or irreducible in the algebraic case).

Proof: We prove this by induction on the number of blowups in π_1. If there are no blowups, let $S \in \mathcal{C}$ be a sheet so that $S \cap V = A_i$ for some $i \in D$ with $i \geq d$. Now $(S, \mathcal{A}_S)$ is a regular tico. Let $D' = D - \{i\}$. For each $j \in D'$ let $E_j \in \mathcal{A}_S$ be the sheet so that $E_j \cap V = S \cap A_j$. Then we must have K be the connected (or irreducible) component of $\bigcap_{j \in D'} E_j$ which contains y.

Now suppose that the assertion is true for π_1. Let L' be an allowable center and let $\pi' = \pi(M_1, L')$ and $M' = \mathcal{B}(M_1, L')$. Pick any $D' \in \mathcal{D}(\pi_1 \pi')$. Let $D = D' - \{a+s+1\}$. Then $D \in \mathcal{D}(\pi_1)$. Let $K \subset M_1$ be the unique fat submanifold so that $K \cap V_1 = \bigcap_{i \in D} A_i'$ and so that K is connected (or irreducible in the algebraic case).

If $D' = D$ we set $K' = \mathcal{B}(K, K \cap L')$. If $D' \neq D$ then set $K' = \mathcal{B}(K, K \cap L') \cap \pi'^{-1}(L')$. Then $K' \cap \pi'^{-1}(V_1) = \bigcap_{i \in D'} A_i''$, where A_i'' is the strict transform of A_i' for $i \leq a+s$ and $A_{a+s+1}'' = \pi'^{-1}(L' \cap V_1)$. $\square$

Assertion 3.3.9.4 Suppose $D \in \mathcal{D}(\pi_1)$. Then for some $i \leq c$ we have $\alpha_{ij} = 0$ for all $j \in D$ with $j < d$.

Proof: If not, then by Lemma 3.2.2, $\emptyset \neq \bigcap_{j \in D}^{j < d} A_j' \subset (f\pi_1)^{-1}(L)$ which contradicts the fact that $\pi_1^{-1}f^{-1}(L) \cap V_1 = \bigcup_{i=d}^{a+s} A_i'$. $\square$

Let $\mathbf{c}$ denote the set $\{1, 2, \dots, c\}$. Pick any $D \in \mathcal{D}(\pi_1)$. Let

$$C(D) = \{i \in \mathbf{c} \mid \alpha_{ij} = 0 \text{ for all } j \in D \text{ with } j < d\}.$$

By Assertion 3.3.9.4 we know that $C(D)$ is nonempty. Let

$$B(D) = \{(i, i') \in C(D) \times \mathbf{c} \mid \alpha_{ij} > \alpha_{i'j} \text{ and } \alpha_{ij'} < \alpha_{i'j'} \text{ for some } j, j' \in D\}.$$

The significance of $B(D)$ comes from the following:

Assertion 3.3.9.5 Suppose $D \in \mathcal{D}(\pi_1)$ and either $B(D) = \emptyset$ or $D \notin \mathcal{D}'(\pi_1)$. Then $\bigcap_{i \in D} A_i' - \bigcup_{i \notin D} A_i' \subset \mathcal{U}(f\pi_1)$. Hence $B(D) \neq \emptyset$ for all $D \in \mathcal{D}'(\pi_1)$. Also, if $\mathcal{D}'(\pi_1) = \emptyset$ then $V_1 \subset \mathcal{U}(f\pi_1)$.

Proof: Pick any $y \in \bigcap_{i \in D} A'_i - \bigcup_{i \notin D} A'_i$. Pick a tico chart $\psi': (\mathbf{R}^m, 0) \to (V_1, y)$ so that $\psi'^{-1}(|\mathcal{A}'_1|) = \bigcup_{i=1}^{e} \mathbf{R}^m_i$. For $i = 1, \ldots, e$ let $\beta(i) \in D$ be defined by $\mathbf{R}^m_i = \psi'^{-1}(A'_{\beta(i)})$. So β is a bijection of $\{1, \ldots, e\}$ with D. Then

$$(f\pi_1\psi')^* \mathcal{J}^\infty(L) = \langle \prod_{j=1}^{e} x_j^{\alpha_{1\beta(j)}}, \ldots, \prod_{j=1}^{e} x_j^{\alpha_{c\beta(j)}} \rangle.$$

If $D \notin \mathcal{D}'(\pi_1)$ then by openness of $\mathcal{U}(f\pi_1)$ we could have picked $y \in \mathcal{U}(f\pi_1)$, in which case we see that the only way for $(f\pi_1\psi')^* \mathcal{J}^\infty(L)$ to be locally principal is to have some i' so that $\alpha_{i'j} \leq \alpha_{ij}$ for all $1 \leq i \leq c$ and $j \in D$. But the exponents α_{ij} do not depend on the point y so this implies that $\bigcap_{i \in D} A'_i - \bigcup_{i \notin D} A'_i \subset \mathcal{U}(f\pi_1)$.

Now suppose $B(D) = \emptyset$. Pick some $i' \in C(D)$ such that in lexicographical order we have $(\alpha_{i'\beta(1)}, \ldots, \alpha_{i'\beta(e)}) \leq (\alpha_{i\beta(1)}, \ldots, \alpha_{i\beta(e)})$ for all $i \in C(D)$. Pick any $i \in \mathbf{c}$. If $\prod_{j=1}^{e} x_j^{\alpha_{i\beta(j)}}$ is not a multiple of $\prod_{j=1}^{e} x_j^{\alpha_{i'\beta(j)}}$ then we may pick the smallest j so that $\alpha_{i'\beta(j)} > \alpha_{i\beta(j)}$. If $i \notin C(D)$ then $\alpha_{ij'} > 0$ for some $j' \in D$ with $j' < d$. But $\alpha_{i'j'} = 0$ so $(i', i) \in B(D)$, contradicting the fact that $B(D) = \emptyset$. So we must have $i \in C(D)$, but then there is a first j' so that $\alpha_{i'\beta(j')} < \alpha_{i\beta(j')}$ and we again get $(i', i) \in B(D) = \emptyset$. So $(f\pi_1\psi')^* \mathcal{J}^\infty(L) = \langle \prod_{j=1}^{e} x_i^{\alpha_{i'\beta(j)}} \rangle$ and thus $y \in \mathcal{U}(f\pi_1)$.   □

In light of Assertion 3.3.9.5 we may as well assume that $\mathcal{D}'(\pi_1) \neq \emptyset$.

We now assign a complexity $\eta_1(D)$ to any $D \in \mathcal{D}'(\pi_1)$. Let e be the number of elements in D and let e' be the number of elements in $D \cap \{1, \ldots, d-1\}$. We let (i, i') be the lexicographical maximum of all elements of $B(D)$ and we let $\eta_1(D) = (e, e', i, i')$. We order the complexities $\eta_1(D)$ using lexicographical order.

Let $L' \subset M_1$ be an allowable center. Let π' be the blowup map $\pi(M_1, L')$. If $L' \cap V_1 = \emptyset$ then $\mathcal{D}(\pi_1) = \mathcal{D}(\pi_1\pi')$. For $D \in \mathcal{D}(\pi_1)$ and $D' \in \mathcal{D}(\pi_1\pi')$ we say in this case that $D' \to D$ if $D' = D$.

If $L' \cap V_1 \neq \emptyset$ then by Assertion 3.3.9.3 there are k and k' with $L' \cap V_1 = A'_k \cap A'_{k'}$ and $k \geq d$. The total transform of $\mathcal{A}'_1$ is the tico $\{A''_j\}$ for $j = 1, \ldots, a+s+1$ where $A''_j = \mathfrak{B}(A'_j, L' \cap A'_j\}$ for $j = 1, \ldots, a+s$ and $A''_{a+s+1} = \pi'^{-1}(L')$. By Lemmas 3.2.4b and 3.3.4 we know that the exponent for A''_j in B_ℓ is $\alpha_{\ell j}$ for $j \neq a+s+1$ and is $\alpha_{\ell,a+s+1} = \alpha_{\ell k} + \alpha_{\ell k'}$ for $j = a+s+1$. Thus we see that $\mathcal{D}(\pi_1\pi') = \mathcal{D}_0 \cup \mathcal{D}_1 \cup \mathcal{D}_2 \cup \mathcal{D}_3$ where

$$
\begin{aligned}
\mathcal{D}_0 &= \{D \in \mathcal{D}(\pi_1) \mid \{k, k'\} \not\subset D\} \\
\mathcal{D}_1 &= \{(D - \{k\}) \cup \{a+s+1\} \mid D \in \mathcal{D}(\pi_1) - \mathcal{D}_0\} \\
\mathcal{D}_2 &= \{(D - \{k'\}) \cup \{a+s+1\} \mid D \in \mathcal{D}(\pi_1) - \mathcal{D}_0\} \\
\mathcal{D}_3 &= \{(D - \{k, k'\}) \cup \{a+s+1\} \mid D \in \mathcal{D}(\pi_1) - \mathcal{D}_0\}
\end{aligned}
$$

Suppose $D \in \mathcal{D}(\pi_1)$ and $D' \in \mathcal{D}(\pi_1\pi')$. Then we say that $D' \to D$ if either $D' = D \in \mathcal{D}_0$ or $a+s+1 \in D'$ and $D - \{k, k'\} = D' - \{k, k', a+s+1\}$.

Note that $D' \to D$ means π' maps $\bigcap_{j \in D'} A_j''$ onto $\bigcap_{j \in D} A_j'$.

Assertion 3.3.9.6 *Let L' be an allowable center. Let π' be the blowup map $\pi(M_1, L')$. Pick any $D' \in \mathcal{D}'(\pi_1 \pi')$. Then $D' \to D$ for some unique $D \in \mathcal{D}(\pi_1)$ and we must have $\eta_1(D') \le \eta_1(D)$.*

Proof: By definition, there is a unique $D \in \mathcal{D}(\pi_1)$ with $D' \to D$. By Assertion 3.3.9.2 and Lemma 3.3.6 we know that $D \in \mathcal{D}'(\pi_1)$.

If $L' \cap V_1 = \emptyset$ then the assertion is trivial so we may assume that $L' \cap V_1 \ne \emptyset$. Let k, k' and $\mathcal{D}_i$ be as above.

If $D' \in \mathcal{D}_0$ then $D = D'$ and the complexity is the same. If $D' \in \mathcal{D}_3$ then D' has fewer elements than D so the complexity decreases. So we may assume $D' \in \mathcal{D}_1 \cup \mathcal{D}_2$. If $k' < d$ and $D' \in \mathcal{D}_2$ then $D' \cap \{1, \dots, d-1\}$ has one less element than $D \cap \{1, \dots, d-1\}$ so the complexity decreases. Hence by symmetry we need only consider the case $D' \in \mathcal{D}_1$.

Note that the first two coordinates of $\eta_1(D)$ and $\eta_1(D')$ are the same, so it suffices to prove that $B(D') \subset B(D)$. We have $C(D') = C(D)$ so letting $D'' = D' - \{a+s+1\} = D - \{k\}$ we have:

$$
\begin{aligned}
B(D') &= \{\, (i, i') \in C(D) \times \mathbf{c} \mid \alpha_{ij} > \alpha_{i'j} \text{ and } \alpha_{ij'} < \alpha_{i'j'} \\
&\qquad \text{for some } j, j' \in D' \,\} \\
&= B_1 \cup B_2 \cup B_3
\end{aligned}
$$

where

$$
\begin{aligned}
B_1 &= \{\, (i, i') \mid \alpha_{ij} > \alpha_{i'j} \text{ and } \alpha_{ij'} < \alpha_{i'j'} \text{ for some } j, j' \in D'' \,\} \\
B_2 &= \{\, (i, i') \mid \alpha_{ik} + \alpha_{ik'} > \alpha_{i'k} + \alpha_{i'k'} \text{ and } \alpha_{ij'} < \alpha_{i'j'} \text{ for some } j' \in D'' \,\} \\
B_3 &= \{\, (i, i') \mid \alpha_{ij} > \alpha_{i'j} \text{ and } \alpha_{ik} + \alpha_{ik'} < \alpha_{i'k} + \alpha_{i'k'} \text{ for some } j \in D'' \,\}
\end{aligned}
$$

The sets B_2 and B_3 correspond to setting $j = a+s+1$ and $j' = a+s+1$ respectively.

Note that $B_1 \subset B(D)$. To see that $B_2 \subset B(D)$, pick any $(i, i') \in B_2$. Then either $\alpha_{ik} > \alpha_{i'k}$ or $\alpha_{ik'} > \alpha_{i'k'}$ so setting $j = k$ or k' we see that $(i, i') \in B(D)$. Likewise $B_3 \subset B(D)$. Hence $B(D') \subset B(D)$.   □

We need finer invariants than just η_1 however. Let (e, e', i, i') be the lexicographical maximum of $\eta_1(D)$ over all $D \in \mathcal{D}'(\pi_1)$. Let $\delta_{jj'} = \alpha_{ij} + \alpha_{i'j'} - \alpha_{i'j} - \alpha_{ij'}$. If $(i, i') \in B(D)$ let

$$\gamma(D) = \max\{\, \delta_{jj'} \mid \alpha_{ij} > \alpha_{i'j} \text{ and } \alpha_{ij'} < \alpha_{i'j'} \text{ with } j, j' \in D \,\}.$$

Let $\lambda(D)$ be the number of pairs $(j, j') \in D \times D$ so that $\alpha_{ij} > \alpha_{i'j'}$, $\alpha_{ij'} < \alpha_{i'j'}$ and $\delta_{jj'} = \gamma(D)$. If $(i, i') \notin B(D)$ set $\gamma(D) = \lambda(D) = 0$.

We define $\eta_2(D) = (\eta_1(D), \gamma(D), \lambda(D))$. Let $(e, e', i, i', \gamma, \lambda)$ be the lexicographical maximum of $\eta_2(D)$ over all $D \in \mathcal{D}'(\pi_1)$. Let μ be the number of

$D \in \mathcal{D}'(\pi_1)$ with $\eta_2(D) = (e, e', i, i', \gamma, \lambda)$. Finally define an invariant of the multiblowup, $\eta(\pi_1) = (e, e', i, i', \gamma, \lambda, \mu)$.

Pick some $D^* \in \mathcal{D}'(\pi_1)$ so that $\eta_2(D^*) = (e, e', i, i', \gamma, \lambda)$. Pick $k, k' \in D^*$ so that $\alpha_{ik} > \alpha_{i'k}$, $\alpha_{ik'} < \alpha_{i'k'}$ and $\delta_{kk'} = \gamma$. Note that $\alpha_{ik} > 0$ and $i \in C(D^*)$ so we must have $k \geq d$. Let $L' \subset M_1$ be the allowable center so that $L' \cap V_1 = A'_k \cap A'_{k'}$. Let $\pi' \colon M' \to M_1$ be the blowup with center L'.

Assertion 3.3.9.7 If $D' \in \mathcal{D}'(\pi_1 \pi')$, $D' \to D$ and $\eta_1(D) = (e, e', i, i')$ then $\eta_2(D') \leq \eta_2(D)$. If $\eta_2(D') = \eta_2(D)$ then $D' = D \neq D^*$.

Proof: If $D = D'$ then $D \neq D^*$ and $\eta_2(D') = \eta_2(D)$. So assume $D \neq D'$. From the proof of Assertion 3.3.9.6 we may assume that $D' \in \mathcal{D}_1$. We may also assume that $(i, i') \in B(D')$, otherwise $\eta_1(D') < \eta_1(D)$. So if $D'' = D' - \{a + s + 1\} = D - \{k\}$ we have:

$$
\begin{aligned}
\gamma(D') &= \max\{\, \delta_{jj'} \mid \alpha_{ij} > \alpha_{i'j} \text{ and } \alpha_{ij'} < \alpha_{i'j'} \text{ with } j, j' \in D' \,\} \\
&= \max(\{\, \delta_{jj'} \mid j, j' \in D'' \text{ and } \alpha_{ij} > \alpha_{i'j'} \text{ and } \alpha_{ij'} < \alpha_{i'j'} \,\} \\
&\quad \cup \{\, \delta_{kj'} + \alpha_{ik'} - \alpha_{i'k'} \mid j' \in D'', \\
&\qquad \alpha_{ik} + \alpha_{ik'} > \alpha_{i'k} + \alpha_{i'k'} \text{ and } \alpha_{ij'} < \alpha_{i'j'} \} \\
&\quad \cup \{\, \delta_{jk'} + \alpha_{i'k} - \alpha_{ik} \mid j \in D'', \\
&\qquad \alpha_{ij} > \alpha_{i'j} \text{ and } \alpha_{ik} + \alpha_{ik'} < \alpha_{i'k} + \alpha_{i'k'} \}).
\end{aligned}
$$

But $\alpha_{ik'} - \alpha_{i'k'} < 0$ and $\alpha_{i'k} - \alpha_{ik} < 0$. So

$$
\gamma(D') = \max\{\, \delta_{jj'} \mid j, j' \in D'' \text{ and } \alpha_{ij} > \alpha_{i'j'} \text{ and } \alpha_{ij'} < \alpha_{i'j'} \,\}.
$$

So $\gamma(D') \leq \gamma(D)$ and if $\gamma(D) = \gamma(D')$ we must have $\lambda(D) < \lambda(D')$ since

$$
\{\, (j, j') \in D' \times D' \mid \alpha_{ij} > \alpha_{ij'}, \ \alpha_{i'j} < \alpha_{ij'} \text{ and } \delta_{jj'} = \gamma(D') \,\}
$$

$$
\subset \{\, (j, j') \in D \times D \mid \alpha_{ij} > \alpha_{ij'}, \ \alpha_{i'j} < \alpha_{ij'} \text{ and } \delta_{jj'} = \gamma(D) \,\} - \{(k, k')\}.
$$

$\square$

It is now easy to finish the proof, since from Assertion 3.3.9.7 we see that either $\mathcal{D}'(\pi_1 \pi') = \emptyset$ or $\eta(\pi_1 \pi') < \eta(\pi_1)$. So there is a multiblowup π_2 with $\mathcal{D}'(\pi_1 \pi_2) = \emptyset$. Now Assertion 3.3.9.5 implies that we are done.

One could also go about this proof in a slightly different way which some readers might prefer. First of all, it suffices to only consider D with cardinality a. To see this, note that any stratum in V_1 with codimension $< a$ has in its frontier a stratum of codimension a. So if we know all the codimension a strata of V_1 are in $\mathcal{U}(f)$ then we know all strata of V_1 are in $\mathcal{U}(f\pi_1)$. This is what we did in our example above, we only considered codimension 4 strata. Thus one could proceed by taking any pair $(i, i') \in \mathbf{c} \times \mathbf{c}$ and blowing up until $(i, i') \notin B(D)$ (i.e., $\gamma(D) = 0$) for all $D \in \mathcal{D}'(\pi_1)$, just as we did in the example. Do this for all pairs (i, i') and eventually you are done. $\square$

Next we have an algebraic analogue of Lemma 3.1.2.

Lemma 3.3.10 *Let* $(V, \mathcal{A})$ *be an algebraic tico and let* K *and* L *be Zclosed subsets of* V *which are unions of strata. Let* $r\colon K \to \mathbf{R}$ *and* $s\colon L \to \mathbf{R}$ *be entire rational functions so that* $r|_{K \cap L} = s|$. *Suppose* L *is nonsingular and* $K \cap L = |L \cap \mathcal{A}|$. *Then* $r \cup s\colon K \cup L \to \mathbf{R}$ *is an entire rational function.*

Proof: Extend r to an entire rational function $r\colon V \to \mathbf{R}$. Define $t\colon K \cup L \to \mathbf{R}$ by $t|_K = 0$ and $t|_L = s - r$. Then $r + t|_K = r|_K$ and $r + t|_L = s$ and we are done if t is an entire rational function. It suffices to show t is entire rational in a neighborhood of $K \cap L$.

So pick any $y \in K \cap L$. Let

$$\{A_1, \dots, A_k\} = \{A \in \mathcal{A} \mid y \in A \text{ and } L \pitchfork A \text{ at } y\}.$$

Let $u_i\colon (V, A_i) \to (\mathbf{R}, 0)$, $i = 1, \dots, k$ be polynomials so that du_i has rank 1 at y. But $K \cap L = |\mathcal{A} \cap L|$ so $U \cap L \cap K = U \cap L \cap \bigcup_{i=1}^k A_i = U \cap L \cap \bigcup_{i=1}^k u_i^{-1}(0)$ for some neighborhood U of y. Also $s - r|_{L \cap K} = 0$, so repeated applications of Lemma 2.2.11 show that there is a Zopen neighborhood U of y in L and an entire rational function $v\colon U \to \mathbf{R}$ so that $s(x) - r(x) = (\prod_{i=1}^k u_i(x))v(x)$ for all $x \in U$. Extend v to any entire rational function $v\colon U' \to \mathbf{R}$ where U' is a Zopen neighborhood of y in $K \cup L$. Then $t = (\prod_{i=1}^k u_i(x))v(x)$ for all x in some Zopen neighborhood of y in $K \cup L$. So t is entire rational. $\qquad\square$

Finally, the following lemma shows that under mild conditions, restricting to fat submanifolds preserves ticoness of a map.

Lemma 3.3.11 *Let* $f\colon (M, \mathcal{A}) \to (N, \mathcal{B})$ *be a tico map between regular ticos. Let* $S \subset M$ *be a fat submanifold and let* $T \subset N$ *be the smallest fat submanifold of* N *so that* $f(S) \subset T$. *Suppose that either*

1) *S is connected or*
2) *f has constant exponents or*
3) *M and N are algebraic, f is an entire rational function and S is irreducible.*

Then $f|\colon (S, S \cap \mathcal{A}) \to (T, T \cap \mathcal{B})$ *is a tico map.*

Proof: For any sheets $A \in \mathcal{A}$ and $B \in \mathcal{B}$ let $\alpha_{AB}\colon A \to \mathbf{Z}$ be the exponent map of A in B for f. Let $\mathcal{S} = \{A \in \mathcal{A} \mid S \subset A\}$. Then S is a union of connected components of $\cap \mathcal{S}$. Let $\mathcal{C} = \{B \in \mathcal{B} \mid \alpha_{AB}(x) > 0 \text{ for some } A \in \mathcal{S} \text{ and for some } x \in S\}$. Either of conditions 1,2 or 3 guarantee that each α_{AB} is constant on S. Hence

$$\mathcal{C} = \{B \in \mathcal{B} \mid \alpha_{AB}(x) > 0 \text{ for some } A \in \mathcal{S} \text{ and all } x \in S\}.$$

Assertion 3.3.11.1 *T is a union of connected components of* $\cap \mathcal{C}$.

Proof: By Lemma 3.2.2, $f(S) \subset \cap \mathcal{C}$. On the other hand, if $\dim T < \dim (\cap \mathcal{C})$ then there is a $B \in \mathcal{B} - \mathcal{C}$ so that $T \subset \cap \mathcal{C} \cap B$, hence $f(S) \subset B$. But then by Lemma 3.2.2 we must have $\alpha_{AB}(x) > 0$ for some $x \in S$, $A \in \mathcal{S}$. So $B \in \mathcal{C}$, a contradiction. So $\dim T = \dim (\cap \mathcal{C})$, so T is a union of components of $\cap \mathcal{C}$. □

Now take any $z \in S$. Pick tico charts $\psi \colon (\mathbf{R}^m, 0) \to (M, z)$ and $\theta \colon (\mathbf{R}^n, 0) \to (N, f(z))$ so $f\psi(\mathbf{R}^m) \subset \theta(\mathbf{R}^n)$, $\psi^{-1}(|\mathcal{A}|) = \bigcup_{i=1}^a \mathbf{R}_i^m$, $\psi^{-1}(|\mathcal{S}|) = \bigcup_{i=1}^d \mathbf{R}_i^m$, $\theta^{-1}(|\mathcal{B}|) = \bigcup_{i=1}^b \mathbf{R}_i^n$ and $\theta^{-1}(|\mathcal{C}|) = \bigcup_{i=1}^c \mathbf{R}_i^n$. Let $f_i(x)$ be the i-th coordinate of $\theta^{-1} f\psi(x)$. Then there are nowhere 0 functions φ_i so that $f_i(x) = \prod_{j=1}^a x_j^{\alpha_{ij}} \varphi_i(x)$ for all $i \leq b$. For each $c < i \leq b$ we must have $\alpha_{ij} = 0$ for all $j \leq d$, otherwise $\mathbf{R}_i^n$ would be in $\theta^{-1}(|\mathcal{C}|)$. Hence $f_i(x) = \prod_{j>d}^a x_j^{\alpha_{ij}} \varphi_i(x)$ for all $c < i \leq b$. So by considering the charts $\psi|_{0 \times \mathbf{R}^{m-d}}$ and $\theta|_{0 \times \mathbf{R}^{n-c}}$ we see that $f|_S$ is a tico map. □

4. Full Ticos

Definitions: Recall a manifold M is full if all of its $\mathbf{Z}/2\mathbf{Z}$ homology is generated by embedded smooth submanifolds (see 2.9). A tico $(M, \mathcal{A})$ is *full* if it is regular and if M, all sheets of $\mathcal{A}$ and all intersections of sheets of $\mathcal{A}$ are full manifolds. That is, $\cap \mathcal{D}$ is full for all $\mathcal{D} \subset \mathcal{A}$. A tico blowup $\pi \colon (M', \mathcal{A}') \to (M, \mathcal{A})$ with center L is *full* if $(L, L \cap \mathcal{A})$ is a full tico. Likewise a tico multiblowup is *full* if each of its blowups is full.

Lemma 3.4.1 *Let $\pi \colon (Y, \mathcal{B}) \to (X, \mathcal{A})$ be a full tico multiblowup of a regular tico $\mathcal{A}$ in a closed manifold X.*

    a) *Suppose $B \in \mathcal{B}$ is such that $(B, \mathcal{B} \cap B)$ is not full. Then B is the strict preimage of some $A \in \mathcal{A}$ so that $(A, \mathcal{A} \cap A)$ is not full.*

    b) *If $(X, \mathcal{A})$ is full then $(Y, \mathcal{B})$ is full.*

Proof: It suffices to prove this in the case where π is a single blowup. Let the center of π be L. Since π is full we know $(L, \mathcal{A} \cap L)$ is full.

If B is the strict preimage of some $A \in \mathcal{A}$ then $\mathcal{B} \cap B$ is the total transform of $\mathcal{A} \cap A$ under the full tico blowup with center $A \cap L$. Hence by induction on dimension we cannot have $(A, \mathcal{A} \cap A)$ full since b) would imply that $(B, \mathcal{B} \cap B)$ is full.

The only other case is where $B = \pi^{-1}(L)$. Let L' be a connected component of L and $B' = \pi^{-1}(L')$. It suffices to show that $(B', B' \cap \mathcal{B})$ is full since B' is just a connected component of B.

For each $A \in \mathcal{A}$ let $A' \in \mathcal{B}$ be the strict preimage of A. Let $\mathcal{C} = \{A \in \mathcal{A} \mid L' \subset A\}$. We need to show that $(B', \mathcal{B} \cap B')$ is full. The sheets of $\mathcal{B} \cap B'$ are just $A' \cap B'$ for sheets $A \in \mathcal{A}$. So pick any $\mathcal{D} \subset \mathcal{B} \cap B'$. Let $\mathcal{E} \subset \mathcal{A}$ be such that $\mathcal{D} = \{A' \cap B' \mid A \in \mathcal{E}\}$. Then $\cap \mathcal{D} = \pi^{-1}(L') \cap \mathcal{B}(\cap \mathcal{E}, \cap \mathcal{E} \cap L)$ is the projectivized normal bundle of $L' \cap (\bigcap \mathcal{E})$ in $\bigcap \mathcal{E}$. By Lemma 2.9.2 and the fact that $L' \cap (\bigcap \mathcal{E})$ is full, we know that $\bigcap \mathcal{D}$ is full. So $(B', \mathcal{B} \cap B')$ is full.

To finish off the proof we must show that if $(X, \mathcal{A})$ is full, then $(Y, \mathcal{B})$ is full. By a) we know that $\cap \mathcal{D}$ is full for all non-empty $\mathcal{D} \subset \mathcal{B}$ so we must only show that Y is full. But this follows from fullness of L and Proposition 2.9.3c. $\square$

Lemma 3.4.2 *Let* $\pi \colon (Y, \mathcal{B}) \to (X, \mathcal{A})$ *be a tico multiblowup. Suppose that* X *is closed and full and* B *is full for every* $B \in \mathcal{B}$. *Then* Y *is full.*

Proof: Let the multiblowup be

$$(Y, \mathcal{B}) = (X_n, \mathcal{A}_n) \xrightarrow{\pi_n} (X_{n-1}, \mathcal{A}_{n-1}) \longrightarrow \cdots \xrightarrow{\pi_1} (X_0, \mathcal{A}_0) = (X, \mathcal{A})$$

where π_i is the blowup with center $L_i \subset X_{i-1}$. Let $\rho_i \colon X_n \to X_i$ be the composition $\pi_{i+1} \circ \pi_{i+2} \circ \cdots \circ \pi_n$. For any manifold Z, recall $H_*^{imb}(Z)$ be the subgroup of the homology $H_*(Z)$ generated by embedded submanifolds of Z (all homology is with $\mathbf{Z}/2\mathbf{Z}$ coefficients).

Suppose $H_*(Y) \neq H_*^{imb}(Y)$. Then take $\alpha \in H_*(Y) - H_*^{imb}(Y)$. By Proposition 2.9.1b we know that $\pi_* \left(H_*^{imb}(Y) \right) \supset H_*^{imb}(X)$. But $H_*^{imb}(X) = H_*(X)$. Hence there is a $\beta \in H_*^{imb}(Y)$ so that $\pi_*(\alpha - \beta) = 0$. Note $\pi = \rho_0$ so there is some smallest $i > 0$ so that $(\rho_i)_*(\alpha - \gamma) \neq 0$ in $H_*(X_i)$ for all $\gamma \in H_*^{imb}(Y)$. Then there is some $\gamma \in H_*^{imb}(Y)$ with $\rho_{i-1_*}(\alpha - \gamma) = 0$. Let $\delta = \rho_{i_*}(\alpha - \gamma)$, then $\pi_{i_*}(\delta) = 0$. So by Proposition 2.9.3d, $\delta \in j_*(H_*(P))$ where P is $\pi_i^{-1}(L_i)$ and $j \colon P \hookrightarrow X_i$ is inclusion.

Let B be the strict transform of P under the multiblowup $\rho_i \colon X_n \to X_i$. We know $(\rho_i|_B)_* \colon H_*(B) \to H_*(P)$ is onto by Proposition 2.9.1a. So $\delta = \rho_{i_*}(\epsilon)$ for some ϵ in the image $H_*(B) \to H_*(Y)$. But $H_*(B) = H_*^{imb}(B)$ since $B \in \mathcal{B}$ so $\epsilon \in H_*^{imb}(Y)$ and $\rho_{i_*}(\alpha - (\gamma + \epsilon)) = 0$ which is a contradiction. So $H_*(Y) = H_*^{imb}(Y)$, i.e., Y is full. $\square$

5. Type N Tico Maps

We will need tico maps with certain stability properties. Submersive tico maps are very stable; this is shown in Chapter V. At the moment, however, we do not know how to transform an entire rational function into a submersive tico map except in low dimensions. Thus we take what we can do and call it type N.

Definition: Let $f \colon (M, \mathcal{A}) \to (N, \mathcal{B})$ be a tico map. Then f has *type N* if for every $z \in M$ there are charts $\psi \colon Z \to M$, $\theta \colon (\mathbf{R}^n, 0) \to (N, f(z))$ so that

   1) Z is an open subset of $\mathbf{R}^m$.
   2) $\psi(Z)$ is neighborhood of x.
   3) $f\psi(Z) \subset \theta(\mathbf{R}^n)$.
   4) $\psi^{-1}(|\mathcal{A}|) = \bigcup_{i=1}^{a} \mathbf{R}_i^m \cap Z$.

5) $\theta^{-1}(|\mathcal{B}|) = \bigcup_{i=1}^{b} \mathbf{R}_i^n$.

6) $f_i(x) = \prod_{j=1}^{c} x_j^{\alpha_{ij}}$ for all $x \in Z$ and $i \le b$ where $f_i(x)$ is the i-th coordinate of $\theta^{-1} f \psi(x)$ and $c \ge a$.

7) The $b \times c$ matrix $\alpha = (\alpha_{ij})$ is onto, i.e., has rank b.

The basic idea is that the φ_i's in the definition of tico map are pure monomials $\prod_{j=a+1}^{c} x_j^{\alpha_{ij}}$ and the matrix of exponents is onto.

Definition: The map f above is *submersive* if in addition $m - c \ge n - b$ and $f_i(x) = x_{i-n+m}$ for all $i > b$ and all $x \in \mathbf{Z}$.

At the moment the difference between submersive and type N is the only thing which stands in the way of a complete topological classification of real algebraic sets in all dimensions. This is explained in Chapter V and Chapter VI.

The goal for this section is Proposition 3.5.3 which says that by doing fat unobtrusive blowups we may transform any tico map to a tico map of type N. Before proving this, we introduce some notation.

Let $(M, \mathcal{A})$ be a regular tico and let S be a stratum of M. Then we may associate to S the vector space $\mathbf{Z}^{\mathcal{A}_S}$ where $\mathcal{A}_S = \{ A \in \mathcal{A} \mid S \subset A \}$. Note that the dimension of this vector space is the codimension of S in M. For convenience, if $\sigma = \sum_{A \in \mathcal{A}_S} \sigma_A \cdot A \in \mathbf{Z}^{\mathcal{A}_S}$ and $A \in \mathcal{A}$ we define $\sigma \cdot A$ to be σ_A if $S \subset A$ and 0 otherwise.

Now let $f \colon (M, \mathcal{A}) \to (N, \mathcal{B})$ be a tico map with $\mathcal{A}$ and $\mathcal{B}$ regular. If T is a stratum of M let S be the stratum of N so that $f(T) \subset S$. Then we may define a linear map $\mu_{f,T} \colon \mathbf{Z}^{\mathcal{B}_S} \to \mathbf{Z}^{\mathcal{A}_T}$ by

$$\mu_{f,T}\Big(\sum_{B \in \mathcal{B}_S} \sigma_B \cdot B \Big) = \sum_{A \in \mathcal{A}_T} \sum_{B \in \mathcal{B}_S} \alpha_{AB}(T) \sigma_B \cdot A.$$

Here $\alpha_{AB}(T)$ is the local exponent for f of A in B at points of T. This is well defined by Lemma 3.2.2 since T is connected. We could interpret $\mathbf{Z}^{\mathcal{B}_S}$ and $\mathbf{Z}^{\mathcal{A}_T}$ as being the space of monomial maps in tico coordinates for S and T and then $\mu_{f,T}$ is just the pullback map f^*.

If f is as above, define $\mathcal{U}(f)$ to be the maximal open subset of M so that $f| \colon (\mathcal{U}(f), \mathcal{A} \cap \mathcal{U}(f)) \to (N, \mathcal{B})$ has type N. For each stratum S of N, let $\mathcal{C}(f, S)$ be the set of strata T of M so that $T \not\subset \mathcal{U}(f)$ and $f(T) \subset S$. Define $\mathcal{K}(f, T)$ to be the kernel of $\mu_{f,T}$.

Lemma 3.5.1 *Suppose* $f \colon (M, \mathcal{A}) \to (N, \mathcal{B})$ *is a tico map with* $\mathcal{A}$ *and* $\mathcal{B}$ *regular. Suppose* T *and* S *are strata so that* $f(T) \subset S$. *Then:*

a) *If* $g \colon (N, \mathcal{B}) \to (P, \mathcal{C})$ *is a tico map (with* $\mathcal{C}$ *regular), then* $\mu_{gf,T} = \mu_{f,T} \circ \mu_{g,S}$

b) *Suppose* $\mu_{f,T}$ *is injective, i.e.,* $\mathcal{K}(f, T) = \{0\}$. *Then* $T \subset \mathcal{U}(f)$.

c) *If* f *is a fat unobtrusive multiblowup then* $\mu_{f,T}$ *is onto. Consequently, if* $\dim T = \dim S$ *we know that* $\mu_{f,T}$ *is an isomorphism.*

d) *Suppose f is a fat unobtrusive multiblowup, T' is a stratum of M, $T' \subset A \in \mathcal{A}_T$, $f(T') \subset S$ and $\dim T' = \dim T$. Then $\mu_{f,T}(\sigma) \cdot A = \mu_{f,T'}(\sigma) \cdot A$ for all $\sigma \in \mathbf{Z}^{\mathcal{B}_S}$.*

Proof: Note that by Lemma 3.2.2 if $A \in \mathcal{A}_T$ then $\alpha_{AB} = 0$ for all $B \in \mathcal{B} - \mathcal{B}_S$. Hence a) follows from Lemma 3.2.4.

Now let us prove b). Pick any $x \in T$. Pick standard tico charts $\psi \colon (\mathbf{R}^m, 0) \to (M, x)$ and $\theta \colon (\mathbf{R}^n, 0) \to (N, f(x))$ for f. So $f\psi(\mathbf{R}^m) \subset \theta(\mathbf{R}^n)$, $\psi^{-1}(|\mathcal{A}|) = \bigcup_{i=1}^{a} \mathbf{R}_i^m$ and $\theta^{-1}(|\mathcal{B}|) = \bigcup_{i=1}^{b} \mathbf{R}_i^n$. Let $A_i \in \mathcal{A}$, $i = 1, \dots, a$ and $B_i \in \mathcal{B}$, $i = 1, \dots, b$ be such that $\psi^{-1}(A_i) = \mathbf{R}_i^m$ and $\theta^{-1}(B_i) = \mathbf{R}_i^n$. Notice $\psi^{-1}(T) = \bigcap_{i=1}^{a} \mathbf{R}_i^m$, $\theta^{-1}(S) = \bigcap_{i=1}^{b} \mathbf{R}_i^n$, $\mathcal{B}_S = \{B_1, \dots, B_b\}$ and $\mathcal{A}_T = \{A_1, \dots, A_a\}$.

Let $f_i(z)$ be the i-th coordinate of $\theta^{-1} f\psi(z)$. Then we know that $f_i(z) = \prod_{j=1}^{a} z_j^{\alpha_{ij}} \varphi_i(z)$ if $i \le b$ where $\varphi_i(\mathbf{R}^n) \subset \mathbf{R} - 0$ and $\alpha_{ij} = \alpha_{A_j B_i}(T)$. After composing θ with reflections through $\mathbf{R}_i^n$ we may also suppose that $\varphi_i(\mathbf{R}^m) \subset (0, \infty)$. Now since the matrix (α_{ij}) of $\mu_{f,T}$ is injective, its transpose is surjective after tensoring with $\mathbf{Q}$. So there are $\gamma_{jk} \in \mathbf{Q}$, $j = 1, \dots, a$, $k = 1, \dots, b$ so that $\sum_{j=1}^{a} \alpha_{ij} \gamma_{jk} = \delta_{ik}$, the Kronecker delta.

Define $\kappa \colon \mathbf{R}^n \to \mathbf{R}^n$ by letting the j-th coordinate of $\kappa(z)$ be z_j if $j > a$ and $z_j \prod_{k=1}^{b} \varphi_k(z)^{\gamma_{jk}}$ if $j \le a$. The Jacobian matrix of κ at 0 is diagonal with positive entries so there is a neighborhood U of 0 in $\mathbf{R}^m$ on which κ^{-1} is defined. We claim that for $i \le b$, the i-th coordinate of $\theta^{-1} f\psi\kappa^{-1}(z)$ is $\prod_{j=1}^{a} z_j^{\alpha_{ij}}$. Hence $x \in \mathcal{U}(f)$ so $T \subset \mathcal{U}(f)$.

To prove the claim, let $\kappa^{-1}(z) = y$. Then $\theta^{-1} f\psi\kappa^{-1}(z) = \theta^{-1} f\psi(y)$ whose i-th coordinate is $\prod_{j=1}^{a} y_j^{\alpha_{ij}} \varphi_i(y)$. But $z = \kappa(y)$ so

$$\prod_{j=1}^{a} z_j^{\alpha_{ij}} = \prod_{j=1}^{a} y_j^{\alpha_{ij}} \prod_{j=1}^{a} \prod_{k=1}^{b} \varphi_k(y)^{\gamma_{jk}\alpha_{ij}}$$

$$= \prod_{j=1}^{a} y_j^{\alpha_{ij}} \prod_{k=1}^{b} \varphi_k(y)^{\delta_{ik}} = \prod_{j=1}^{a} y_j^{\alpha_{ij}} \varphi_i(y).$$

Now let us prove c). By part a) and Lemma 3.3.6 it suffices to consider the case where f is a single blowup with fat unobtrusive center L. If $f(T) \not\subset L$ then $\mathcal{A}_T = \{\mathcal{B}(B, B \cap L) \mid B \in \mathcal{B}_S\}$ so $\mu_{f,T}$ is an isomorphism by Lemma 3.3.4. In fact we have $\mu_{f,T}(\sum \sigma_B \cdot B) = \sum \sigma_B \cdot \mathcal{B}(B, B \cap L)$. Now suppose $f(T) \subset L$. Let $\mathcal{L} \subset \mathcal{B}$ be such that L is a union of components of $\bigcap \mathcal{L}$. Let $\mathcal{D} \subset \mathcal{B}$ be such that $\mathcal{A}_T = \{f^{-1}(L)\} \cup \{\mathcal{B}(B, B \cap L) \mid B \in \mathcal{D}\}$. Then $\mathcal{D} \cup \mathcal{L} \subset \mathcal{B}_S$. Note that $\mathcal{L} \ne \mathcal{D} \cap \mathcal{L}$ since otherwise, $T \subset f^{-1}(L) \cap \bigcap_{B \in \mathcal{L}} \mathcal{B}(B, L) = f^{-1}(L) \cap \mathcal{B}(\bigcap \mathcal{L}, L) = \emptyset$. Take $B_0 \in \mathcal{L} - \mathcal{D}$. Then by Lemma 3.3.4 $\mu_{f,T}(B_0) = f^{-1}(L)$, $\mu_{f,T}(B - B_0) = \mathcal{B}(B, L)$ for $B \in \mathcal{D} \cap \mathcal{L}$ and $\mu_{f,T}(B) = \mathcal{B}(B, B \cap L)$ for $B \in \mathcal{D} - \mathcal{L}$. So $\mu_{f,T}$ is onto. If $\dim T = \dim S$ then $\mu_{f,T}$ is injective because $\mathbf{Z}^{\mathcal{A}_T}$ and $\mathbf{Z}^{\mathcal{B}_S}$ have the same rank.

Now let us prove d). Suppose we have two fat unobtrusive multiblowups
$\pi\colon (M,\mathcal{A}) \to (N',\mathcal{B}')$ and $\pi'\colon (N',\mathcal{B}') \to (N,\mathcal{B})$ so that $f = \pi'\pi$ and π consists
of a single blowup with fat unobtrusive center L. Let Q and Q' be the strata of
N' with $\pi(T) \subset Q$ and $\pi(T') \subset Q'$.

First suppose $A = \pi^{-1}(L)$. Let $\mathcal{L} \subset \mathcal{B}'$ be the sheets of $\mathcal{B}'$ which contain L.
Then $\pi(T') \subset L$ and $\pi(T) \subset L$ so we know that $Q \cup Q' \subset C$ for each $C \in \mathcal{L}$.
Hence $\mu_{\pi',Q}(\sigma) \cdot C = \mu_{\pi',Q'}(\sigma) \cdot C$ for all $C \in \mathcal{L}$, $\sigma \in \mathbf{Z}^{\mathcal{B}_S}$ by induction. But by
Lemma 3.3.4 and part a),

$$
\begin{aligned}
\mu_{f,T}(\sigma) \cdot A &= \mu_{\pi,T}\mu_{\pi',Q}(\sigma) \cdot A \\
&= \sum_{C \in \mathcal{L}} \mu_{\pi',Q}(\sigma) \cdot C \\
&= \sum_{C \in \mathcal{L}} \mu_{\pi',Q'}(\sigma) \cdot C = \mu_{f,T'}(\sigma) \cdot A.
\end{aligned}
$$

So d) is true in this case. Now suppose $A = \mathcal{B}(C, C \cap L)$ for some $C \in \mathcal{B}'$. Then
$Q \cup Q' \subset C$ so

$$
\mu_{f,T}(\sigma) \cdot A = \mu_{\pi,T}\mu_{\pi',Q}(\sigma) \cdot A = \mu_{\pi',Q}(\sigma) \cdot C = \mu_{\pi',Q'}(\sigma) \cdot C = \mu_{f,T'}(\sigma) \cdot A.
$$

So d) is true in all cases. $\square$

Lemma 3.5.2 Let $f\colon (M,\mathcal{A}) \to (N,\mathcal{B})$ be a tico map. Suppose that we have fat
unobtrusive tico multiblowups $\rho\colon (M',\mathcal{A}') \to (M,\mathcal{A})$ and $\pi\colon (N',\mathcal{B}') \to (N,\mathcal{B})$
and suppose $f'\colon M' \to N'$ is a smooth function so that $\pi f' = f\rho$. (Note f' is a
tico map by Lemma 3.2.7.) Suppose also that $\mathcal{A}$ and $\mathcal{B}$ are regular. Then

   a) $\mathcal{U}(f') \supset \rho^{-1}(\mathcal{U}(f))$
   b) Suppose S, S', T and T' are strata for N, N', M and M' respectively and
      $\pi(S') \subset S$, $\rho(T') \subset T$ and $T' \in \mathcal{C}(f',S')$. Then $T \in \mathcal{C}(f,S)$.
   c) Suppose S and S' are strata of N and N' respectively and $\pi(S') \subset S$.
      Suppose T and T' are strata of M and M' respectively, $\dim T = \dim T'$,
      $\rho(T') \subset T$, and $f'(T') \subset S'$. Then $\mathcal{K}(f',T') = \mu_{\pi,S'}\mathcal{K}(f,T)$.

Proof: Let us prove a). It suffices to consider the cases where one of ρ or π is the
identity and the other is a single blowup with fat unobtrusive center L. First
suppose that $\pi = $ identity. Since the restriction $\rho|\colon \rho^{-1}(\mathcal{U}(f) - L) \to \mathcal{U}(f) - L$
is a tico isomorphism and $f' = f\rho$ we easily see that $\rho^{-1}(\mathcal{U}(f) - L) \subset \mathcal{U}(f')$. So
pick $x \in \mathcal{U}(f) \cap L$. We may pick coordinate charts $\psi\colon Z \to M$ and $\theta\colon (\mathbf{R}^n,0) \to$
$(N, f(x))$ so that $f\psi(Z) \subset \theta(Z^n)$, $Z \subset \mathbf{R}^m$ is open, $\psi(Z)$ is a neighborhood
of x, $\psi^{-1}(|\mathcal{A}|) = \bigcup_{i=1}^{a} \mathbf{R}_i^m \cap Z$, $\theta^{-1}(|\mathcal{B}|) = \bigcup_{i=1}^{b} \mathbf{R}_i^n$ and if $f_i(z)$ is the i-th
coordinate of $\theta^{-1}f\psi(z)$ then $f_i(z) = \prod_{j=1}^{d} z_j^{\alpha_{ij}}$ for all $i \leq b$ and some $d \geq a$.
We may also assume after reordering that $\psi^{-1}(L) = \bigcap_{i=1}^{c} \mathbf{R}_i^m \cap Z$ for some
$c \leq a$. By Lemma 2.5.1 we have charts $\psi_k\colon Z_k \to M'$, $k = 1,\dots,c$ so that
the j-th coordinate of $\psi^{-1}\rho\psi_k(z)$ is z_j if $j > c$ or $j = k$ and is $z_k z_j$ if $j \leq c$

and $j \neq k$. Also these charts cover a neighborhood of $\rho^{-1}(x)$. Then the i-th coordinate of $\theta^{-1} f' \psi_k(z)$ is $\prod_{j=1}^{d} z_j^{\alpha_{ij}^k}$ for all $i \leq b$ where $\alpha_{ij}^k = \alpha_{ij}$ for $j \neq k$ and $\alpha_{ik}^k = \sum_{j=1}^{c} \alpha_{ij}$. So $\rho^{-1}(x) \subset \mathcal{U}(f')$. So $\rho^{-1}(\mathcal{U}(f)) \subset \mathcal{U}(f')$.

Now suppose $\rho = $ identity and π is a single blowup with center L. Since $\pi^{-1} \colon N - L \to N' - \pi^{-1}(L)$ is a tico isomorphism and $f'| \colon M - f^{-1}(L) \to N' - \pi^{-1}(L)$ is $\pi^{-1} f$ we know that $\mathcal{U}(f) - f^{-1}(L) \subset \mathcal{U}(f')$. So pick $x \in \mathcal{U}(f) \cap f^{-1}(L)$. Pick ψ and θ as above, only now we have $\theta^{-1}(L) = \bigcap_{i=1}^{c} \mathbf{R}_i^n$ for some $c \leq b$. We have charts $\theta_k \colon \mathbf{R}^n \to N'$, $k = 1, \dots, c$ covering $\pi^{-1}\theta(\mathbf{R}^n)$ so that the j-th coordinate of $\theta^{-1}\pi\theta_k(z)$ is z_j if $j > c$ or $j = k$ and $z_k z_j$ if $j \leq c$ and $j \neq k$. Then the i-th coordinate of $\theta_k^{-1} f' \psi(z)$ is $\prod_{j=1}^{a} z_j^{\alpha_{ij}^k}$ for all $i \leq b$ where $\alpha_{ij}^k = \alpha_{ij}$ if $i > c$ or $i = k$ and $\alpha_{ij}^k = \alpha_{ij} - \alpha_{kj}$ if $i \leq c$ and $i \neq k$. Since $\theta_k^{-1}(|\mathcal{B}'|) = \bigcup_{i=1}^{b} \mathbf{R}_i^n$ we have thus shown that $\mathcal{U}(f') \supset \mathcal{U}(f)$. So a) is proven.

Let us now prove b). Pick $T' \in \mathcal{C}(f', S')$. Now $f(T) \supset f\rho(T') = \pi f'(T')$ and $\pi f'(T') \subset \pi(S') \subset S$. Hence $f(T)$ intersects S, so $f(T) \subset S$ by Lemma 3.2.6. Also $\emptyset \neq T' - \mathcal{U}(f') \subset T' - \rho^{-1}\mathcal{U}(f)$. So $\emptyset \neq \rho(T' - \rho^{-1}(\mathcal{U}(f))) = \rho(T') - \mathcal{U}(f) \subset T - \mathcal{U}(f)$. So $T \not\subset \mathcal{U}(f)$ so $T \in \mathcal{C}(f, S)$.

We will now prove c). By Lemma 3.5.1 we know that $\mu_{\rho, T'}$ is an isomorphism, $\mu_{\pi, S'}$ is onto and $\mu_{f', T'} \mu_{\pi, S'} = \mu_{\rho, T'} \mu_{f, T}$. Consequently, c) is true. $\qquad\square$

Proposition 3.5.3 below shows that a tico map may be made type N by blowing up. Before proving this we will show this for two examples which help illustrate the ideas of the proof.

The first example is the map $f \colon M \to N$ where $M = \mathbf{R}^2$, $N = \mathbf{R}^2$ and f is given by $f(x, y) = (x^2 y, x^2 y(1 + y^2))$. We have the ticos $\{A_1, A_2\}$ and $\{B_1, B_2\}$ in M and N respectively where $A_i = \mathbf{R}_i^2$ and $B_i = \mathbf{R}_i^2$. So $A_1 = \{x = 0\}$, $A_2 = \{y = 0\}$, etc. Then $\mathcal{U}(f) = \mathbf{R}^2 - A_2$. To see this note that if $y \neq 0$ the coordinate change $x' = x|y|^{1/2}$, $y' = 1 + y^2$ in the source makes the map $(\pm x'^2, \pm x'^2 y')$. The $\pm$ sign may be eliminated by doing reflections in the target.

Exercise: If $y = 0$, show that f is not locally of type N. $\qquad\diamond$

There are four types of strata, the connected components of $\mathbf{R}^2 - (A_1 \cup A_2)$, $A_1 - A_2$, $A_2 - A_1$ and $A_1 \cap A_2$. If T is a component of $\mathbf{R}^2 - (A_1 \cup A_2)$ then $\mathcal{K}(f, T) = 0$. Otherwise, $\mathcal{K}(f, T)$ is generated by $B_1 - B_2$. We now blow up the point 0 in N. This gives two charts N_1 and N_2 above N, both diffeomorphic to $\mathbf{R}^2$. The blowup maps $\pi \colon N_1 \to N$ and $\pi \colon N_2 \to N$ are given by $(x, y) \mapsto (x, xy)$ and $(x, y) \to (xy, y)$ respectively. The map f lifts to a map to either chart. The map to N_1 is $(x^2 y, 1 + y^2)$. Thus the lifted f is type N since it maps to a codimension one stratum.

The next example is the map $f \colon M \to N$ with $M = N = \mathbf{R}^3$ and $f(x, y, z) = (x^2, y^2 z^2, xyz(1 + z^2))$. We have ticos $\{A_i\}$ and $\{B_i\}$ where $A_i = \mathbf{R}_i^3$ and $B_i = \mathbf{R}_i^3$ for $i = 1, 2, 3$.

Exercise: Show that $\mathfrak{U}(f) = \mathbf{R}^3 - \{z = 0, x = 0\} = M - A_3 \cap A_1$.   ◇

Exercise: Show that $\mathcal{K}(f, T) = 0$ if T is a component of the big stratum $M - (A_1 \cup A_2 \cup A_3)$, $\mathcal{K}(f, T) = (B_1 - 2B_3)$ for T a component of $A_1 - (A_2 \cup A_3)$, $\mathcal{K}(f, T) = (B_2 - 2B_3)$ for T a component of $A_3 - A_1$, $A_2 - A_1$ or $A_2 \cap A_3 - A_1$ and $\mathcal{K}(f, T) = (B_1 + B_2 - 2B_3)$ for all other strata T.   ◇

Notice that $\mathcal{K}(f, T) \neq 0$ even for some T's where f has type N.

We will want to blow up the line $B_2 \cap B_3$ in N. We choose $B_2 \cap B_3$ since for T a stratum in $A_3 \cap A_1$, the coefficients of B_2 and B_3 in $\mathcal{K}(f, T)$ have opposite sign and largest magnitude. The map f does not lift to the blowup since $f^*(\mathfrak{I}(B_2 \cap B_3)) = \langle y^2z^2, xyz \rangle$ is not locally principal. So we must blow up the source as in Proposition 3.3.9. We first blow up $A_1 \cap A_2 = \{x = y = 0\}$. This gives us two charts M_1 and M_2 with blowup maps (x, xy, z) and (xy, y, z) respectively. The maps from M_i to N are given by:

$$
\begin{aligned}
M_1 \to N & \quad \left(x^2, x^2y^2z^2, x^2yz\left(1 + z^2\right)\right) \\
M_2 \to N & \quad \left(x^2y^2, y^2z^2, xy^2z\left(1 + z^2\right)\right)
\end{aligned}
$$

The map from M_1 lifts but the map from M_2 does not, since the pullback of the ideal is $y^2z\langle x, z \rangle$. So we blow up $x = z = 0$ in M_2 obtaining two charts M_3 and M_4 with projection maps (x, y, xz) and (xz, y, z) respectively. The maps from M_i to N for $i = 3, 4$ are given by:

$$
\begin{aligned}
M_3 \to N & \quad \left(x^2y^2, x^2y^2z^2, x^2y^2z\left(1 + x^2z^2\right)\right) \\
M_2 \to N & \quad \left(x^2y^2z^2, y^2z^2, xy^2z^2\left(1 + z^2\right)\right)
\end{aligned}
$$

Both of these maps now lift.

Let us look more carefully at the lifted maps. We have charts N_1 and N_2 in the blowup of N with blowup maps given by (x, y, yz) and (x, yz, z) respectively. The tico in the blowup of N has four sheets. Three of them are the strict transforms of B_1, B_2 and B_3; we will just call them B_1, B_2 and B_3 also and hope no confusion will ensue. The fourth sheet B_4 is the inverse image of the center. The sheets B_1, B_4 and B_3 are in N_1 and B_1, B_2 and B_4 are in N_2. The liftings of the maps from the M_i's to the N_j's are given as follows. The last column gives generators for $\mathcal{K}$ of the zero stratum in M_i.

$$
\begin{array}{lll}
M_1 \to N_2 & \left(x^2, yz/\left(1 + z^2\right), x^2yz\left(1 + z^2\right)\right) & B_1 + B_2 - B_4 \\
M_3 \to N_2 & \left(x^2y^2, z/\left(1 + x^2z^2\right), x^2y^2z\left(1 + x^2z^2\right)\right) & B_1 + B_2 - B_4 \\
M_4 \to N_1 & \left(x^2y^2z^2, y^2z^2, x\left(1 + x^2\right)\right) & B_1 - B_4 - 2B_3
\end{array}
$$

We now blow up $B_1 \cap B_3$ which only lies in N_1 and misses N_2. Thus N_1 is replaced by two charts N_3 and N_4. Meanwhile, N_2 remains the same. The blowup map from N_3 and N_4 to N_1 is given by (x, y, zx) and (xz, y, z) respectively. The map from M_4 lifts to N_4 and becomes the map

$$
\begin{array}{ll}
M_4 \to N_4 & \left(xy^2z^2/\left(1 + z^2\right), y^2z^2, x\left(1 + z^2\right)\right) \qquad B_1 - B_4 - B_5
\end{array}
$$

where B_5 is the new sheet, the inverse image of the center.

We now blow up $B_1 \cap B_4$ which lies in both charts N_2 and N_4. We thus replace N_2 and N_4 with two charts each. However, as with N_3 above, we only need one of the charts over each and can ignore the other two. Let N_5 be the chart over N_2 with blowup map (x, y, xz) and let N_6 be the chart over N_4 with blowup map (xy, y, z). Our lifted maps are now as follows:

$$
\begin{array}{lll}
M_1 \to N_5 & \left(x^2, yz/\left(1+z^2\right), yz\left(1+z^2\right)\right) & B_2 - B_4 \\
M_3 \to N_5 & \left(x^2y^2, z/\left(1+x^2z^2\right), z\left(1+x^2z^2\right)\right) & B_2 - B_4 \\
M_4 \to N_6 & \left(x/\left(1+z^2\right), y^2z^2, x\left(1+z^2\right)\right) & B_1 - B_5
\end{array}
$$

We now blow up $B_2 \cap B_4$. Let N_7 be the chart over N_5 with blowup map (x, y, yz). Then the maps from M_1 and M_3 both lift and now have type N (and $\mathcal{K} = 0$).

$$
\begin{array}{ll}
M_1 \to N_7 & \left(x^2, yz/\left(1+z^2\right), \left(1+z^2\right)^2\right) \\
M_3 \to N_7 & \left(x^2y^2, z/\left(1+x^2z^2\right), \left(1+x^2z^2\right)^2\right)
\end{array}
$$

Note they now map to a neighborhood of the codimension two stratum $B_6 \cap B_7$, since the last coordinate is never zero.

To finish off, we blow up $B_1 \cap B_5$. Let N_8 be the chart over N_6 with blowup map (x, y, xz). Then the map $M_4 \to N_8$ is given by

$$
M_4 \to N_8 \qquad \left(x/\left(1+z^2\right), y^2z^2, \left(1+z^2\right)^2\right)
$$

which is type N (and has $K = 0$).

Now that we have seen some examples, let us prove this result in general.

Proposition 3.5.3 Let $f\colon (M, \mathcal{A}) \to (N, \mathcal{B})$ be a tico map with $\mathcal{A}$ and $\mathcal{B}$ finite. Then there are tico multiblowups $\pi\colon N' \to N$ and $\rho\colon M' \to M$ with fat unobtrusive centers and a tico map $g\colon M' \to N'$ so that $f\rho = \pi g$ and g has type N. Furthermore, if $\mathcal{A}$ is regular we may guarantee that the centers of ρ lie over $f^{-1}(|\mathcal{B}|)$. If M, N, $\mathcal{A}$ and $\mathcal{B}$ are algebraic and f is entire rational we may also guarantee that π and ρ are algebraic multiblowups and g is entire rational.

Proof: By Proposition 3.3.8 and 3.3.9 we may as well assume that $\mathcal{A}$ and $\mathcal{B}$ are regular. Let M_c and N_c denote the union of all strata of M and N with dimension $\leq c$. Suppose we know that for some integers a and b, $\mathcal{U}(f) \supset M_{a-1} \cup \left(M_a \cap f^{-1}(N_{b-1})\right)$. What we will do is show that there are tico multiblowups $\rho\colon M' \to M$ and $\pi\colon N' \to N$ with fat unobtrusive centers so that the centers of ρ lie over $f^{-1}(|\mathcal{B}|)$, there is a smooth map $f'\colon M' \to N'$ with $\pi f' = f\rho$ and so that $\mathcal{U}(f') \supset M'_{a-1} \cup \left(M'_a \cap f'^{-1}(N'_b)\right)$. Then by induction we will be done.

Assertion 3.5.3.1 If $\rho\colon M' \to M$ and $\pi\colon N' \to N$ are any tico multiblowups with fat unobtrusive centers and there is a smooth $f'\colon M' \to N'$ with $\pi f' = f\rho$ then $\mathcal{U}(f') \supset M'_{a-1} \cup \left(M'_a \cap f'^{-1}(N'_{b-1})\right)$.

Proof: Let T' be a stratum of M' with dimension $< a$. Let T be the stratum of M with $\rho(T') \subset T$. Then $\dim T \leq \dim T' < a$ by Lemma 3.3.6 so $T \subset \mathcal{U}(f)$ so $T' \subset \rho^{-1}\mathcal{U}(f) \subset \mathcal{U}(f')$ by Lemma 3.5.2. Hence $M'_{a-1} \subset \mathcal{U}(f')$. Now suppose T' is a stratum of M' with dimension a and S' is the stratum of N' with $f'(T') \subset S'$ and $\dim S' < b$. Let T and S be the strata of M and N with $\rho(T') \subset T$ and $\pi(S') \subset S$. Now $\dim T \leq \dim T' = a$ and $\dim S \leq \dim S' < b$ and $f(T) \subset S$ so $T \subset M_a \cap f^{-1}(N_{b-1}) \subset \mathcal{U}(f)$. Hence $T' \subset \rho^{-1}(T) \subset \rho^{-1}\mathcal{U}(f) \subset \mathcal{U}(f')$. So we have shown $\mathcal{U}(f') \supset M'_{a-1} \cup \left(M'_a \cap f'^{-1}(N'_{b-1})\right)$.   □

Thus if we find π, ρ and f' with $\mathcal{U}(f') \supset \left(M'_a - M'_{a-1}\right) \cap f'^{-1}\left(N'_b - N'_{b-1}\right)$, the rest is automatic.

So suppose we have tico multiblowups $\rho \colon M' \to M$ and $\pi \colon N' \to N$ with fat unobtrusive centers, the centers of ρ lie over $f^{-1}(|\mathcal{B}|)$ and there is a smooth map $g \colon M' \to N'$ with $\pi g = f\rho$. Let $\mathcal{A}'$ and $\mathcal{B}'$ be the total transforms of $\mathcal{A}$ and $\mathcal{B}$ respectively. For each stratum S of N' let

$$\mathcal{D}(g, S) = \{\, T \in \mathcal{C}(g, S) \mid T \text{ has dimension } a \,\}.$$

So $\mathcal{D}(g, S)$ is the set of a dimensional strata T of $g^{-1}(S)$ which contain points where g is not locally of type N. Let

$$\mathcal{J}(g, S) = \{\, \mathcal{K}(g, T) \mid T \in \mathcal{D}(g, S) \,\}.$$

Note that by taking ρ and π to be identities, we have defined $\mathcal{D}(f, R)$ and $\mathcal{J}(f, R)$ for any stratum R of N.

For each dimension b stratum R of N let $\mathcal{E}(g, R)$ be the set of dimension b strata S of N' so that $\pi(S) \subset R$ and $\mathcal{D}(g, S) \neq \emptyset$, i.e., $g(M'_a - \mathcal{U}(g)) \cap S \neq \emptyset$.

Let R be a dimension b stratum of N and suppose $\mathcal{J}(f, R) \neq \emptyset$. Define $\gamma(g, R)$ to be the number of $K \in \mathcal{J}(f, R)$ so that $\mu_{\pi, S}(K) \in \mathcal{J}(g, S)$ for some $S \in \mathcal{E}(g, R)$.

Suppose π is factored as $\pi''\pi'$ and ρ is factored as $\rho''\rho'$ where the maps $\pi'' \colon (N'', \mathcal{B}'') \to (N, \mathcal{B})$, $\pi' \colon (N', \mathcal{B}') \to (N'', \mathcal{B}'')$, $\rho'' \colon (M'', \mathcal{A}'') \to (M, \mathcal{A})$ and $\rho' \colon (M', \mathcal{A}') \to (M'', \mathcal{A}'')$ are fat unobtrusive multiblowups and there is a map $f' \colon (M'', \mathcal{A}'') \to (N'', \mathcal{B}'')$ so $\pi''f' = f\pi''$ and $\pi'g = f'\rho'$.

Assertion 3.5.3.2 *For all* $S \in \mathcal{E}(g, R)$

$$\mathcal{J}(g, S) \subset \{\, \mu_{\pi', S}(K) \mid K \in \mathcal{J}(f', S') \,\}$$

where S' *is the stratum of* N'' *such that* $S' \supset \pi'(S)$.

Proof: Suppose $\mathcal{K}(g, T') \in \mathcal{J}(g, S)$ with $T' \in \mathcal{D}(g, S)$. Let T be the stratum of M containing $\rho(T')$ and let T'' be the stratum of M'' containing $\rho'(T')$. If $\dim T < \dim T' = a$ then $T \subset M_{a-1} \subset \mathcal{U}(f)$ so $T' \subset \mathcal{U}(g)$, a contradiction. So $\dim T = \dim T'' = \dim T' = a$ so $\mathcal{K}(g, T') = \mu_{\pi', S}\mathcal{K}(f', T'')$ by Lemma 3.5.2c. But $T'' \not\subset \mathcal{U}(f')$ (otherwise $T' \subset \mathcal{U}(g)$) and $f'(T'') \subset S'$. Then $T'' \in \mathcal{D}(f', S')$ so $\mathcal{K}(f', T'') \in \mathcal{J}(f', S')$.   □

As a consequence of Assertion 3.5.3.2 and Lemma 3.5.1a and c, we see that as we blow up more and more then $\gamma(g, R)$ cannot increase. We will show below that if $\gamma(g, R) > 0$ then we may choose multiblowups so that $\gamma(g, R)$ decreases. Consequently we may as well assume that all $\gamma(g, R)$'s are 0, in other words each $\mathcal{J}(g, S)$ is empty. But that means each $\mathcal{D}(g, S)$ is empty so $\left(M'_a - M'_{a-1}\right) \cap g^{-1}\left(N'_b - N'_{b-1}\right) \subset \mathcal{U}(g)$ which is all we needed to prove.

So it only remains to prove the claim that we can make $\gamma(g, R)$ decrease. So suppose $\gamma(g, R) > 0$. Then we may pick $K_0 \in \mathcal{J}(f, R)$ so that $\mu_{\pi, S}(K_0) \in \mathcal{J}(g, S)$ for some $S \in \mathcal{E}(g, R)$. Notice $K_0 \neq 0$ by Lemma 3.5.1b. So we may pick $\sigma \in K_0 - 0$. For each $S \in \mathcal{E}(g, R)$ let σ_S denote $\mu_{\pi, S}(\sigma)$ and $K_S = \mu_{\pi, S}(K_0)$. For each $B_0, B_1 \in \mathcal{B}'$ define

$$
\lambda(g, R, S, B_0, B_1) = \begin{cases} \sigma_S \cdot B_0 - \sigma_S \cdot B_1 & \text{if } S \subset B_0 \cap B_1 \\ & \text{and } \sigma_S \cdot B_0 > 0 > \sigma_S \cdot B_1 \\ 0 & \text{otherwise} \end{cases}
$$

Note that $\lambda(g, R, S, B_0, B_1)$ is independent of S as long as $S \subset B_0 \cap B_1$, $S \in \mathcal{E}(g, R)$ and $K_S \in \mathcal{J}(g, S)$. This is because if $S' \subset B_0 \cap B_1$, $S' \in \mathcal{E}(g, R)$ and $K_{S'} \in \mathcal{J}(g, S')$ then by Lemma 3.5.1d:

$$
\sigma_{S'} \cdot B_i = \mu_{\pi, S'}(\sigma) \cdot B_i = \mu_{\pi, S}(\sigma) \cdot B_i = \sigma_S \cdot B_i \quad \text{for } i = 0, 1.
$$

So we can define $\lambda(g, R, B_0, B_1) = \lambda(g, R, S, B_0, B_1)$ if there is an $S \in \mathcal{E}(g, R)$ with $S \subset B_0 \cap B_1$, $K_S \in \mathcal{J}(g, S)$ and $\sigma_S \cdot B_0 > 0 > \sigma_S \cdot B_1$. If no such S exists we set $\lambda(g, R, B_0, B_1) = 0$. Now let

$$
\begin{aligned}
\lambda(g, R) &= \max\{\lambda(g, R, B_0, B_1) \mid B_i \in \mathcal{B}'\} \quad \text{and} \\
\eta(g, R) &= \text{the number of pairs } (B_0, B_1) \in \mathcal{B}' \times \mathcal{B}' \\
& \qquad \text{such that } \lambda(g, R, B_0, B_1) = \lambda(g, R).
\end{aligned}
$$

Suppose $\lambda(g, R) > 0$. Pick $L_0, L_1 \in \mathcal{B}'$ so that $\lambda(g, R, L_0, L_1) = \lambda(g, R)$. Let $\pi' \colon N'' \to N'$ be the blowup with center $L_0 \cap L_1$. By Proposition 3.3.9 there is a tico multiblowup $\rho' \colon M'' \to M'$ with fat unobtrusive centers lying over $g^{-1}(|\mathcal{B}'|)$ so that there is a smooth $g' \colon M'' \to N''$ with $\pi' g' = g \rho'$. Note that $g^{-1}(|\mathcal{B}'|) = g^{-1}\pi^{-1}(|\mathcal{B}|) = \rho^{-1}f^{-1}(|\mathcal{B}|)$ so the centers of $\rho \rho'$ lie over $f^{-1}(|\mathcal{B}|)$.

Assertion 3.5.3.3 $(\lambda(g', R), \eta(g', R)) < (\lambda(g, R), \eta(g, R))$ with lexicographical ordering.

Proof: Let us calculate $\lambda(g', R)$. Take $S' \in \mathcal{E}(g', R)$ with $K_{S'} \in \mathcal{J}(g', S')$. Let S be the stratum of N' containing $\pi'(S')$. Pick $T' \in \mathcal{D}(g', S')$ so that $K_{S'} = \mathcal{K}(g', T')$. Let T be the stratum of M' with $\rho'(T') \subset T$. If $\dim T < \dim T' = a$ then $T \subset \mathcal{U}(g)$ so $T' \subset \mathcal{U}(g')$, contradiction. So $\dim T = a$. Hence $\mathcal{K}(g', T') = \mu_{\pi', S'}\mathcal{K}(g, T)$ by Lemma 3.5.2. If $\dim S < b$ then $T \subset \mathcal{U}(g)$ so

$T' \subset \mathcal{U}(g')$, a contradiction. So $\dim S = b$ so $\mu_{\pi',S'}$ is injective by Lemma 3.5.1c. But

$$\mathcal{K}(g',T') = K_{S'} = \mu_{\pi\pi',S'}(K_0) = \mu_{\pi',S'}(\mu_{\pi,S}(K_0)) = \mu_{\pi',S'}(K_S)$$

so $\mathcal{K}(g,T) = K_S$ by the injectivity of $\mu_{\pi',S'}$. Because $K_{S'} = \mu_{\pi',S'}(K_S)$, we have $\sigma_{S'} \cdot \mathfrak{B}(B, B \cap L) = \sigma_S \cdot B$ for all $B \in \mathcal{B}'$ with $S' \subset \mathfrak{B}(B, B \cap L)$. Also $\sigma_{S'} \cdot \pi'^{-1}(L) = \sigma_S \cdot L_0 + \sigma_S \cdot L_1$ if $S \subset L$ and $\sigma_{S'} \cdot \pi'^{-1}(L) = 0$ if $S \not\subset L$.

Now suppose $\eta(g,R) = 1$. Then $\lambda(g,R,B_0,B_1) < \lambda(g,R)$ unless $B_0 = L_0$ and $B_1 = L_1$. Suppose $\lambda(g',R,B_0',B_1') > 0$ for B_i' in the transform $\mathcal{B}''$ of $\mathcal{B}'$. Then there is an $S' \in \mathcal{E}(g',R)$ with $S' \subset B_0' \cap B_1'$, $K_{S'} \in \mathcal{J}(g',S')$ and $\sigma_{S'} \cdot B_0' > 0 > \sigma_{S'} \cdot B_1'$. Let S be the stratum containing $\pi'(S')$. First suppose that $B_i' = \mathfrak{B}(B_i, B_i \cap L)$ for some $B_i \in \mathcal{B}'$, $i = 0,1$. Note we cannot have $B_i = L_i$, $i = 0,1$ since we would have $S' \subset B_0' \cap B_1' = \mathfrak{B}(B_0 \cap B_1, L) = \mathfrak{B}(L,L) = \emptyset$. We have seen above that $\sigma_{S'} \cdot B_i' = \sigma_S \cdot B_i$. Hence

$$\lambda(g',R,B_0',B_1') = \sigma_S \cdot B_0 - \sigma_S \cdot B_1 = \lambda(g,R,B_0,B_1) < \lambda(g,R).$$

Now suppose $B_0' = \pi'^{-1}(L)$ and $B_1' = \mathfrak{B}(B_1, B_1 \cap L)$. Then

$$\begin{aligned}\lambda(g',R,B_0',B_1') &= \sigma_S \cdot L_0 + \sigma_S \cdot L_1 - \sigma_S \cdot B_1 \\ &< \sigma_S \cdot L_0 - \sigma_S \cdot B_1 \\ &= \lambda(g,R,L_0,B_1) \le \lambda(g,R).\end{aligned}$$

Likewise $\lambda(g',R,B_0',B_1') < \lambda(g,R,B_0,L_1) \le \lambda(g,R)$ if $B_0' = \mathfrak{B}(B_0, B_0 \cap L)$ and $B_1' = \pi'^{-1}(L)$. In any case we have seen $\lambda(g',R) < \lambda(g,R)$.

If $\eta(g,R) > 1$ then the above calculations show that $\lambda(g',R) \le \lambda(g,R)$ but $\eta(g',R) < \eta(g,R)$ if $\lambda(g',R) = \lambda(g,R)$, since $\lambda(g',R,B_0',B_1') = \lambda(g,R)$ implies $B_i' = \mathfrak{B}(B_i, B_i \cap L)$, $\lambda(g,R,B_0,B_1) = \lambda(g,R)$ and $(B_0,B_1) \ne (L_0,L_1)$. $\square$

By Assertion 3.5.3.3, we may as well assume that $\lambda(g,R) = 0$. Then for each $S \in \mathcal{E}(g,R)$ either $K_S \notin \mathcal{J}(g,S)$ or $\sigma_S \cdot B \ge 0$ for all $B \in \mathcal{B}'$ or $\sigma_S \cdot B \le 0$ for all $B \in \mathcal{B}'$. (Recall that $\sigma_S \cdot B = 0$ if $S \not\subset B$.)

If $A \in \mathcal{A}'$ and $B \in \mathcal{B}'$ let α_{AB} be the exponent of A in B for g.

Suppose $K_S \in \mathcal{J}(g,S)$ and $\sigma_S \cdot B \le 0$ for all $B \in \mathcal{B}'$. Pick any $T \in \mathcal{D}(g,S)$ with $\mathcal{K}(g,T) = K_S$. In particular, $\sigma_S \in \mathcal{K}(g,T)$. If $T \subset A \in \mathcal{A}'$ then $\mu_{g,T}(\sigma_S) = \sum_{B \in \mathcal{B}'}(\sigma_S \cdot B)\alpha_{AB}(T) = 0$ so since $\alpha_{AB}(T) \ge 0$ for all B and $\sigma_S \cdot B \le 0$ for all B we must have $(\sigma_S \cdot B)\alpha_{AB}(T) = 0$ for each $B \in \mathcal{B}'$. By Lemma 3.5.1c, $\mu_{\pi,S}$ is injective so $\sigma_S \ne 0$. So we must have $\sigma_S \cdot B < 0$ for some $B \in \mathcal{B}'$. But then we know $g(T) \subset S \subset B$. Hence $\alpha_{AB}(T) > 0$ for some $A \in \mathcal{A}'$ with $T \subset A$ by Lemma 3.2.2. This is a contradiction since $(\sigma_S \cdot B)\alpha_{AB}(T) = 0$.

Likewise we cannot have $K_S \in \mathcal{J}(g,S)$ and $\sigma_S \cdot B \ge 0$ for all $B \in \mathcal{B}'$. Consequently we must have $K_S \notin \mathcal{J}(g,S)$ for each $S \in \mathcal{E}(g,R)$. If we do this for each $K_0 \in \mathcal{J}(f,R)$ we are finished since then $\mathcal{J}(g,S) = \emptyset$ for each $S \in \mathcal{E}(g,R)$. $\square$

6. Submersive Tico Maps

Let us recall the definition of submersive tico map in a slightly different form; in particular we leave out the assumption of surjection of the matrix (β_{ij}). However we prove in Lemma 3.6.1 that this property will hold anyway, so our definitions are equivalent.

Definition: A tico map $f: (M, \mathcal{A}) \to (N, \mathcal{B})$ is *submersive* if for each $q \in M$ there are charts $\psi: Z \to M$ and $\theta: (\mathbf{R}^n, 0) \to (N, f(q))$ where Z is an open subset of $\mathbf{R}^m$ so that

1) $\psi(Z)$ is a neighborhood of q.
2) $f\psi(Z) \subset \theta(\mathbf{R}^n)$.
3) $\psi^{-1}(|\mathcal{A}|) = \bigcup_{j>c}^m \mathbf{R}_j^m \cap Z$.
4) $\theta^{-1}(|\mathcal{B}|) = \bigcup_{i>d}^n \mathbf{R}_i^n$, $d \le c$.
5) If $f_i(x)$ is the i-th coordinate of $\theta^{-1}f\psi(x)$ then $f_i(x) = x_i$ for $i \le d$.
6) For some nonnegative integers β_{ij}, $f_i(x) = \prod_{j>d}^m x_j^{\beta_{ij}}$ for $i > d$ and for all $x \in Z$.

We should emphasize the case where $f(q) \notin |\mathcal{B}|$, for instance when $q \notin |\mathcal{A}|$. Then f is locally a submersion. So $f|: M - f^{-1}(|\mathcal{B}|) \to N - |\mathcal{B}|$ is a smooth submersion.

The basic idea is that by 5), f submerses each stratum of M to its target stratum and by 6) the φ_i's in the definition of tico map have a particularly simple form. Recall, since f is a tico map, $f_i(x) = \prod_{j>c}^m x_j^{\beta_{ji}} \varphi_i(x)$ (with φ_i nowhere zero). So $\varphi_i(x) = \prod_{j>d}^c x_j^{b_{ij}}$. In particular, we must have $x_j \ne 0$ for $j = d+1, \ldots, c$ and x near $\psi^{-1}(q)$.

Lemma 3.6.1 *The matrix (β_{ij}) formed by the exponents in 6) is onto.*

Proof: Take the Jacobian matrix of $\theta^{-1}f\psi(x)$ at some point x with $x_j \ne 0$ for all j. This matrix is onto since $x \in \psi^{-1}(M - |\mathcal{A}|)$ so $\theta^{-1}f\psi$ is a submersion at x. Multiply each j-th column by x_j. Divide each i-th row by f_i. The resulting matrix is still onto. It is $\begin{pmatrix} I & 0 \\ 0 & \beta \end{pmatrix}$. So $\beta = (\beta_{ij})$ is onto. $\square$

Notice that the integers d and c are independent of all choices of θ and ψ since $n - d$ is the number of local sheets of $\mathcal{B}$ containing $f(q)$ and $m - c$ is the number of local sheets of $\mathcal{A}$ containing q. However the next lemma shows that there are essentially no restrictions on θ, since if θ satisfies 4) we may find a ψ so that 1)-6) are satisfied after perhaps composing θ with some hyperplane reflections.

Lemma 3.6.2 *Let $f: (M, \mathcal{A}) \to (N, \mathcal{B})$ be a submersive tico map. Pick any $q \in M$ and any chart $\theta: (\mathbf{R}^n, 0) \to (N, f(q))$ so that $\theta^{-1}(|\mathcal{B}|) = \bigcup_{i>d}^n \mathbf{R}_i^n$. Then we may pick a chart $\psi: Z \to M$ so that 1)-5) in the definition of submersive tico map hold. In addition a slightly weakened 6) holds, we have $f_i(x) = \pm \prod_{j>d}^m x_j^{\beta_{ij}}$ for $i > d$.*

Proof: Since this is a local result, we may as well assume that M is an open subset of $\mathbf{R}^m$, $N = \mathbf{R}^n$, $|\mathcal{A}| = \bigcup_{j>c}^m \mathbf{R}_j^m \cap M$, $|\mathcal{B}| = \bigcup_{i>d}^n \mathbf{R}_i^m$ and

$$f(x) = \left(x_1, \dots, x_d, \prod_{j>d}^m x_j^{\beta_{d+1,j}}, \dots, \prod_{j>d}^m x_j^{\beta_{nj}}\right).$$

By permuting coordinates we may as well assume that $\theta^{-1}(\mathbf{R}_i^n) = \mathbf{R}_i^n$ for all $i > d$. It suffices to find a coordinate chart $\psi\colon Z \to M$ with $Z \subset \mathbf{R}^m$ so that $q \in \psi(Z)$, $\psi^{-1}(\mathbf{R}_j^m) = \mathbf{R}_j^m \cap Z$ for all $j > c$ and $\theta f = f\psi$.

Let $\theta_i(x)$ and $\psi_i(x)$ denote the i-th coordinates of $\theta(x)$ and $\psi(x)$ respectively. Since $\theta(\mathbf{R}_i^n) \subset \mathbf{R}_i^n$ for $i > d$ we know $\theta_i(\mathbf{R}_i^n) = 0$ so by Lemma 3.1.1, $\theta_i(x) = x_i\theta_i'(x)$ for some smooth function θ_i', $i > d$. Notice that the i-th row of the Jacobian of θ is $(0, 0, \dots, \theta_i'(x), \dots, 0)$ at any point $x \in \mathbf{R}_i^n$. Hence $\theta_i'(x) \neq 0$ for all $x \in \mathbf{R}_i^n$. But $\mathbf{R}_i^n = \theta^{-1}(\mathbf{R}_i^n) = \theta_i^{-1}(0) = \mathbf{R}_i^n \cup \theta_i'^{-1}(0)$ so θ_i' is nowhere zero for $i > d$. After composing θ with reflections about the hyperplane $\mathbf{R}_i^n$ we may as well assume that each θ_i' is positive, $i > d$. (This is the source of the $\pm$ signs.)

We wish to solve $\theta f = f\psi$. In other words,

   i) $\theta_i f(x) = \psi_i(x)$ for $i \le d$.
   ii) $\theta_i f(x) = \prod_{j>d}^m \psi_j(x)^{\beta_{ij}}$ for $i > d$.

So equation i) defines $\psi_i(x)$ for $i \le d$. For ii) we know that if $i > d$ then $\theta_i(f(x)) = \prod_{j>d}^m x_j^{\beta_{ij}}\theta_i'(f(x))$. By Lemma 3.6.1 the matrix (β_{ij}) is onto so we can find rational numbers γ_{jk}, $d < j \le m$, $d < k \le n$ so that $\sum_{j>d}^m \beta_{ij}\gamma_{jk} = \delta_{ik}$, the Kronecker delta.

Given this, observe that $\psi_j(x) = x_j \prod_{k>d}^n \theta_k'(f(x))^{\gamma_{jk}}$ will solve equation ii). So we have a solution ψ to $\theta f = f\psi$, we must only show that ψ is a local diffeomorphism. But by computing the Jacobian we can see that it is. In particular, the Jacobian of ψ at q is of the form $\begin{pmatrix} A & 0 & B \\ C & D & E \\ 0 & 0 & F \end{pmatrix}$ where D and F are diagonal with nonzero i-th entry $\prod_{k>d}^n \theta_k'(f(q))^{\gamma_{ik}}$ and A is nonsingular since the Jacobian of θ at $f(q)$ is of the form $\begin{pmatrix} A & G \\ 0 & H \end{pmatrix}$ and is nonsingular. $\qquad\square$

Finally, we have the following result about submersive tico maps. Its proof is very similar to the proof of Lemma 3.5.2a, so we leave it as an exercise.

Lemma 3.6.3 *Suppose $\rho\colon (M', \mathcal{A}') \to (M, \mathcal{A})$ and $\pi\colon (N', \mathcal{B}') \to (N, \mathcal{B})$ are fat unobtrusive tico multiblowups and $f\colon (M, \mathcal{A}) \to (N, \mathcal{B})$ is a submersive tico map. Suppose $f'\colon M' \to N'$ is a smooth map so that $\pi f' = f\rho$. Then f' is a submersive tico map.*

7. Micos

We will need some notion of a tico map which is defined on only part of a manifold.

Definition: Let $(M, \mathcal{A})$ and $(N, \mathcal{B})$ be ticos and suppose $\mathcal{C} \subset \mathcal{A}$. We say $f \colon |\mathcal{C}| \to N$ is a *mico* if f is the restriction of a tico map, i.e., there is a neighborhood U of $|\mathcal{C}|$ in M and a tico map $f' \colon (U, U \cap (\mathcal{A} - \mathcal{C})) \to (N, \mathcal{B})$ so that $f = f'|$. We say the mico has type N if the extension f' can be taken so that $f' \colon (U, U \cap \mathcal{A}) \to (N, \mathcal{B})$ has type N. Likewise f is submersive if $f' \colon (U, U \cap \mathcal{A}) \to (N, \mathcal{B})$ is submersive.

Lemma 3.7.1 *Let* $(M, \mathcal{A})$ *and* $(N, \mathcal{B})$ *be ticos. Suppose* $\mathcal{C} \subset \mathcal{A}$ *and* $f \colon |\mathcal{C}| \to N$ *is a smooth map. For every* $S \in \mathcal{C}$ *let* $g_S \colon S' \to M$ *be the immersion associated to* S. *Then*

  a) f *is a mico if and only if* $f g_S \colon (S', g_S^* (\mathcal{A} - \mathcal{C})) \to (N, \mathcal{B})$ *is a tico map for each* $S \in \mathcal{C}$.

  b) *If* f *is a mico then* f *is a mico of type* N *if* $f g_S \colon (S', g_S^* (\mathcal{A})) \to (N, \mathcal{B})$ *has type* N *for all* $S \in \mathcal{C}$.

  c) *If* f *is a mico then* f *is a submersive mico if* $f g_S \colon (S', g_S^* (\mathcal{A})) \to (N, \mathcal{B})$ *is submersive for all* $S \in \mathcal{C}$.

Proof: Suppose that f is a mico. Then f extends to a tico map f' on a neighborhood U of $|\mathcal{C}|$. Pick any sheet $S \in \mathcal{C}$, let $g_S \colon S' \to M$ be the immersion associated to S. By Lemma 3.3.2 we know that $g_S \colon (S', g_S^*(\mathcal{A} - \mathcal{C})) \to (U, U \cap (\mathcal{A} - \mathcal{C}))$ is a tico map and hence $f g_S$ is a tico map also.

Now suppose that $f g_S \colon (S', g_S^* (\mathcal{A} - \mathcal{C})) \to (N, \mathcal{B})$ is a tico map for each $S \in \mathcal{C}$. Pick any $z \in |\mathcal{C}|$. Pick tico charts $\psi \colon (\mathbf{R}^m, 0) \to (M, z)$ and $\theta \colon (\mathbf{R}^n, 0) \to (N, f(z))$ so that $\psi^{-1} (|\mathcal{A}|) = \bigcup_{i=1}^a \mathbf{R}_i^m$, $\psi^{-1} (|\mathcal{C}|) = \bigcup_{i=1}^c \mathbf{R}_i^m$, $f \psi (\mathbf{R}^m) \subset \theta (\mathbf{R}^n)$ and $\theta^{-1} (|\mathcal{B}|) = \bigcup_{i=1}^b \mathbf{R}_i^n$. For $y \in \psi^{-1} (|\mathcal{C}|)$ let $\lambda_i (y)$ be the i-th coordinate of $\theta^{-1} f \psi (y)$.

Since $f g_S \colon (S', g_S^* (\mathcal{A} - \mathcal{C})) \to (N, \mathcal{B})$ is a tico map for each $S \in \mathcal{C}$ we know that there are integers α_{ij}^k and smooth functions $\varphi_i^k \colon \mathbf{R}^m \to \mathbf{R} - 0$ for $1 \leq k \leq c$, $1 \leq i \leq b$ and $c < j \leq a$ so that $\lambda_i (y) = \prod_{j=c+1}^a y_j^{\alpha_{ij}^k} \varphi_i^k (y)$ for all $y \in \mathbf{R}_k^m$. But if $y \in \mathbf{R}_k^m \cap \mathbf{R}_\ell^m$, we have $\prod_{j=c+1}^a y_j^{\alpha_{ij}^k} \varphi_i^k (y) = \prod_{j=c+1}^a y_j^{\alpha_{ij}^\ell} \varphi_i^\ell (y)$. Consequently, we know that $\alpha_{ij}^k = \alpha_{ij}^\ell$ for all $i = 1, \dots, b$, $j = c+1, \dots, a$ and $\varphi_i^k (y) = \varphi_i^\ell (y)$ for all $i = 1, \dots, b$, $y \in \mathbf{R}_k^m \cap \mathbf{R}_\ell^m$. Hence by Lemma 3.1.2 we may find smooth functions $\varphi_i \colon \mathbf{R}^m \to \mathbf{R}$ so that $\varphi_i|_{\mathbf{R}_k^m} = \varphi_i^k$ for all $i = 1, \dots, b$ and $k = 1, \dots, c$.

Now we may extend each λ_i to $\mathbf{R}^m$ by letting

$$\lambda_i (y) = \prod_{j=c+1}^a y_j^{\alpha_{ij}^1} \varphi_i (y)$$

for $i \le b$ and taking any extension we please for $i > b$. So we can extend f to $U = \psi \left(\mathbf{R}^m - \bigcup_{i=1}^b \varphi_i^{-1}(0) \right)$ by setting

$$f'(x) = \theta \left(\lambda_1 \psi^{-1}(x), \ldots, \lambda_n \psi^{-1}(x) \right).$$

Furthermore, there is a relative version. Suppose we already have a tico map $f'': (U', (\mathcal{A} - \mathcal{C}) \cap U') \to (N, \mathcal{B})$ so that $f''|_{U' \cap |\mathcal{C}|} = f|$ and U' is some open set. Let $K \subset U'$ be closed. Then we claim there is a neighborhood U'' of $(|\mathcal{C}| \cap \psi(\mathbf{R}^m)) \cup K$ and a tico map $f''': (U'', (\mathcal{A} - \mathcal{C}) \cap U'') \to (N, \mathcal{B})$ so that $f'''|_{U'' \cap |\mathcal{C}|} = f|$ and $f'''|_K = f''|_K$.

To see this relative version, note that on $\psi^{-1}(U')$ we already have an extension of the λ_i's and the φ_i's. So we need only pick our extensions of the λ_i's and φ_i's to $\mathbf{R}^m$ so that they agree with the previous extensions on $\psi^{-1}(K')$ for some closed neighborhood K' of K in U'. We may then let $U'' = \operatorname{Int}(K' \cup U)$ and let $f'''|_{\operatorname{Int} K'} = f''|$ and $f'''|_U = f'$. Since f'' and f' agree on the open set $U \cap \operatorname{Int} K'$, we know that f''' is smooth and is clearly a tico map.

Now that we have the relative version we easily get the compact global version. More specifically, let $L \subset M$ be compact and suppose $f': (U', (\mathcal{A} - \mathcal{C}) \cap U') \to (N, \mathcal{B})$ is an extension where U' is open and $K \subset U'$ is closed. Then there is a closed neighborhood U'' of $K \cup (L \cap |\mathcal{C}|)$ and an extension f'' so that $f''|_K = f'|$. We prove this by covering $L \cap |\mathcal{C}|$ with a finite number of charts and applying the above relative version a finite number of times. In particular, if M is compact we may set $L = M$, $U' = \emptyset$ and the lemma is proven.

So suppose M is not compact. Pick a sequence of open sets $W_i \subset M$ so that for each i, $\operatorname{Cl}(W_i)$ is compact, $\operatorname{Cl}(W_i) \subset W_{i+1}$ and $\bigcup_{i=1}^{\infty} W_i = M$. Let $W_0 = \emptyset$. Since $\operatorname{Cl}(W_i)$ is compact for all i we know that for each $i = 0, 1, \ldots$ there is a neighborhood U_i of $\operatorname{Cl}(W_{4i+3}) \cap |\mathcal{C}|$ and a tico map $f_i: (U_i, (\mathcal{A} - \mathcal{C}) \cap U_i) \to (N, \mathcal{B})$ extending $f|$. Pick closed neighborhoods K_i of $\operatorname{Cl}(W_{4i+3}) \cap |\mathcal{C}|$ in U_i. We know that for each $i = 0, 1, \ldots$ there is a neighborhood U_i' of $\operatorname{Cl}(W_{4i+6}) \cap |\mathcal{C}|$ and a tico map $f_i': (U_i', (\mathcal{A} - \mathcal{C}) \cap U_i') \to (N, \mathcal{B})$ extending $f|$ so that

i) f_i' agrees with f_i on $K_i \cap U_i' \cap W_{4i+3}$.
ii) f_i' agrees with f_{i+1} on $K_{i+1} \cap U_i' - W_{4i+4}$.

Let

$$U = \bigcup_{i=0}^{\infty} \left(\operatorname{Int} K_i \cap (W_{4i+3} - Cl(W_{4i})) \right) \cup \left(U_i' \cap (W_{4i+5} - \operatorname{Cl}(W_{4i+2})) \right).$$

Then certainly U is a neighborhood of $|\mathcal{C}|$. We define $f': U \to N$ as follows. If $x \in \operatorname{Int} K_i \cap (W_{4i+3} - \operatorname{Cl}(W_{4i}))$ let $f'(x) = f_i(x)$. If $x \in U_i' \cap (W_{4i+5} - \operatorname{Cl}(W_{4i+2}))$ let $f(x) = f_i'(x)$. This is well defined by i) and ii). It is certainly a tico map since it is locally either f_i or f_i'. Hence the lemma is proven, since $f': (U, (\mathcal{A} - \mathcal{C}) \cap U) \to (N, \mathcal{B})$ being a tico map implies that $f': (U, \mathcal{A} \cap U) \to (N, \mathcal{B})$ is a tico map also by Lemma 3.2.5.

Now suppose that f is a mico and $fg_S \colon (S', g_S^* (\mathcal{A})) \to (N, \mathcal{B})$ has type N for all $S \in \mathcal{C}$. Pick any $z \in |\mathcal{C}|$. Pick $S \in \mathcal{C}$ so that $z \in S$. We may pick charts $\psi' \colon Z' \to S'$ and $\theta \colon (\mathbf{R}^n, 0) \to (N, f(z))$ where Z' is a connected open subset of $\mathbf{R}_1^m$, $\psi'(Z')$ is a neighborhood of some point $z' \in g_S^{-1}(z)$, $fg_S\psi'(Z) \subset \theta(\mathbf{R}^n)$, $\psi'^{-1}(|g_S^*(\mathcal{A})|) = \bigcup_{i=2}^a \mathbf{R}_i^m \cap Z'$, $\theta^{-1}(|\mathcal{B}|) = \bigcup_{i=1}^b \mathbf{R}_i^n$, $f_i(x) = \prod_{j=2}^c x_j^{\alpha_{ij}}$ for $i \leq b$ and the matrix (α_{ij}) is onto, where $f_i(x)$ is the i-th coordinate of $\theta^{-1}fg_S\psi'(x)$. We may also assume $Z' \cap \mathbf{R}_i^m$ is empty for $a < i \leq c$.

Pick any chart $\psi \colon Z \to M$ where Z is some connected neighborhood of Z' in $\mathbf{R}^m$ so that $Z \cap \mathbf{R}_1^m = Z'$, $g_S\psi = \psi|$, $\psi^{-1}(|\mathcal{A}|) = \bigcup_{i=1}^a \mathbf{R}_i^m \cap Z$ and $Z \cap \mathbf{R}_i^m$ is empty for all $a < i \leq c$. Let $h_i(x)$ be the i-th coordinate of $\theta^{-1}f'\psi(x)$ where f' is any extension of f to a tico map on a neighborhood of x. Then for $i \leq b$, we have $h_i(x) = \prod_{j=1}^a x_j^{\beta_{ij}} \varphi_i(x)$ for some smooth maps $\varphi_i \colon Z \to \mathbf{R} - 0$. Since $h_i|_{Z'} = f_i$ we know that $\beta_{i1} = 0$ for all $i \leq b$, $\beta_{ij} = \alpha_{ij}$ for all $i \leq b$, $1 < j \leq a$ and $\varphi_i(x) = \prod_{j>a}^c x_j^{\alpha_{ij}}$ for $x \in Z'$.

Since (α_{ij}) is onto we may pick rational numbers γ_{jk} for $2 \leq j \leq c$, $1 \leq k \leq b$ so that $\sum_{j=2}^c \alpha_{ij}\gamma_{jk} = \delta_{ik}$, the Kronecker delta. Define $\mu_j \colon Z \to \mathbf{R}$ for $1 \leq j \leq m$ by letting $\mu_j(x) = x_j$ if $j = 1$, or $j > c$ and $\mu_j(x) = x_j \prod_{k=1}^b \left(\varphi_k(x) \prod_{i>a}^c x_i^{-\alpha_{ki}} \right)^{\gamma_{jk}}$ for $2 \leq j \leq c$. Define $\mu \colon Z \to \mathbf{R}^m$ by $\mu(x) = (\mu_1(x), \dots, \mu_m(x))$. Note that $\mu|_{Z'}$ is the identity. Then the Jacobian matrix of μ at any point of Z' is the identity (except for some garbage in the first columns of rows 2 to c). Hence there is a neighborhood Z'' of $\psi^{-1}(z)$ in Z on which μ^{-1} is well defined.

Let $g_i(x)$ be the i-th coordinate of $\theta^{-1}f'\psi\mu^{-1}(x)$ for $x \in Z''$. We claim that $g_i(x) = \prod_{j=2}^c x_j^{\alpha_{ij}}$ for all $i \leq b$. To see this, let $y = \mu^{-1}(x)$. Then $g_i(x) = h_i(y) = \prod_{j=2}^a y_j^{\alpha_{ij}} \varphi_i(y)$. But

$$\prod_{j=2}^c x_j^{\alpha_{ij}} = \prod_{j=2}^c \mu_j(y)^{\alpha_{ij}}$$

$$= \prod_{j=2}^c y_j^{\alpha_{ij}} \prod_{j=2}^c \prod_{k=1}^b \left(\varphi_k(y) \prod_{d>a}^c y_d^{-\alpha_{kd}} \right)^{\gamma_{jk}\alpha_{ij}}$$

$$= \prod_{j=2}^c y_j^{\alpha_{ij}} \prod_{k=1}^b \left(\varphi_k(y) \prod_{d>a}^c y_d^{-\alpha_{kd}} \right)^{\delta_{ik}}$$

$$= \prod_{j=2}^c y_j^{\alpha_{ij}} \varphi_i(y) \prod_{d>a}^c y_d^{-\alpha_{id}} = \prod_{j=2}^a y_j^{\alpha_{ij}} \varphi_i(y) = g_i(x).$$

So f' has type N. The proof for submersive micos is similar to the above type N proof. $\qquad\square$

CHAPTER IV

RESOLUTION TOWERS

1. Definition of Resolution Towers

Finally we are in a position to define resolution towers. The main result of this chapter is Theorem 4.2.8 which says that once we have a resolution tower, we can blow it up to make it nicer.

Definition: A *resolution tower* $\mathfrak{T} = \{V_i, \mathcal{A}_i, p_i\}_\mathfrak{I}$ is a partially ordered index set $\mathfrak{I}$, a collection of ticos $(V_i, \mathcal{A}_i)$, $i \in \mathfrak{I}$ and collections of proper maps $p_i = \{p_{ji}\}_{j<i}$ where $p_{ji} \colon V_{ji} \to V_j$ so that for each $j < i$, $V_{ji} = |\mathcal{A}_{ji}|$ for some $\mathcal{A}_{ji} \subset \mathcal{A}_i$ and:

    I. If $V_{ji} \cap V_{ki} \neq \emptyset$ then either $j < k$ or $k < j$ or $k = j$. Moreover, $p_{ji}(V_{ji} \cap V_{ki}) \subset V_{kj}$ if $k < j < i$.

    II. $p_{kj} \circ p_{ji}| = p_{ki}|_{V_{ji} \cap V_{ki}}$ if $k < j < i$.

    III. $p_{ji}^{-1}\left(\bigcup_{k<m} V_{kj}\right) = \bigcup_{k<m} V_{ki} \cap V_{ji}$ if $m \leq j < i$.

    IV. $\mathcal{A}_i$ is the disjoint union $\bigcup_{j<i} \mathcal{A}_{ji}$.

    V. $p_{ji}^{-1}(\partial V_j) = \partial V_i \cap V_{ji}$ if $j < i$.

Note that although $\mathcal{A}_{ji} \cap \mathcal{A}_{ki}$ is empty if $j \neq k$, $|\mathcal{A}_{ji}| \cap |\mathcal{A}_{ki}|$ is not usually empty.

We normally take the index set $\mathfrak{I}$ to be $\{0, \dots, n\}$, indeed if there are $n + 1$ elements of $\mathfrak{I}$ there is a one to one order preserving map of $\mathfrak{I}$ onto $\{0, \dots, n\}$ so we might as well do so. However we will sometimes find it convenient to use other indexing sets.

Definitions: A subset $\mathfrak{I}' \subset \mathfrak{I}$ is called an *upper index subset* if $i < j$ and $i \in \mathfrak{I}'$ implies $j \in \mathfrak{I}'$. A subset $\mathfrak{I}' \subset \mathfrak{I}$ is called a *lower index subset* if $i > j$ and $i \in \mathfrak{I}'$ implies $j \in \mathfrak{I}'$.

Definition: A resolution tower is *finite* if each tico $(V_i, \mathcal{A}_i)$ is finite.

To make a resolution tower useful we will usually need some additional properties. The ones we will use here are:

    R - Each tico $\mathcal{A}_i$ is regular

M - Each p_{ji} is a mico

E - Each tico $\mathcal{A}_i$ is regular and also each p_{ji} is a mico with constant exponents

S - Each p_{ji} is a submersive mico

N - Each p_{ji} has type N

F - Each $(V_i, \mathcal{A}_i)$ is full

U - $p_{ji}|_S$ is a submersion for every stratum S of $(V_i, \mathcal{A}_i)$ with $S \subset V_{ji} - \bigcup_{k<j} V_{ki}$

We can combine these properties. Thus a resolution tower for which properties R and M hold is said to be of type RM (or type MR). A resolution tower of type E is automatically of type MR.

Definition: If $\mathfrak{T} = \{V_i, \mathcal{A}_i, p_i\}_{\mathfrak{I}}$ is a resolution tower, then $\partial\mathfrak{T}$ is the tower $\{V_i', \mathcal{A}_i', p_i'\}_{\mathfrak{I}}$ where $V_i' = \partial V_i$, $\mathcal{A}_{ji} = \mathcal{A}_{ji} \cap \partial V_i$ and $p_{ji}' = p_{ji}|_{V_{ji}'}$.

Exercise: $\partial\mathfrak{T}$ is a resolution tower. ◇

As for the case of ticos, we will tend to ignore boundary points in proofs. Taking them into account would only complicate the proofs and obscure the ideas.

Definition: If $\mathfrak{T} = \{V_i, \mathcal{A}_i, p_i\}_{\mathfrak{I}}$ is a resolution tower and $\mathfrak{I}' \subset \mathfrak{I}$ is a lower index subset then the *truncated resolution tower* $\mathfrak{T}[\mathfrak{I}']$ is $\{V_i, \mathcal{A}_i, p_i\}_{\mathfrak{I}'}$. In other words we just forget V_i, $\mathcal{A}_i$ and p_i for $i \notin \mathfrak{I}'$. In case $\mathfrak{I} = \{0, \ldots, n\}$ we make the convention that $\mathfrak{T}[m] = \mathfrak{T}[\{0, 1, \ldots, m\}]$.

Definition: If $\mathfrak{T} = \{V_i, \mathcal{A}_i, p_i\}_{\mathfrak{I}}$ is a resolution tower then $|\mathfrak{T}|$ denotes the *realization* of $\mathfrak{T}$. This is the quotient space obtained from the disjoint union $\bigcup_{i \in \mathfrak{I}} V_i$ by identifying every $x \in V_{ji}$ with $p_{ji}(x) \in V_j$.

Exercise: $|\mathfrak{T}|$ is a locally compact Hausdorff space. (Hint: you need the properness of the p_i's). ◇

Notice that $|\mathfrak{T}|$ is a stratified set with strata $V_i - |\mathcal{A}_i|$. The reader should think of V_i as being a resolution of the singularities of the i-skeleton of $|\mathfrak{T}|$ and think of $V_{ji} \subset V_i$ as the closure of the inverse image of the j strata. The maps p_{ji} glue all these different resolutions together. We normally think of V_i as having dimension i but for convenience we do not actually require this.

For example, if $n = 1$, V_1 is a circle, $\mathcal{A}_{01}$ is 3 points, V_0 is 2 points and $p_{01}(|\mathcal{A}_{01}|)$ is one point in V_0, then the realization will be as in Figure IV.1.1, a three-leaf rose union a point.

The following gives another example of a resolution tower. Consider the resolution tower $\mathfrak{T} = \{V_i, \mathcal{A}_i, p_i\}_{i=0}^2$ where $V_2 = S^1 \times S^1$, $\mathcal{A}_{02} = \{S^1 \times a, S^1 \times b\}$, $\mathcal{A}_{12} = \{a \times S^1\}$, $V_1 = S^1$, $\mathcal{A}_{01} = \{a, b\}$ where $a \neq b$ are points on S^1,

FIGURE IV.1.1. A realization of a resolution tower

$V_0 = S^0 = \{a', b'\}$, p_{02} collapses $S^1 \times a$ and $S^1 \times b$ to a' and b' respectively, p_{01} takes a, b to a', b' respectively, p_{12} is the obvious homeomorphism $a \times S^1 \to S^1$ induced by projection. This tower is shown in Figure IV.1.2. The realization of

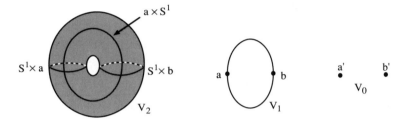

FIGURE IV.1.2. A resolution tower

this resolution tower is the suspension of two circles shown in Figure IV.1.3.

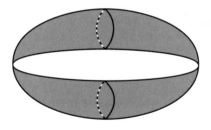

FIGURE IV.1.3. The realization of the resolution tower

Throughout this book, when we mention a resolution tower $\mathfrak{T}$ then unless we state otherwise, we automatically mean that $\mathfrak{T} = \{V_i, \mathcal{A}_i, p_i\}_{\mathfrak{J}}$, for some $\mathfrak{J}$, p_{ji} is an element of the collection p_i, V_{ji} is the domain of p_{ji} and $\mathcal{A}_{ji} \subset \mathcal{A}_i$ is such that $V_{ji} = |\mathcal{A}_{ji}|$. Likewise if we have another resolution tower $\mathfrak{T}'$ we will mean that $\mathfrak{T}' = \{V_i', \mathcal{A}_i', p_i'\}_{\mathfrak{J}'}$, p_{ji}' will be an element of the collection p_j', V_{ji}' will be the domain of p_{ji}' and $|\mathcal{A}_{ji}'| = V_{ji}'$. A similar convention holds for $\mathfrak{T}''$ etc. Thus, naming $\mathfrak{T}$ automatically names each V_i, $\mathcal{A}_i$, p_i, V_{ji}, $\mathcal{A}_{ji}$ and p_{ji} and we may begin using these symbols without explicitly defining them. By default, unless we state otherwise, the index set $\mathfrak{J}$ will be $\{0, \ldots, n\}$.

Definition: We call a resolution tower $\mathfrak{T} = \{V_i, \mathcal{A}_i, p_i\}_{\mathfrak{I}}$ an *algebraic resolution tower* if each $(V_i, \mathcal{A}_i)$ is an algebraic tico and all p_{ji} are entire rational functions. In particular all algebraic resolution towers are of type R. Sometimes to emphasize the fact that resolution towers are topological objects we will call them *topological resolution towers*.

Definitions: Let $\mathfrak{T} = \{V_i, \mathcal{A}_i, p_i\}_{\mathfrak{I}}$ and $\mathfrak{T}' = \{V_i', \mathcal{A}_i', p_i'\}_{\mathfrak{I}'}$ be resolution towers. Let $\alpha: \mathfrak{I} \to \mathfrak{I}'$ be an order preserving map and let $f = \{f_i\}_{\mathfrak{I}}$ be a collection of smooth maps $f_i: V_i \to V_{\alpha(i)}'$. Then we say that $f: \mathfrak{T} \to \mathfrak{T}'$ is a *tower morphism* if

1) $f_i\left(|\mathcal{A}_{ji}|\right) \subset |\mathcal{A}_{\alpha(j)\alpha(i)}'|$ for all $j < i$.
2) $f_i^{-1}\left(\bigcup_{j \leq b} |\mathcal{A}_{j\alpha(i)}'|\right) = \bigcup_{\alpha(k) \leq b} |\mathcal{A}_{ki}|$ for all $b < \alpha(i)$.
3) $p_{\alpha(j)\alpha(i)}' f_i|_{|\mathcal{A}_{ji}|} = f_j p_{ji}$ for all $j < i$.

We say that f is a *tico tower morphism* if in addition, each $f_i: (V_i, \mathcal{A}_i) \to (V_i', \mathcal{A}_i')$ is a tico map. If $\mathfrak{T}$ and $\mathfrak{T}'$ are algebraic, each f_i is an entire rational function and $\{f_i\}_{\mathfrak{I}}$ is a tower morphism we say that $\{f_i\}_{\mathfrak{I}}$ is a *rational tower morphism*.

Definition: An *isomorphism* $h = (h_i)_{\mathfrak{I}}$ between two resolution towers $\mathfrak{T} = \{V_i, \mathcal{A}_i, p_i\}_{\mathfrak{I}}$ and $\mathfrak{T}' = \{V_i', \mathcal{A}_i', p_i'\}_{\mathfrak{I}'}$ is an invertible tower morphism. In other words, it is an isomorphism $\alpha: \mathfrak{I} \to \mathfrak{I}'$ of ordered sets and a set of diffeomorphisms $h_i: V_i \to V_{\alpha(i)}'$, $i \in \mathfrak{I}$, so that $h_i^* \mathcal{A}_{\alpha(j)\alpha(i)}' = \mathcal{A}_{ji}$ for all $j < i$ and $h_j p_{ji} = p_{\alpha(j)\alpha(i)}' h_i|$ for all $j < i$.

Definition: In Chapter VII we will find a weaker notion useful. A *weak isomorphism* $h = (h_i)_{\mathfrak{I}}$ between two resolution towers $\mathfrak{T} = \{V_i, \mathcal{A}_i, p_i\}_{\mathfrak{I}}$ and $\mathfrak{T}' = \{V_i', \mathcal{A}_i', p_i'\}_{\mathfrak{I}'}$ is an isomorphism $\alpha: \mathfrak{I} \to \mathfrak{I}'$ of ordered sets and a set of homeomorphisms $h_i: V_i \to V_{\alpha(i)}'$, $i \in \mathfrak{I}$, so that $h_i^* \mathcal{A}_{\alpha(j)\alpha(i)}' = \mathcal{A}_{ji}$ for all $j < i$ and $h_j p_{ji} = p_{\alpha(j)\alpha(i)}' h_i|$ for all $j < i$ and so that each h_i restricts to a diffeomorphism of $V_i - |\mathcal{A}_i|$ to $V_{\alpha(i)}' - |\mathcal{A}_{\alpha(i)}'|$. Notice that a weak isomorphism induces an isomorphism of stratified sets from the realization $|\mathfrak{T}|$ to $|\mathfrak{T}'|$.

We often will say that a collection of functions $\{f_i\}_{i \in \mathfrak{I}}$ is a tower morphism, tico tower morphism, etc., without mentioning the map α since α can be recovered from $\{f_i\}_{i \in \mathfrak{I}}$ anyway. But sometimes it is convenient to mention α.

Let $\mathfrak{TR}$ and $\mathcal{AR}$ be the set of all topological and algebraic resolution towers respectively. When we put subscripts R, M, E, S, N, F, U to $\mathfrak{TR}$ or $\mathcal{AR}$ they will denote the subsets of $\mathfrak{TR}$ or $\mathcal{AR}$ of that type. For example, $\mathfrak{TR}_{RM}$ is the set of resolution towers of type RM. Clearly we can map $\mathcal{AR} \to \mathfrak{TR}$ by the forgetful map. Notice $\mathfrak{TR}_S \subset \mathfrak{TR}_{NU} \subset \mathfrak{TR}_{MU}$, $\mathcal{AR}_S \subset \mathcal{AR}_{NU} \subset \mathcal{AR}_{MU}$ and $\mathcal{AR} = \mathcal{AR}_R$.

Let us make the convention that if $\mathcal{D}$ is any subset of $\mathfrak{TR}$ or $\mathcal{AR}$ then $|\mathcal{D}| = \{|\mathfrak{T}| \mid \mathfrak{T} \in \mathcal{D}\}$. We think of $|\mathcal{D}|$ as lying in the collection of smooth stratified sets up to P.L. isomorphism, i.e., we identify two stratified sets if they have isomorphic refinements.

Theorem 4.28 and Proposition 4.27 below have the following corollary.

Theorem $|\mathfrak{TR}| = |\mathfrak{TR}_R|$ and $|\mathfrak{TR}_M| = |\mathfrak{TR}_{ERN}|$. *In fact* $|\mathcal{AR}_{*M}| = |\mathcal{AR}_{*ERN}|$, $|\mathfrak{TR}_*| = |\mathfrak{TR}_{*R}|$ *and* $|\mathfrak{TR}_{*M}| = |\mathfrak{TR}_{*ERN}|$ *where* * *is any string of symbols from* $\{F, U, N, S\}$.

Let us now prove some results about resolution towers.

Lemma 4.1.1 *Let* $\mathfrak{T} = \{V_i, \mathcal{A}_i, p_i\}_{\mathfrak{I}}$ *be a resolution tower of type* M. *Pick any extension* p'_{ji} *of* p_{ji} *to a neighborhood* U *of* V_{ij} *so that* $p'_{ji}\colon (U, U \cap \mathcal{A}_i) \to (V_j, \mathcal{A}_j)$ *is a tico map. Then for any* m, *there is a neighborhood* U' *of* V_{ji} *so that*

$$p'_{ji}\colon \big(U',\, U' \cap \bigcup_{k \leq m} \mathcal{A}_{ki}\big) \to \big(V_j,\, \bigcup_{k \leq m} \mathcal{A}_{kj}\big)$$

is a tico map.

Proof: By Lemma 3.2.5 we may as well assume that $m < j$. We know that

$$p'^{-1}_{ji}\big(\bigcup_{k \leq m} V_{kj}\big) \cap V_{ji} = p_{ji}^{-1}\big(\bigcup_{k \leq m} V_{kj}\big) = V_{ji} \cap \bigcup_{k \leq m} V_{ki}.$$

Since $p'^{-1}_{ji}(\bigcup_{k \leq m} V_{kj}) \subset |U \cap \mathcal{A}_i|$, Lemma 3.2.2 implies that we can find a neighborhood U' of V_{ji} so that $U' \cap p'^{-1}_{ji}(\bigcup_{k \leq m} V_{kj}) = U' \cap \bigcup_{k \leq m} V_{ki}$. The result now follows from Lemma 3.2.5. $\qquad\square$

Definition: A resolution tower $\mathfrak{T} = \{V_i, \mathcal{A}_i, p_i\}_{i=a}^{n}$ is called *well indexed with bias* m if $\dim V_i = i + m$ for all $i = a, \dots, n$. It is simply called *well indexed* if it is well indexed with some bias.

Proposition 4.1.2 *Suppose* $\mathfrak{T} = \{V_i, \mathcal{A}_i, p_i\}_{\mathfrak{I}}$ *is a resolution tower and* $\dim V_j < \dim V_i$ *whenever* $V_{ji} \neq \emptyset$. *Then there is a canonical way of changing* $\mathfrak{T}$ *into a well indexed resolution tower* $\mathfrak{T}' = \{V'_i, \mathcal{A}'_i, p'_i\}_{i=0}^{m}$ *with bias 0 so that* $|\mathfrak{T}| = |\mathfrak{T}'|$. *Furthermore, if* $\mathfrak{T}$ *is of type* R, M, E, S, N, F *or* U *or is algebraic, so is* $\mathfrak{T}'$.

Proof: Let $m = \max \dim V_i$ and let $D_k = \{\, i \in \mathfrak{I} \mid \dim V_i = k \,\}$. Let $V'_k = \bigcup_{i \in D_k} V_i$, $\mathcal{A}'_k = \bigcup_{i \in D_k} \mathcal{A}_i$ and $\mathcal{A}'_{bk} = \bigcup_{j \in D_b} \bigcup_{i \in D_k} \mathcal{A}_{ji}$. Our hypothesis implies that:

Assertion 4.1.2.1 $\mathcal{A}'_k = \bigcup_{b < k} \mathcal{A}'_{bk}$.

We also know that:

Assertion 4.1.2.2 $V_{ji} \cap V_{ki} = \emptyset$ if $k < j$ and $\dim V_j \leq \dim V_k$.

Proof: Suppose not. Then $\emptyset \neq p_{ji}(V_{ji} \cap V_{ki}) \subset V_{kj}$. So $\dim V_k < \dim V_j$, a contradiction. $\qquad\square$

So we may define $p'_{bc}\colon |\mathcal{A}'_{bc}| \to V'_b$ by $p'_{bc}|_{\mathcal{A}_{ji}} = p_{ji}$ if $j \in D_b$ and $i \in D_c$. By Assertion 4.1.2.2, this is well defined since if $k \neq j$ and $\dim V_j = \dim V_k$ we must have $V_{ji} \cap V_{ki} = \emptyset$.

Assertion 4.1.2.3 $\mathfrak{T}'$ is a resolution tower.

Proof: Note that Assertion 4.1.2.1 implies condition IV in the definition of resolution tower. The first part of condition I is trivial. For the last part of condition I, if $0 \leq a < b < c \leq m$ then

$$p'_{bc}(V'_{bc} \cap V'_{ac}) = p'_{bc}\Big(\bigcup_{k \in D_a} \bigcup_{j \in D_b} \bigcup_{i \in D_c} (V_{ki} \cap V_{ji}) \Big) = \bigcup_{k \in D_a} \bigcup_{j \in D_b} \bigcup_{i \in D_c} p_{ji}(V_{ki} \cap V_{ji}).$$

But by Assertion 4.1.2.2, $V_{ki} \cap V_{ji} \neq \emptyset$ for $k \in D_a$ and $j \in D_b$ implies $k < j$ so

$$p'_{bc}(V'_{bc} \cap V'_{ac}) = \bigcup_{j \in D_b} \bigcup_{i \in D_c} \overset{k<j}{\bigcup_{k \in D_a}} p_{ji}(V_{ki} \cap V_{ji}) \subset \bigcup_{k \in D_a} \bigcup_{j \in D_b} V_{kj} = |\mathcal{A}'_{ab}|.$$

Now condition II is immediate from condition I and the definition of p'_{bc}. For condition III, pick integers $d \leq b < c$ then by condition I we know that

$$p'_{bc}{}^{-1}\Big(\bigcup_{a<d} V'_{ab} \Big) \supset \bigcup_{a<d} (V'_{ac} \cap V'_{bc}).$$

So take any $x \in p'_{bc}{}^{-1}(\bigcup_{a<d} V'_{ab})$. Then $x \in p_{ji}^{-1}(V_{kj})$ for some $i \in D_c$, $j \in D_b$, $k \in D_a$ with $k < j < i$ and $a < d$. Choose a minimal m so that $k < m \leq j$. Then by condition III we know that $x \in V_{\ell i} \cap V_{ji}$ for some $\ell < m$. Then $p_{ji}(x) \in V_{\ell j} \cap V_{kj}$ so we know from condition I that either $\ell < k$ or $\ell > k$ or $\ell = k$. We can not have $\ell > k$ since the minimality of m would be violated. So $\ell \leq k$ and thus by Assertion 4.1.2 $\dim V_\ell \leq \dim V_k < d$ so $x \in V_{\ell i} \cap V_{ji} \subset \bigcup_{a<d}(V'_{ac} \cap V'_{bc})$.

□

We leave the verification that the types don't change as an exercise. □

Lemma 4.1.3 *Suppose* $\mathfrak{T} = \{V_i, \mathcal{A}_i, p_i\}_{\mathfrak{I}}$ *is a resolution tower of type U. Then* $\dim V_j < \dim V_i$ *whenever* $V_{ji} \neq \emptyset$.

Proof: Pick i and j so $V_{ji} \neq \emptyset$. Pick a stratum S of V_i so that $S \subset V_{ji} - \bigcup_{b<j} V_{bi}$. Then $p_{ji}|_S$ submerses S to V_j. Hence $\dim V_j \leq \dim S < \dim V_i$. □

2. Blowing up Resolution Towers

We need to develop a notion of blowing up resolution towers. The first step is to blow up a single level of a resolution tower.

Definition: Let $\mathfrak{T} = \{V_i, \mathcal{A}_i, p_i\}_{\mathfrak{I}}$ be a resolution tower and let $\pi \colon (V'_k, \mathcal{A}'_k) \to (V_k, \mathcal{A}_k)$ be a tico multiblowup with unobtrusive, unskewed centers. We will define the *tower transforms* $\mathcal{A}'_{jk}, p'_{jk}$ and p'_k of $\mathcal{A}_{jk}, p_{jk}$ and p_k with $\mathcal{A}'_{jk} \subset \mathcal{A}'_k$, $p'_{jk} \colon |\mathcal{A}'_{jk}| \to V_j$ and $p'_k = \{p'_{jk}\}_{j<k}$. Factor π into a composition of a single blowup $\pi' \colon (V'_k, \mathcal{A}'_k) \to (V''_k, \mathcal{A}''_k)$ with center L and a multiblowup $\pi'' \colon (V''_k, \mathcal{A}''_k) \to (V_k, \mathcal{A}_k)$. We may suppose we have already defined the tower transforms $\mathcal{A}''_{jk}$ of $\mathcal{A}_{jk}$, $0 \leq j < k$ (via the multiblowup π''), so that $\mathcal{A}''_k$ is the disjoint union $\bigcup_{j<k} \mathcal{A}''_{jk}$ and so that $\mathcal{A}''_{jk} = \mathcal{A}_{jk}$ in case π'' is the identity. Let m be the smallest index with $L \subset |\mathcal{A}''_{mk}|$. We define $\mathcal{A}'_{jk}$ to be the strict transform of $\mathcal{A}''_{jk}$ (via the blowup π') if $j \neq m$ and let $\mathcal{A}'_{mk}$ be the total transform of $\mathcal{A}''_{mk}$. Let $p'_{jk} \colon |\mathcal{A}'_{jk}| \to V_j$ be $p_{jk}\pi|$ and let $p'_k = \{p'_{jk}\}_{j<k}$.

Lemma 4.2.1 *With the notation as above,*

$$\pi\left(|\mathcal{A}'_{ik}|\right) = |\mathcal{A}_{ik}| \quad \text{and}$$

$$\pi^{-1}\left(\bigcup_{j\leq i} |\mathcal{A}_{jk}|\right) = \bigcup_{j\leq i} |\mathcal{A}'_{jk}|$$

for all $i < k$.

Proof: We may assume π is a single blowup with center L. That $\pi\left(|\mathcal{A}'_{ik}|\right) = |\mathcal{A}_{ik}|$ follows from the definition of strict transform and the fact that L is nowhere dense in $|\mathcal{A}_{ik}|$ if $i \neq m$. As usual, let V'_{jk} denote $|\mathcal{A}'_{jk}|$. Pick any $x \in \pi^{-1}(V_{jk})$ with $j \leq i$. Then either $x \in V'_{jk}$ or $\pi(x) \in L$. If $\pi(x) \in L$ suppose first that L is transverse to V_{jk} at $\pi(x)$. Then V'_{jk} coincides with $\pi^{-1}(V_{jk})$ near x so $x \in V'_{jk}$. If L is not transverse, then $L \subset V_{jk}$ so $\pi^{-1}(L) \in \mathcal{A}'_{mk}$ for some $m \leq j$. So in any case $x \in \bigcup_{b\leq j} V'_{bk}$. So $\pi^{-1}(V_{jk}) \subset \bigcup_{b\leq j} V'_{bk}$. But $\pi(V'_{jk}) \subset V_{jk}$, so we are done. $\qquad\square$

In order for a blowup of a single level to fit into the resolution tower, we need liftings of maps from all the higher levels. The liftings may not exist but the following lemma says that if they do, a resolution tower is obtained.

Lemma 4.2.2 *Let $\mathfrak{T} = \{V_i, \mathcal{A}_i, p_i\}_{\mathfrak{I}}$ be a resolution tower and suppose that $\pi \colon (V'_k, \mathcal{A}'_k) \to (V_k, \mathcal{A}_k)$ is a tico blowup with unobtrusive, unskewed center L. Suppose further that there are maps $p'_{ki} \colon V_{ki} \to V'_k$ for all $i > k$ so that $\pi p'_{ki} = p_{ki}$. Let $\mathcal{A}'_{jk}$, p'_k, p'_{jk} be the tower transforms of $\mathcal{A}_{jk}$, p_k and p_{jk}, $j < k$. For $i \neq k$ let $V'_i = V_i$, $\mathcal{A}'_i = \mathcal{A}_i$ and $p'_i = \{p'_{ji}\}_{j<i}$ where $p'_{ji} = p_{ji}$ for $i \neq k \neq j$. Then $\mathfrak{T}' = \{V'_i, \mathcal{A}'_i, p'_i\}_{\mathfrak{I}}$ is a resolution tower. Also $\mathfrak{T}'$ has type U if $\mathfrak{T}$ has type U and if $p_{jk}|$ submerses all strata of $L \cap V_{jk} - \bigcup_{m<j} V_{mk}$ for all $j < k$.*

Proof: First let us show that the maps p'_{ji} are proper. This is trivial for $j \neq k \neq i$ since $p'_{ji} = p_{ji}$. Now $p'_{jk} = p_{jk}\pi$ which is proper since p_{jk} and π are. So let us consider p'_{ki}. Take a compact set $K \subset V'_k$. Then $\pi(K)$ is compact so $p_{ki}^{-1}\pi(K)$ is compact by properness of p_{ki}. But $p'^{-1}_{ki}(K)$ is a closed subset of $p'^{-1}_{ki}\pi^{-1}\pi(K) = p_{ki}^{-1}\pi(K)$ which is compact. So p'_{ki} is proper.

Let us prove condition I in the definition of resolution tower. If $V'_{jk} \cap V'_{ik} \neq \emptyset$ then $\emptyset \neq \pi(V'_{jk} \cap V'_{ik}) \subset V_{jk} \cap V_{ik}$ so either $i < j$, $j < i$ or $i = j$. Next, if $j < i < k$, then

$$p'_{ik}(V'_{ik} \cap V'_{jk}) = p_{ik}\pi(V'_{ik} \cap V'_{jk}) \subset p_{ik}(V_{ik} \cap V_{jk}) \subset V_{ji} = V'_{ji}.$$

If $j < k < i$ then pick any $x \in V_{ki} \cap V_{ji} - \bigcup_{m<j} V_{mi}$. Then

$$p'_{ki}(x) \in \pi^{-1} p_{ki}(x) \subset \pi^{-1}\left(V_{jk} - \bigcup_{m<j} V_{mk}\right) = V'_{jk} - \bigcup_{m<j} V'_{mk}$$

by Lemma 4.2.1. So $p'_{ki}(V_{ji} \cap V_{ki}) \subset V'_{jk}$ by continuity. So condition I is true.

The only nontrivial case in condition II is to show that if $k < j < i$ then $p'_{ki}(x) = p'_{kj} p_{ji}(x)$ for all $x \in V_{ji} \cap V_{ki}$. First take $x \in V_{ji} \cap V_{ki} - \bigcup_{m<k} V_{mi}$. Then $p_{ki}(x) \in V_k - |\mathcal{A}_k| \subset V_k - L$. But $\pi| \colon V'_k - \pi^{-1}(L) \to V_k - L$ is a diffeomorphism. So $p'_{ki}(x) = \pi^{-1} p_{ki}(x) = \pi^{-1} p_{kj} p_{ji}(x) = p'_{kj} p_{ji}(x)$. Since $\bigcup_{m<k} V_{mi}$ is nowhere dense in $V_{ji} \cap V_{ki}$ we know that $p'_{ki}(x) = p'_{kj} p_{ji}(x)$ for all $x \in V_{ji} \cap V_{ki}$ by continuity.

To see condition III, note that if $m \le i < k$ then

$$
\begin{aligned}
p'_{ik}{}^{-1}\left(\bigcup_{j<m} V_{ji}\right) &= V'_{ik} \cap \pi^{-1} p_{ik}^{-1}\left(\bigcup_{j<m} V_{ji}\right) \\
&= V'_{ik} \cap \pi^{-1}\left(V_{ik} \cap \bigcup_{j<m} V_{jk}\right) = V'_{ik} \cap \bigcup_{j<m} V'_{jk}
\end{aligned}
$$

by Lemma 4.2.1. If $m \le k < i$ then

$$p'_{ki}{}^{-1}\left(\bigcup_{j<m} V'_{jk}\right) = p'_{ki}{}^{-1}\pi^{-1}\left(\bigcup_{j<m} V_{jk}\right) = p_{ki}^{-1}\left(\bigcup_{j<m} V_{jk}\right) = V_{ki} \cap \bigcup_{j<m} V_{ji}$$

So III is shown. Condition IV is immediate, and condition V follows since π preserves the boundary. Hence $\mathfrak{T}'$ is a resolution tower.

Now suppose that $\mathfrak{T}$ is of type U. We must show that $\mathfrak{T}'$ is of type U also. Pick a stratum T of $V_{ki} - \bigcup_{j<k} V_{ji}$. Since $p_{ki}| \colon T \to V_k - |\mathcal{A}_k|$ is a submersion and $\pi^{-1}| \colon V_k - |\mathcal{A}_k| \to V'_k - |\mathcal{A}'_k|$ is a diffeomorphism, we know that $p'_{ki}| = \pi^{-1} p_{ki}| \colon T \to V'_k - |\mathcal{A}'_k|$ is submersion. Now pick a connected component S of a stratum of $V'_{ik} - \bigcup_{j<i} V'_{jk}$. If $\pi(S) \not\subset L$ then $\pi(S) \cap L$ is empty so $\pi|_S$ is an embedding onto an open subset of a stratum of $V_{ik} - \bigcup_{j<i} V_{jk}$. Hence $p'_{ik}|_S = p_{ik} \pi|_S$ is a submersion. If $\pi(S) \subset L$ then $\pi| \colon S \to S'$ is a bundle over a stratum S' of $L \cap V_{ik} - \bigcup_{j<i} V_{jk}$. By assumption, $p_{ik}| \colon S' \to V_i$ is a submersion. Hence $p'_{ik}|_S$ is a submersion. So $\mathfrak{T}'$ is of type U. $\qquad \square$

We can now blow up a resolution tower.

Definition: A *multiblowup* of a resolution tower $\mathfrak{T} = \{V_i, \mathcal{A}_i, p_i\}_{\mathfrak{I}}$ is a collection $\{\pi_i\}_{\mathfrak{I}}$ of unobtrusive unskewed tico multiblowups

$$\pi_i \colon (V'_i, \mathcal{A}'_i) \to (V_i, \mathcal{A}_i)$$

and maps

$$p'_{ji} \colon |\mathcal{A}'_{ji}| \to V'_j$$

where each $\mathcal{A}'_{ji}$ is the tower transform of $\mathcal{A}_{ji}$ so that

$$\pi_j p'_{ji} = p_{ji} \pi_i|.$$

Lemma 4.2.3 below shows that $\mathfrak{T}' = \{V_i', \mathcal{A}_i', p_i'\}_{\mathfrak{I}}$ is a resolution tower.

Lemma 4.2.3 *If $\{\pi_i\}_{\mathfrak{I}} \colon \mathfrak{T}' \to \mathfrak{T}$ is a multiblowup then:*

  a) $\mathfrak{T}'$ *is a resolution tower.*
  b) $|\mathfrak{T}'| = |\mathfrak{T}|$.

Proof: Note that Lemma 4.2.2 implies that a) is true if all but one of the π_i's are the identity. We may as well assume that $\mathfrak{I} = \{0, \dots, n\}$. But $\{\pi_i\}_{i=0}^n$ may be factored into $n + 1$ such multiblowups, first do $\{id, id, \dots, id, \pi_n\}$ then $\{id, id, \dots, \pi_{n-1}, id\}$ and so forth.

To prove b), note that $\{\pi_i\}_{i=0}^n$ induces a continuous proper map $\pi \colon |\mathfrak{T}'| \to |\mathfrak{T}|$. By unobtrusiveness of each π_i, π restricts to a diffeomorphism on each stratum since $\pi_i| \colon V_i' - |\mathcal{A}_i'| \to V_i - |\mathcal{A}_i|$ is a diffeomorphism. So π is one to one and onto. Hence π^{-1} exists since $|\mathfrak{T}'|$ and $|\mathfrak{T}|$ are both locally compact Hausdorff. $\square$

Exercise: Any multiblowup is a tico tower morphism.

Lemma 4.2.4 *Let $\{\pi_i\}_{\mathfrak{I}} \colon \mathfrak{T}' \to \mathfrak{T}$ be a multiblowup and let $\{f_i\}_{\mathfrak{I}''} \colon \mathfrak{T}'' \to \mathfrak{T}$, $\alpha \colon \mathfrak{I}'' \to \mathfrak{I}$ be a tower morphism. Suppose for each $i \in \mathfrak{I}''$, we have a map $g_i \colon V_i'' \to V_{\alpha(i)}'$ so that $\pi_{\alpha(i)} g_i = f_i$. Then $\{g_i\}_{\mathfrak{I}''} \colon \mathfrak{T}'' \to \mathfrak{T}'$ is a tower morphism. If $\{f_i\}_{\mathfrak{I}''}$ is a tico tower morphism then $\{g_i\}_{\mathfrak{I}''}$ is a tico tower morphism also.*

Proof: Let $\mathfrak{T} = \{V_i, \mathcal{A}_i, p_i\}_{\mathfrak{I}}$, let $\mathfrak{T}' = \{V_i', \mathcal{A}_i', p_i'\}_{\mathfrak{I}}$ and let $\mathfrak{T}'' = \{V_i'', \mathcal{A}_i'', p_i''\}_{\mathfrak{I}''}$. Then Lemma 3.2.7 implies that $g_i \colon (V_i'', \mathcal{A}_i'') \to \left(V_{\alpha(i)}', \mathcal{A}_{\alpha(i)}'\right)$ is a tico map if $f_i \colon (V_i'', \mathcal{A}_i'') \to \left(V_{\alpha(i)}, \mathcal{A}_{\alpha(i)}\right)$ is a tico map. So we must only show that

$$g_i \left(|\mathcal{A}_{ji}''|\right) \;\subset\; |\mathcal{A}_{\alpha(j)\alpha(i)}'|$$

$$g_i^{-1}\Big(\bigcup_{d \leq b} |\mathcal{A}_{d\alpha(i)}'|\Big) \;=\; \bigcup_{\alpha(c) \leq b} |\mathcal{A}_{ci}''| \ \text{ and }$$

$$g_j p_{ji}'' \;=\; p_{\alpha(j)\alpha(i)}' g_i|$$

for all $j < i$ and $b < \alpha(i)$.

It suffices to consider the case where for some k, $\pi_i =$ identity for $i \neq k$ and π_k is a single blowup with center L. This is because, as in Lemma 4.2.3, we may factorize the multiblowup into such simple multiblowups.

If $\alpha(i) \neq k$ then $g_i = f_i$. Hence for all $i \in \alpha^{-1}(k)$, we only need show that:

$$g_i \left(|\mathcal{A}_{ji}''|\right) \;\subset\; |\mathcal{A}_{\alpha(j)k}'| \ \text{ for all } \ j < i$$

$$g_i^{-1}\Big(\bigcup_{d \leq b} |\mathcal{A}_{dk}'|\Big) \;=\; \bigcup_{\alpha(c) \leq b} |\mathcal{A}_{ci}''| \ \text{ for all } \ b < k$$

$$g_j p_{ji}'' \;=\; p_{\alpha(j)k}' g_i| \ \text{ for all } \ j < i$$

$$g_i p_{ij}'' \;=\; p_{k\alpha(j)}' g_j| \ \text{ for all } \ j > i$$

Pick the smallest m so that $L \subset |\mathcal{A}_{mk}|$. Pick any $i \in \alpha^{-1}(k)$. By definition of the tower transform, we know that $\pi_k^{-1}(|\mathcal{A}_{jk}|) = |\mathcal{A}'_{jk}|$ for all $j \leq m$. Hence if $j < i$ and $\alpha(j) \leq m$,

$$g_i\left(|\mathcal{A}''_{ji}|\right) \subset \pi_k^{-1} f_i\left(|\mathcal{A}''_{ji}|\right) \subset \pi_k^{-1}\left(|\mathcal{A}_{\alpha(j)k}|\right) = |\mathcal{A}'_{\alpha(j)k}|.$$

If $j < i$ and $\alpha(j) > m$ then take any $x \in |\mathcal{A}''_{ji}| - \bigcup_{b<j} |\mathcal{A}''_{bi}|$. Then

$$f_i(x) \in |\mathcal{A}_{\alpha(j)k}| - \bigcup_{b<\alpha(j)} |\mathcal{A}_{bk}| \subset |\mathcal{A}_{\alpha(j)k}| - L.$$

But $\pi_k^{-1}\left(|\mathcal{A}_{\alpha(j)k}| - L\right) = |\mathcal{A}'_{\alpha(j)k}| - \pi_k^{-1}(L)$ so $g_i(x) \in |\mathcal{A}'_{\alpha(j)k}|$. So by continuity, $g_i\left(|\mathcal{A}''_{ji}|\right) \subset |\mathcal{A}'_{\alpha(j)k}|$.

We also know that for any $b < k$, $\bigcup_{d \leq b} |\mathcal{A}'_{dk}| = \pi_k^{-1}(\bigcup_{d \leq b} |\mathcal{A}_{dk}|)$ by Lemma 4.2.1 so

$$g_i^{-1}(\bigcup_{d \leq b} |\mathcal{A}'_{dk}|) = f_i^{-1}(\bigcup_{d \leq b} |\mathcal{A}_{dk}|) = \bigcup_{\alpha(c) \leq b} |\mathcal{A}''_{ci}|.$$

Next, for $x \in |\mathcal{A}''_{ji}|$ we know that

$$p'_{\alpha(j)k} g_i(x) = p_{\alpha(j)k} \pi_k g_i(x) = p_{\alpha(j)k} f_i(x) = f_j p''_{ji}(x) = g_j p''_{ji}(x).$$

Finally, for any $j > i$, pick any $x \in |\mathcal{A}''_{ij}| - \bigcup_{b<i} |\mathcal{A}''_{bj}|$. Then

$$\pi_k p'_{k\alpha(j)} g_j(x) = p_{k\alpha(j)} f_j(x) = f_i p''_{ij}(x) = \pi_k g_i p''_{ij}(x).$$

But $p''_{ij}(x) \in V''_i - |\mathcal{A}''_i|$ so $f_i p''_{ij}(x) \in V_i - |\mathcal{A}_i| \subset V_i - L$. Consequently $\pi_k p'_{k\alpha(j)} g_j(x) = \pi_k g_i p''_{ij}(x) \in V_k - L$ means that $p'_{k\alpha(j)} g_j(x) = g_i p''_{ij}(x)$. Since $|\mathcal{A}''_{ij}| - \bigcup_{b<i} |\mathcal{A}''_{bj}|$ is dense in $|\mathcal{A}''_{ij}|$ we know by continuity that $g_i p''_{ij} = p'_{k\alpha(j)} g_j$. So we are finished. $\qquad\square$

The following technical lemma allows one to lift a map - something you must do if you wish to construct multiblowups of resolution towers.

Lemma 4.2.5 *Let* $\mathfrak{T} = \{V_i, \mathcal{A}_i, p_i\}_{\mathfrak{J}}$ *be a finite resolution tower of type* M. *Let* $(N, \mathcal{B})$ *be a tico and let* $\rho: (N', \mathcal{B}') \to (N, \mathcal{B})$ *be a fat unobtrusive tico multiblowup. Let* $q: V_{ji} \to N$ *be a mico. Suppose that* $\mathcal{A}_i$ *is a regular tico and for some* $k \leq j$, $q|: \left(S, S \cap (\bigcup_{m<k} \mathcal{A}_{mi})\right) \to (N, \mathcal{B})$ *is a tico map for all* $S \in \mathcal{A}_{ji}$. *Then there is a fat unobtrusive tico multiblowup* $\pi: (V'_i, \mathcal{A}'_i) \to (V_i, \mathcal{A}_i)$ *with centers lying over* $\bigcup_{m<k} V_{mi}$ *and there is a mico* $r: |\mathcal{A}'_{ji}| \to N'$ *so that* $\rho r = q\pi|$ *where* $\mathcal{A}'_{ji}$ *is the tower transform of* $\mathcal{A}_{ji}$.

Furthermore, in the algebraic case where $\mathfrak{T}$, $(N, \mathcal{B})$, $(N', \mathcal{B}')$ *and* ρ *are algebraic and* q *is an entire rational function, we may guarantee that* π *is an algebraic multiblowup and* r *is an entire rational function.*

Proof: Suppose we have a mico $f\colon |\mathcal{C}| \to N'$ so that $\rho f = q|$ for some $\mathcal{C} \subset \mathcal{A}_{ji}$. Furthermore, suppose f is an entire rational function if we are in the algebraic case. If $\mathcal{C} = \mathcal{A}_{ji}$ we are done, so we may as well assume that $C \neq \mathcal{A}_{ji}$.

Pick $S \in \mathcal{A}_{ji} - \mathcal{C}$. By Proposition 3.3.9 there is a fat unobtrusive tico multiblowup $\pi\colon (S', \mathcal{E}) \to \big(S, S \cap \big(\bigcup_{m<k} \mathcal{A}_{mi}\big)\big)$ and a tico map $g\colon (S', \mathcal{E}) \to (N', \mathcal{B}')$ so that $\rho g = f\pi$. By Lemma 3.3.7, π gives a fat unobtrusive tico multiblowup of $(S, S \cap \mathcal{A}_i)$. This induces a fat unobtrusive tico multiblowup $\pi\colon (V_i', \mathcal{A}_i') \to (V_i, \mathcal{A}_i)$ with the same centers so that $S' \subset V_i'$. Notice that the centers of π all lie over $\bigcup_{m<k} V_{mi}$, hence the tower transform of $\mathcal{A}_{mi}$ is the strict transform of $\mathcal{A}_{mi}$ for all $m \geq k$, in particular for $m = j$. Let $\mathcal{C}'$ be the strict transform of $\mathcal{C}$ and let $\mathcal{A}_{mi}'$ be the tower transform of each $\mathcal{A}_{mi}$.

Assertion 4.2.5.1 $g|_{S' \cap |\mathcal{C}'|} = f\pi|_{S' \cap |\mathcal{C}'|}$.

Proof: Since $S \cap |\mathcal{C}| \cap \bigcup_{m<k} V_{mi}$ is nowhere dense in $S \cap |\mathcal{C}|$, $S' \cap |\mathcal{C}'| \cap \bigcup_{m<k} |\mathcal{A}_{mi}'|$ is nowhere dense in $S' \cap |\mathcal{C}'|$. Hence if $g| \neq f\pi|$ there is an $x \in S' \cap |\mathcal{C}'| - \bigcup_{m<k} |\mathcal{A}_{mi}'|$ such that $g(x) \neq f\pi(x)$. But then $\pi(x) \in S \cap |\mathcal{C}| - \bigcup_{m<k} V_{mi}$ so $q\pi(x) \in N - |\mathcal{B}|$. But ρ^{-1} is single valued on $N - |\mathcal{B}|$ so $g(x) = f\pi(x)$ since $\rho g(x) = q\pi(x) = \rho f\pi(x)$. So $g| = f\pi|$. $\qquad\square$

By Lemma 3.1.2 there is a smooth $h\colon S' \cup |\mathcal{C}'| \to N'$ so that $h|_{S'} = g$ and $h|_{|\mathcal{C}'|} = f\pi|$. By Lemma 3.3.11, h is an entire rational function in the algebraic case.

Assertion 4.2.5.2 h is a mico.

Proof: Pick any $T' \in \mathcal{C}' \cup \{S'\}$. Pick $T \in \mathcal{A}_{ji}$ so $T' \subset V_i'$ is the strict transform of T. We know that $\pi|\colon \big(T', T' \cap \bigcup_{m<k} \mathcal{A}_{mi}'\big) \to \big(T, T \cap \bigcup_{m<k} \mathcal{A}_{mi}\big)$ is a tico map by Lemmas 3.2.5, 4.2.1 and 3.3.4. Hence by Lemma 3.2.4 we know $\rho h| = q\pi|\colon \big(T', T' \cap \bigcup_{m<k} \mathcal{A}_{mk}'\big) \to (N, \mathcal{B})$ is a tico map. Then Lemma 3.2.7 implies that $h|\colon \big(T', T' \cap \bigcup_{m\leq k} \mathcal{A}_{mi}'\big) \to (N', \mathcal{B}')$ is a tico map. But then Lemmas 3.7.1 and 3.2.5 imply that h is a mico. $\qquad\square$

So we may as well replace V_i by V_i', $\mathcal{C}$ by $\mathcal{C}' \cup \{S'\}$ and f by h. But then $\mathcal{A}_{ji} - \mathcal{C}$ has fewer elements so eventually we are finished. $\qquad\square$

Definition: Let $\mathfrak{T} = \{V_i, \mathcal{A}_i, p_i\}_\mathfrak{I}$ be a resolution tower and let $\pi\colon V_j' \to V_j$ be a tico multiblowup. Then we say that π *extends* to a multiblowup $\{\pi_i\}_\mathfrak{I}\colon \mathfrak{T}' \to \mathfrak{T}$ if $\pi_j = \pi$, $V_i' = V_i$ and $\pi_i = $ the identity unless $i \geq j$. Note that in particular, π must have unobtrusive, unskewed centers.

Proposition 4.2.6 *Let* $\mathfrak{T} = \{V_i, \mathcal{A}_i, p_i\}_\mathfrak{I}$ *be a finite resolution tower of type M and let* $\pi\colon (V_k', \mathcal{A}_k') \to (V_k, \mathcal{A}_k)$ *be a tico multiblowup with fat unobtrusive centers. Then* π *extends to a multiblowup* $\{\pi_i\}_\mathfrak{I}\colon \mathfrak{T}' \to \mathfrak{T}$ *with fat unobtrusive centers. Furthermore, if* $\mathfrak{T}$ *is of type R and the centers of* π *lie over* $\bigcup_{j\leq m} V_{jk}$ *we may guarantee that the centers of each* π_i *lie over* $\bigcup_{j\leq m} V_{ji}$. *In addition, if*

$\mathfrak{T}$ and π are algebraic we may guarantee that each multiblowup π_i is algebraic also.

Proof: We will prove this by induction on the number of indices i with $i \geq k$. If k is a maximal element of $\mathfrak{J}$ then we may take $\pi_i =$ the identity for $i \neq k$ and $\pi_k = \pi$. So we may assume that k is not maximal.

Pick a maximal $b > k$ so that for all $k < i < b$ there is a map $p'_{ki} : V_{ki} \to V'_k$ such that $\pi p'_{ki} = p_{ki}$. Furthermore, suppose p'_{ki} is entire rational in the algebraic case. By Proposition 3.3.8 there is a tico multiblowup $\lambda : V'_b \to V_b$ with fat unobtrusive centers so that the total transform of $\mathcal{A}_b$ is regular. By induction λ extends to a multiblowup, hence we may as well assume that $\mathcal{A}_b$ is regular. Notice that for every sheet $S \in \mathcal{A}_{kb}$, $p_{kb}| : (S, S \cap \bigcup_{j \leq m} \mathcal{A}_{jb}) \to (V_k, \bigcup_{j \leq m} \mathcal{A}_{jk})$ is a tico map by Lemmas 3.3.2 and 3.2.5.

So by Lemma 4.2.5 there is a fat unobtrusive tico multiblowup $\mu : V'_b \to V_b$ and a mico $r : |\mathcal{A}'_{kb}| \to V'_k$ so that $\pi r = p_{kb}\mu|$ where $\mathcal{A}'_{kb}$ is the tower transform of $\mathcal{A}_{kb}$. Furthermore, μ is algebraic and r is entire rational in the algebraic case. Also the centers of μ lie over $\bigcup_{j \leq m} V_{jb}$. By induction we may extend μ to a multiblowup. Hence we may as well assume there is a mico $r : V_{kb} \to V'_k$ so that $\pi r = p_{kb}$. But then setting $r = p'_{kb}$ we have improved matters. So we may as well assume that for all $i > k$ there is a map $p'_{ki} : V_{ki} \to V'_k$ such that $\pi p'_{ki} = p_{ki}$.

But then this proposition is implied by Lemma 4.2.3 and the multiblowup $\{\pi_i\}_{\mathfrak{J}}$ where $\pi_i =$ the identity for $i \neq k$ and $\pi_k = \pi$. □

Proposition 4.2.7 *Let* $\mathfrak{T}' \to \mathfrak{T}$ *be a multiblowup of resolution towers. Then* $\mathfrak{T}'$ *has type* R *if* $\mathfrak{T}$ *has type* R *and* $\mathfrak{T}'$ *has type* M *if* $\mathfrak{T}$ *has type* M. *If the centers of the multiblowup are all fat, then* $\mathfrak{T}'$ *has type* N *if* $\mathfrak{T}$ *has type* N, $\mathfrak{T}'$ *has type* F *if* $\mathfrak{T}$ *is compact and has type* F, $\mathfrak{T}$ *has type* U *if* $\mathfrak{T}$ *has type* U *and* $\mathfrak{T}'$ *has type* S *if* $\mathfrak{T}$ *has type* S.

Proof: We may as well assume that $V'_i = V_i$ for $i \neq m$ and $\pi : V'_m \to V_m$ is a single blowup with unobtrusive unskewed center L.

If $\mathfrak{T}$ has type R then $\mathfrak{T}'$ has type R by Lemma 3.3.3. Now suppose $\mathfrak{T}$ has type M. We must show that $p'_{mi} : V_{mi} \to V'_m$ is a mico for all $i > m$ and $p'_{im} = p_{im}\pi| : V'_{im} \to V_i$ is a mico for all $i < m$.

Pick any $S \in \mathcal{A}_{mi}$ for $i > m$. Let $g : S' \to V_i$ be the immersion associated to S. We know that $p_{mi}g : (S', g^*(\mathcal{A}_i - \mathcal{A}_{mi})) \to (V_m, \mathcal{A}_m)$ is a tico map by Lemmas 3.3.2, 4.1.1, 3.2.4 and 3.2.5. But then $p'_{mi}g : (S', g^*(\mathcal{A}_i - \mathcal{A}_{mi})) \to (V'_m, \mathcal{A}'_m)$ is a tico map by Lemma 3.2.7. Hence by Lemma 3.7.1 we know that p'_{mi} is a mico. Notice that if $\mathfrak{T}$ has type N then each $p_{mi}g : (S', g^*(\mathcal{A}_i)) \to (V_m, \mathcal{A}_m)$ has type N. But then if L is fat, we know that $p'_{mi}g : (S', g^*(\mathcal{A}_i)) \to (V'_m, \mathcal{A}'_m)$ has type N by Lemma 3.5.2a. So Lemma 3.7.1 implies that p'_{mi} is a mico of type N.

For $i < m$ pick a neighborhood U of V_{im} and pick an extension of p_{im} to a tico map $p''_{im} : (U, U \cap (\mathcal{A}_m - \mathcal{A}_{im})) \to (V_i, \mathcal{A}_i)$. Since $V'_{im} \subset \pi^{-1}(V_{im})$ we may find a neighborhood U' of V'_{im} in V'_m so that $\pi(U') \subset U$. Then by Lemmas 3.2.4 and 3.3.4, $p''_{im}\pi|_{U'}$ is a tico map. Hence $p'_{im} = p''_{im}\pi|_{V'_{im}}$ is a mico. If $\mathfrak{T}$ is of type N and the centers are fat, we may assume $p''_{im} \colon (U, U \cap \mathcal{A}_m) \to (V_i, \mathcal{A}_i)$ has type N so $p''_{im}\pi|_{U'}$ will have type N by Lemma 3.5.2a. So $\mathfrak{T}'$ has type N.

The proof that $\mathfrak{T}'$ has type S if $\mathfrak{T}$ has type S is very similar to the proof for type N, using Lemma 3.5.4 in place of Lemma 3.5.2a.

If $\mathfrak{T}$ has type F and L is fat then $(L, L \cap \mathcal{A}_m)$ is a full tico, so $(V'_m, \mathcal{A}'_m)$ is full by Lemma 3.4.1. So $\mathfrak{T}'$ has type F.

If $\mathfrak{T}$ has type U and L is fat, then each stratum of $L \cap V_{im} - \bigcup_{j<i} V_{jm}$ is a stratum of $V_{im} - \bigcup_{j<i} V_{jm}$. Hence $\mathfrak{T}'$ has type U by Lemma 4.2.2. $\square$

Theorem 4.2.8 *Let $\mathfrak{T}$ be a finite resolution tower. Then there is a multiblowup $\mathfrak{T}' \to \mathfrak{T}$ with fat centers so that $\mathfrak{T}'$ is of type R. Moreover, if $\mathfrak{T}$ is of type M we may also guarantee that $\mathfrak{T}'$ is of type NE. If $\mathfrak{T}$ is algebraic then we can take the multiblowup to be algebraic also.*

Proof: Suppose some $\mathcal{A}_i$ is not regular. By Proposition 3.3.8 there is a tico multiblowup $\pi_i \colon V'_i \to V_i$ with fat unobtrusive centers so that the total transform of $\mathcal{A}_i$ is regular. By Proposition 4.2.6 we may extend π_i to a multiblowup of $\mathfrak{T}$ with fat unobtrusive centers. Continuing in this way, we have a multiblowup $\mathfrak{T}' \to \mathfrak{T}$ with fat unobtrusive centers so that $\mathfrak{T}'$ has type R.

Now suppose $\mathfrak{T}$ is of type M. We know by Proposition 4.2.7 that $\mathfrak{T}'$ is still of type M. We will blow up some more to make it type N. We will then blow up even more to make it type E also.

For convenience, we may as well assume that $\mathfrak{I} = \{0, \ldots, n\}$. Let $\mathcal{N}'_{ji} \subset \mathcal{A}'_{ji}$ be the set of $S \in \mathcal{A}'_{ji}$ such that $p'_{ji}|\colon (S, S \cap \mathcal{A}'_i) \to (V'_j, \mathcal{A}'_j)$ is a tico map of type N. If $\mathcal{N}'_{ji} = \mathcal{A}'_{ji}$ for all $j < i$ then Lemma 3.7.1 implies that each p'_{ji} is a mico of type N, hence $\mathfrak{T}'$ has type N. So if $\mathfrak{T}'$ is not of type N we may pick k and m so that $\mathcal{N}'_{ji} = \mathcal{A}'_{ji}$ if $i < m$ or $i = m$ and $j > k$ but $\mathcal{N}'_{km} \neq \mathcal{A}'_{km}$. Pick $S' \in \mathcal{A}'_{km} - \mathcal{N}'_{km}$. By Proposition 3.5.3 there are fat unobtrusive tico multiblowups $\pi \colon V''_k \to V'_k$ and $\rho \colon S'' \to S'$ and a tico map $g \colon S'' \to V''_k$ so that $p'_{km}\rho = \pi g$, g has type N and the centers of ρ lie over $S' \cap p'_{km}{}^{-1}(|\mathcal{A}'_k|) = S' \cap \bigcup_{j<k} V'_{jm}$. The multiblowup ρ induces a fat unobtrusive multiblowup $\rho' \colon V''_m \to V'_m$ with the same centers as ρ (so $S'' \subset V''_m$ and $\rho = \rho'|$). By Proposition 4.2.6, ρ' extends to a multiblowup $\{\pi'_i\} \colon \mathfrak{T}'' \to \mathfrak{T}'$ and then π extends to a multiblowup $\{\pi''_i\} \colon \mathfrak{T}''' \to \mathfrak{T}''$. Furthermore, each π'_i and π''_i has fat unobtrusive centers and the centers of π''_i all lie over $\bigcup_{j<k} V''_{ji}$.

Recall $\mathfrak{T}'[m-1]$ is the truncated tower $\{V'_i, \mathcal{A}'_i, p'_i\}_{i=0}^{m-1}$. Note that $\mathfrak{T}'[m-1] = \mathfrak{T}'''[m-1]$. By Lemma 3.7.1 we know that $\mathfrak{T}'[m-1]$ has type N, then Proposition 4.2.7 implies that $\mathfrak{T}'''[m-1]$ has type N. Thus $\mathcal{N}'''_{ji} = \mathcal{A}'''_{ji}$ for all $0 \leq j < i < m$ where $\mathcal{N}'''_{ji}$ is defined similarly to $\mathcal{N}'_{ji}$. Since the centers of π''_m lie over $\bigcup_{j<k} V''_{jm}$

and the centers of $\pi'_m = \rho'$ lie over $\bigcup_{j<k} V'_{jm}$ we know that $\mathcal{A}'''_{jm}$ is the strict transform of $\mathcal{A}''_{jm}$ and $\mathcal{A}''_{jm}$ is the strict transform of $\mathcal{A}'_{jm}$ for all $j \geq k$.

Assertion 4.2.8.1 $\mathcal{N}'''_{jm} = \mathcal{A}'''_{jm}$ for all $k < j < m$. Also, $\mathcal{N}'''_{km}$ contains the strict transform of $\mathcal{N}'_{km} \cup \{S'\}$.

Proof: Pick any $T''' \in \mathcal{A}'''_{jm}$ for $j \geq k$. Then T''' is the strict transform of some $T' \in \mathcal{A}'_{jm}$. If $T' \in \mathcal{N}'_{jm}$ then $p'_{jm}|: (T', T' \cap \mathcal{A}'_m) \to (V'_j, \mathcal{A}'_j)$ is a tico map of type N. Hence by Lemma 3.5.2a, $p'''_{jm}|: (T''', \mathcal{E}) \to (V'''_j, \mathcal{A}'''_j)$ is a tico map of type N where $\mathcal{E}$ is the total transform of $T' \cap \mathcal{A}'_m$ via the multiblowup on T' induced by $\pi'_m \pi''_m$. But $\mathcal{E} = T''' \cap \mathcal{A}'''_m$ so $T''' \in \mathcal{N}'''_{jm}$. If $T' = S'$ then T''' is the strict transform of S'' (via π''_m). So again Lemma 3.5.2 implies that $T''' \in \mathcal{N}'''_{jm}$. $\square$

By Assertion 4.2.8.1 we have improved matters, so by induction we may as well assume that $\mathcal{N}'''_{ji} = \mathcal{A}'''_{ji}$ for all $0 \leq j < i \leq n$. But then Lemma 3.7.1 implies that $\mathfrak{T}'''$ has type N.

So we have proven that we may as well assume that $\mathfrak{T}'$ has type RN. We will now blow up some more to get type E. By Proposition 4.2.7 $\mathfrak{T}'$ will still remain type RN after blowing up, so we will be finished.

Pick $0 \leq k < m \leq n$ so that $p'_{ji}: V'_{ji} \to V'_j$ has constant exponents for all j and i satisfying $0 \leq j < i < m$ or $k < j < i = m$. What we will do is find a tico multiblowup $\pi: V''_m \to V'_m$ with fat unobtrusive connected centers so that $p'_{jm}\pi|: |\mathcal{A}''_{jm}| \to V'_j$ has constant exponents, for all $k \leq j < m$, where $\mathcal{A}''_{jm}$ is the tower transform of $\mathcal{A}'_{jm}$. By Proposition 4.2.6 we may extend π to a multiblowup of $\mathfrak{T}'$ with fat unobtrusive centers. Hence by induction we eventually find a multiblowup of type E. So let us find the promised multiblowup π. For $T \in \mathcal{A}'_{im}$ with $i < k$ let

$$\beta(T) = \text{the number of connected components of } T \cap V'_{km}$$

$$\gamma(T) = \sum_{S \in \mathcal{A}'_{km}} (\text{the number of connected components of } S \cap T)$$

$$\beta = \max\{\,\beta(T) \mid T \in \bigcup_{i<k} \mathcal{A}'_{im}\,\}$$

$$\gamma = \max\{\,\gamma(T) \mid T \in \bigcup_{i<k} \mathcal{A}'_{im} \text{ and } \beta(T) = \beta\,\}$$

$$\lambda = \sharp\{\,T \in \bigcup_{i<k} \mathcal{A}'_{im} \mid \beta(T) = \beta \text{ and } \gamma(T) = \gamma\,\}$$

where $\sharp$ means 'the cardinality of '.

If $\beta = 1$ then $T \cap V'_{km}$ is connected for all $T \in \mathcal{A}'_{im}$ with $i < k$. Consequently the exponent of T in any sheet $U \in \mathcal{A}'_k$ for p'_{km} is constant by continuity. But if $T \in \mathcal{A}'_{im}$ with $i \geq k$ the exponent of T in any sheet $U \in \mathcal{A}'_k$ is zero by Lemmas

4.1.1 and 3.2.2 hence it is constant. So p'_{km} has constant exponents and no blowup is necessary, we are done.

So suppose $\beta > 1$. Pick $T \in \mathcal{A}'_{im}$ for some $i < k$ so that $\beta(T) = \beta$ and $\gamma(T) = \gamma$. Pick $S \in \mathcal{A}'_{km}$ so that $S \cap T \neq \emptyset$ and let L be a connected component of $S \cap T$.

Assertion 4.2.8.2 *After blowing up V'_m with center L the corresponding triple (β, γ, λ) is smaller (with lexicographical order).*

Proof: Let $\pi \colon V''_m \to V'_m$ be the blowup with center L. Let $\mathcal{A}''_{jm}$ be the tower transform of $\mathcal{A}'_{jm}$. So $\mathcal{A}''_{jm}$ is the strict transform of $\mathcal{A}'_{jm}$ if $j \neq i$ and $\mathcal{A}''_{im}$ is the total transform of $\mathcal{A}'_{im}$. Let β', γ' and λ' be the analogues of β, γ and λ for this blowup. Let us calculate β', γ' and λ'.

Pick a sheet $T'' \in \mathcal{A}''_{jm}$ with $j < k$. First suppose $T'' \neq \pi^{-1}(L)$. Then there is a $T' \in \mathcal{A}'_{jm}$ so that $T'' = \mathfrak{B}(T', T' \cap L) =$ the strict transform of T'. If $T' \neq T$ then L is transverse to T' so it is easy to see that $\beta'(T'') = \beta(T')$ and $\gamma'(T'') = \gamma(T')$. If $T' = T$ then $L \subset T'$. If L is an isolated component of $T \cap V'_{km}$ then $\beta'(T'') = \beta(T') - 1$. Otherwise, $\beta'(T'') = \beta(T')$ and $\gamma'(T'') = \gamma(T') - 1$. Now suppose $T'' = \pi^{-1}(L)$. Notice $\pi| \colon T'' \to L$ is a circle bundle. Now L intersects $|\mathcal{A}'_{km} - S|$ transversally and $L \subset S$ so

$$T'' \cap |\mathcal{A}''_{km}| = \pi^{-1}(|\mathcal{A}'_{km} \cap L|) \cup (\mathfrak{B}(S, L) \cap \pi^{-1}(L)).$$

So $T'' \cap |\mathcal{A}''_{km}|$ is the section $\mathfrak{B}(S, L) \cap \pi^{-1}(L)$ union a bunch of circle fibers. So $T'' \cap |\mathcal{A}''_{km}|$ is connected since L is connected. Hence $\beta'(T'') = 1$. Consequently we have $(\beta', \gamma', \lambda') < (\beta, \gamma, \lambda)$.   □

By Assertion 4.2.8.2, after a multiblowup $\pi' \colon V'''_m \to V'_m$ we reach the stage where $\beta = 1$. But then $p'_{km}\pi'|$ has constant exponents. For $i > k$ let U_i be a neighborhood of V_{im} so that there is an extension of p'_{im} to $q_{im} \colon (U_i, U_i \cap \mathcal{A}'_m) \to (V'_i, \mathcal{A}'_i)$, a tico map with constant exponents. Let $U'_i = \pi'^{-1}(U_i)$, then the tico map $q_{im}\pi'| \colon (U'_i, U'_i \cap \mathcal{A}'''_m) \to (V'_i, \mathcal{A}'_i)$ has constant exponents by Lemmas 3.2.4 and 3.3.4. Hence $p'_{im}\pi'|$ has constant exponents for all $i \geq k$ and so we are done by induction.

The algebraic case is proven in the analogous way. For the most part, one just substitutes 'irreducible component' for 'connected component'. The only exception is in the definition of $\beta(T)$. We put an equivalence relation on the connected components of $T \cap V'_{km}$, the one generated by saying two connected components K_1 and K_2 are equivalent if for all $U \in \mathcal{A}'_k$, $\alpha_{TU}|_{K_1} = \alpha_{TU}|_{K_2}$ where α_{TU} is the exponent for T in U for p'_{km}. Then we let $\beta(T)$ be the number of equivalence classes. This will guarantee that $\beta'(\pi^{-1}(L)) = 1$, using Lemmas 3.2.3, 3.2.4, 3.3.3 and 3.3.4.   □

3. Realizations of Resolution Towers

In this section we make a detailed study of the topology of the realization of a resolution tower of type UM. It turns out that the stratified set we obtain has some nice properties, it is not only locally conelike, but each stratum has a sort of tubular neighborhood and the various tubular neighborhoods are compatible. Furthermore, the link of any stratum is itself the realization of a resolution tower. First some definitions.

Definitions: A *Thom stratified set* is a smooth stratified set X such that for each stratum U of X there is a neighborhood X_U of U in X and continuous maps $\pi_U \colon X_U \to U$ and $\rho_U \colon X_U \to [0, \infty)$ so that:

1) $\pi_U(x) = x$ for all $x \in U$.
2) $\rho_U^{-1}(0) = U$.
3) For each stratum V of $X - U$, $(\pi_U, \rho_U)| \colon X_U \cap V \to U \times (0, \infty)$ is a smooth submersion.
4) If V is a stratum of X with $U \cap \operatorname{Cl} V \neq \emptyset$ then $\pi_U(x) = \pi_U \circ \pi_V(x)$ and $\rho_U(x) = \rho_U \circ \pi_V(x)$ for all $x \in X_U \cap X_V$.

The maps $\pi_U \colon X_U \to U$ and $\rho_U \colon X_U \to [0, \infty)$ are called *tubular data* for X.

Definition: A *stratified cone bundle* $\pi \colon E \to X$ is a fibre bundle over a smooth manifold X with fibre the cone on a stratified set. More precisely, E is a smooth stratified set and there is a smooth stratified set L (called the link of the cone bundle) and there are stratified set isomorphisms $h_U \colon U \times \mathfrak{c} L \to \pi^{-1}(U)$ for certain open subsets U of X so that the U's cover X and so if $U \cap V \neq \emptyset$ then $h_V^{-1} h_U| \colon (U \cap V) \times \mathfrak{c} L \to (U \cap V) \times \mathfrak{c} L$ is given by $h_V^{-1} h_U(x, (y, t)) = (x, (g_{UV}^x(y), t))$ where g_{UV}^x is a family of stratified isomorphisms of L. In other words, E is the open mapping cylinder of a fibre bundle over X with fibre L.

Proposition 4.3.1 (*Tubular neighborhood theorem*) *Let X be a locally compact Thom stratified set with tubular data $\pi_U \colon X_U \to U$ and $\rho_U \colon X_U \to [0, \infty)$. For each connected component T of a stratum U pick a point $x \in T$ and a sufficiently small $\epsilon > 0$ and let $L_T = \pi_U^{-1}(x) \cap \rho_U^{-1}(\epsilon)$. Then there is a stratified cone bundle $\theta_T \colon E_T \to T$ with link L_T, a neighborhood E_T' of its zero section and an embedding $h_T \colon E_T' \to X_U$ so that:*

a) *h_T is the identity on the zero section.*
b) *$\pi_U h_T = \theta_T|$.*
c) *Cone levels go to ρ_U levels.*

Proof: This is the usual controlled vector field argument, c.f., [**GWPL**]. They assume that X is realized as a Whitney stratified set, but this is not really needed. □

One consequence of Proposition 4.3.1 is that for each connected component T of a stratum U there is a canonical link of T obtained by taking any $x \in T$

and small enough $\epsilon > 0$ and taking the link $\pi_U^{-1}(x) \cap \rho_U^{-1}(\epsilon)$. Proposition 4.3.1 implies that up to isomorphism, this link is independent of the choice of x and ϵ. We call it the *canonical link*. Note that by restriction the canonical link inherits a Thom stratification. In particular, for any stratum V of $L = \pi_U^{-1}(x) \cap \rho_U^{-1}(\epsilon)$ there is a unique stratum W of X so that $V = W \cap L$. We then let $L_V = L \cap X_W$, $\pi_V = \pi_W|_{L_V}$ and $\rho_V = \rho_W|_{L_V}$.

Exercise: Check that L_V, π_V and ρ_V are tubular data. (Note that $\pi_W^{-1}(L) = L_V$.)   ◇

We will now show that the realization of any resolution tower of type UM has a Thom stratification. It all comes from the following Lemma.

Lemma 4.3.2 *Let $\mathfrak{T} = \{V_i, \mathcal{A}_i, p_i\}_{\mathfrak{I}}$ be a resolution tower of type UM. For each $i \in \mathfrak{I}$, let $W_i = V_i - |\mathcal{A}_i|$ and for each $j < i$, let $W_{ji} = V_{ji} - \bigcup_{m<j} V_{mi}$. Then for each $j < i$ there is a neighborhood U_{ji} of W_{ji} in V_i, a smooth submersion $\pi_{ji} \colon U_{ji} \to W_j$ and a smooth map $\rho_{ji} \colon U_{ji} \to [0, \infty)$ so that:*

    a) $\pi_{ji}(x) = p_{ji}(x)$ for all $x \in W_{ji}$.
    b) $\rho_{ji}^{-1}(0) = W_{ji}$.
    c) $(\pi_{ji}, \rho_{ji})| \colon U_{ji} - W_{ji} \to W_j \times (0, \infty)$ is a smooth submersion.
    d) *If $k < j < i$ then $U_{ji} \cap U_{ki} = \pi_{ji}^{-1}(U_{kj})$ and for all $x \in U_{ji} \cap U_{ki}$, $\pi_{kj} \circ \pi_{ji}(x) = \pi_{ki}(x)$ and $\rho_{kj} \circ \pi_{ji}(x) = \rho_{ki}(x)$.*

Proof: Pick some $n \in \mathfrak{I}$ so that $\dim V_n \geq \dim V_i$ for all $i \in \mathfrak{I}$. By induction on the size of $\mathfrak{I}$ we may apply the Lemma to the truncation $\mathfrak{T}[\mathfrak{I} - \{n\}]$ and obtain $\pi_{ji} \colon U_{ji} \to W_j$ and $\rho_{ji} \colon U_{ji} \to [0, \infty)$ satisfying a)-d) for all $j < i \neq n$. Furthermore, we ask that condition e) below be satisfied.

    e) Let $\theta \colon \mathbf{R}^a \to U_{ji}$ be a tico chart with $\theta(0) \in W_{ji}$. Let $\theta^{-1}(V_{ki}) = \bigcup_{t \in B_k} \mathbf{R}_t^a$. Let $\mathfrak{J} = \{k \in \mathfrak{I} \mid B_k \neq \emptyset\}$. Note that $\mathfrak{J}$ is linearly ordered. For $k \in \mathfrak{J}$ let $C_k = \bigcup_{\ell \leq k} B_\ell$. Then we ask that for x in a neighborhood of 0,
$$\rho_{ji}\theta(x) = \sum_{k \in \mathfrak{J}} \prod_{s \in C_k} x_s^{2\alpha_{ks}} \lambda_k(x)$$
where the λ_k are smooth functions with $\lambda_k(0) > 0$ and where the positive integers α_{ks} are determined in some locally constant way from $\mathfrak{J}$ and the sheet containing $\theta(\mathbf{R}_s^a)$. (It actually turns out that $\alpha_{ks} = 1$ if k is a maximal element of $\mathfrak{J}$ and $\alpha_{ks} = $ the sum of the exponents of $\mathbf{R}_s^a$ in sheets of $\mathcal{A}_{k\ell}$ for $p_{\ell i}\theta$ where ℓ is the smallest element of $\mathfrak{J}$ which is bigger than k).

Pick a maximal lower index subset $\mathfrak{J}' \subset \mathfrak{I}$ so that for all $j < i$ with $j \in \mathfrak{J}'$ or $i \neq n$ we have $\pi_{ji} \colon U_{ji} \to W_j$ and $\rho_{ji} \colon U_{ji} \to [0, \infty)$ satisfying a)-e) and in addition, if $k < j < n$ and $k \in \mathfrak{J}'$ then for all $x \in V_{jn} \cap U_{kn}$, $\pi_{kj} \circ p_{jn}(x) = \pi_{kn}(x)$ and $\rho_{kj} \circ p_{jn}(x) = \rho_{kn}(x)$. After shrinking the U_{ji}'s if necessary we may assume that $U_{ji} \cap U_{ki} \neq \emptyset$ implies that $V_{ji} \cap V_{ki} \neq \emptyset$.

If $\mathfrak{I}' = \mathfrak{I}$ we are done, so choose a minimal element m of $\mathfrak{I} - \mathfrak{I}'$.

Assertion 4.3.2.1 *We may pick a neighborhood U of W_{mn} in $V_n - \bigcup_{k<m} V_{kn}$ and a smooth map $\pi\colon U \to W_m$ so that:*

i) $\pi(x) = p_{mn}(x)$ for all $x \in W_{mn}$.

ii) *If* $m < i < n$ *then* $p_{in}(U \cap V_{in}) \subset U_{mi}$.

iii) *If* $m < i < n$ *then* $\pi(x) = \pi_{mi} \circ p_{in}(x)$ *for all* $x \in U \cap W_{in}$.

iv) *For each stratum S of the tico stratification of V_n, $\pi|\colon U \cap S \to W_m$ is a smooth submersion.*

Proof: Note that if $m < i < n$ then $p_{in}(W_{mn} \cap V_{in}) \subset W_{mi}$, so any sufficiently small U will satisfy ii). Also note that since $\mathfrak{T}$ has type U, iii) and i) imply iv) except when S has codimension 0. But since the smaller strata are submersed, π will certainly be a submersion nearby, so a small enough U will satisfy iv).

Pick a proper embedding of W_m in some $\mathbf{R}^a$. Let $\lambda\colon O \to W_m$ be the closest point map, $\lambda(x)$ is the closest point to x in W_m, O is a neighborhood of W_m in $\mathbf{R}^a$. Then λ is smooth if O is sufficiently small. We may pick a smooth map $\mu\colon U' \to \mathbf{R}^a$ from some neighborhood U' of W_{mn} so that $\mu(x) = p_{mn}(x)$ for all $x \in W_{mn}$ and if $m < i < n$ then $p_{in}(U' \cap W_{in}) \subset U_{mi}$ and $\mu(x) = \pi_{mi} \circ p_{in}(x)$ for all $x \in U \cap W_{in}$. (Just extend locally and piece together with a partition of unity). Now let $U = U' \cap \mu^{-1}(O)$ and $\pi = \lambda\mu$. □

Now pick a maximal upper index subset $\mathfrak{I}'' \subset \mathfrak{I}'$ so that there exists a $\pi\colon U \to W_m$ satisfying i)-iv) above so that in addition for each $i \in \mathfrak{I}''$ with $i < m$ there is a neighborhood U'_{in} of W_{in} in U_{in} so that for all $x \in \pi^{-1}(U_{im}) \cap U'_{in}$, $\pi_{im} \circ \pi(x) = \pi_{in}(x)$ and $\rho_{im} \circ \pi(x) = \rho_{in}(x)$.

Assertion 4.3.2.2 *If $\mathfrak{I}'' = \mathfrak{I}'$ we have a contradiction to the maximality of $\mathfrak{I}'$.*

Proof: Set $U_{mn} = U$ and $\pi_{mn} = \pi$. We must now define ρ_{mn}. To have ρ_{mn} satisfy conditions b),d) and e) it suffices to find ρ_{mn} locally and piece together with a partition of unity. We will then show that e) implies c).

So let us construct ρ_{mn} locally. Pick any tico chart $\theta\colon \mathbf{R}^a \to U$ so $\theta(0) \in W_{mn}$. Let $\mathfrak{J}$, B_k and C_k be as in condition e). If $\mathfrak{J} = \{m\}$ then we may just let $\rho_{mn}\theta(x) = \prod_{s \in B_m} x_s^2$. If $\mathfrak{J} \neq \{m\}$, we know we want $\rho_{mk}p_{kn}\theta(x) = \rho_{mn}\theta(x)$ for all $k \in \mathfrak{J} - \{m\}$. But by condition e) and the fact that p_{kn} is a mico we know that

$$\rho_{mk}p_{kn}\theta(x) = \sum_{\ell \in \mathfrak{J}}^{\ell < k} \prod_{s \in C_\ell} x_s^{2\alpha_{\ell s}} \lambda_\ell(x)$$

$$= \sum_{\ell \in \mathfrak{J}}^{\ell < k} \prod_{s \in C_\ell} x_s^{2\alpha_{\ell s}} \lambda_\ell(x)$$

for all $x \in \theta^{-1}(V_{kn})$ where $\lambda_\ell(0) > 0$. So to define ρ_{mn} locally, we just extend each λ_ℓ defined on $\theta^{-1}\left(\bigcup_{k \in \mathfrak{J}}^{\ell < k} V_{kn}\right)$ to a neighborhood of 0 in $\mathbf{R}^a$ and add in the term $\sum_{s \in C_\ell} x_s^2$ for the maximal $\ell \in \mathfrak{J}$.

Now after perhaps shrinking U_{mn} we may piece these local expressions to-gether with a partition of unity to obtain ρ_{mn}. We will now show that (π_{mn}, ρ_{mn}) satisfies c), after perhaps shrinking U_{mn} some more.

Take a chart θ as above. Let $\psi \colon (\mathbf{R}^b, 0) \to (V_m, p_{mn}\theta(0))$ be a smooth chart. Then by type U and the inverse function theorem we may pick θ so that $\psi^{-1}\pi_{mn}\theta(x) = (x_1, \dots, x_b)$ for all x and also $s \notin B_k$ for any $s \leq b$ and $k \in \mathcal{J}$. Then the Jacobian matrix of $(\psi^{-1}\pi_{mn}\theta(x), \rho_{mn}\theta(x))$ has maximal rank $b + 1$ if and only if the vector $(\partial \rho_{mn}\theta(x)/\partial x_{b+1}, \dots, \partial \rho_{mn}\theta(x)/\partial x_a)$ is nonzero. But if $t \in B_m$ and $x \notin \theta^{-1}(W_{mn}) = \bigcup_{s \in B_m} \mathbf{R}_s^a$, then

$$\partial \rho_{mn} \circ \theta / \partial x_t(x) = (1/x_t) \sum_{\ell \in \mathcal{J}} \prod_{s \in C_\ell} x_s^{2\alpha_{\ell s}} \left(2\alpha_{\ell t} \lambda_\ell(x) + x_t \partial \lambda_\ell / \partial x_t(x) \right).$$

We know that $2\alpha_{\ell t}\lambda_\ell(x) + x_t \partial \lambda_\ell/\partial x_t(x) > 0$ for x near 0, so we only need find some ℓ so that $\prod_{s \in C_\ell} x_s^{2\alpha_{\ell s}} \neq 0$. But since $\rho_{mn}(x) > 0$ we know that such an ℓ exists. So $\partial \rho_{mn} \circ \theta / \partial x_t(x) \neq 0$ for x near 0. So after perhaps shrinking U_{mn}, c) is satisfied.

Now we just shrink the U_{ji}'s so that $U_{ji} \cap U_{ki} = \pi_{ji}^{-1}(U_{kj})$ and see that the maximality of $\mathcal{J}'$ is violated.   □

So we may pick a maximal element k of $\mathcal{J}' - \mathcal{J}''$. Pick a smooth $\beta \colon U \to [0, 1]$ so that $\mathrm{Cl}(\beta^{-1}((0, 1])) \subset U_{kn}$ and for some neighborhood U_{kn}' of W_{kn} in U_{kn}, $\beta(x) = 1$ for all $x \in U \cap U_{kn}'$.

Assertion 4.3.2.3 *For some open neighborhood Z of the graph G of the re-striction $(\pi_{km}, \rho_{km})|_{U_{km} - W_{km}}$ in $(U_{km} - W_{km}) \times W_k \times (0, \infty)$ there is a smooth map $\mu \colon Z \times [0, 1] \to U_{km} - W_{km}$ so that $\mu(x, y, t, 0) = x$, $\pi_{km}\mu(x, y, t, 1) = y$, $\rho_{km}\mu(x, y, t, 1) = t$ and $\mu(x, \pi_{km}(x), \rho_{km}(x), s) = x$.*

Proof: Put a Riemannian metric on U_{km} (or optionally embed it properly in some Euclidean space). Define $\lambda \colon Z \to U_{km} - W_{km}$ for a small neighborhood Z of G by letting $\lambda(x, y, t)$ be the closest point to x in the submanifold $\pi_{km}^{-1}(y) \cap \rho_{km}^{-1}(t)$. Now let $\mu(x, y, t, s)$ be the point which is s of the way along the geodesic from x to $\lambda(x, y, t)$ (or optionally $\mu(x, y, t, s) = \kappa((1 - s)x + s\lambda(x, y, t))$ where κ is the closest point map for $U_{km} \subset \mathbf{R}^b$).   □

Define $\pi' \colon U \to W_m$ by

$$\pi'(x) = \begin{cases} \pi(x) & \text{if} \quad x \in U - U_{kn} \\ \mu(\pi(x), \pi_{kn}(x), \rho_{kn}(x), \beta(x)) & \text{if} \quad x \in U \cap U_{kn} \end{cases}$$

Let us show that after perhaps shrinking U, $\pi' \colon U \to W_m$ satisfies conditions i)-iv) above. If $x \in W_{mn} \cap U_{kn}$ then

$$\begin{aligned} \pi'(x) &= \mu(\pi(x), \pi_{kn}(x), \rho_{kn}(x), \beta(x)) \\ &= \mu(p_{mn}(x), \pi_{km}p_{mn}(x), \rho_{km}p_{mn}(x), \beta(x)) = p_{mn}(x). \end{aligned}$$

(This also shows that after perhaps shrinking U, π' is defined.) If $m < i < n$ and $x \in U \cap W_{in} \cap U_{kn}$ then

$$
\begin{aligned}
\pi'(x) &= \mu(\pi(x), \pi_{kn}(x), \rho_{kn}(x), \beta(x)) \\
&= \mu(\pi_{mi}p_{in}(x), \pi_{km}\pi_{mi}p_{in}(x), \rho_{km}\pi_{mi}p_{in}(x), \beta(x)) \\
&= \pi_{mi}p_{in}(x).
\end{aligned}
$$

So after perhaps shrinking U to satisfy iv), π' satisfies conditions i)-iv).

Now pick $i \in \mathcal{J}''$ with $i < m$. Pick any $x \in U \cap \pi'^{-1}(U_{im}) \cap U'_{in} \cap U_{kn}$. If U is small enough, we know that $\pi(x) \in U_{im}$ since π' approximates π. Now since $U_{in} \cap U_{kn} \neq \emptyset$ we know that $k < i$. Then

$$
\begin{aligned}
\pi'(x) &= \mu(\pi(x), \pi_{kn}(x), \rho_{kn}(x), \beta(x)) \\
&= \mu(\pi(x), \pi_{ki}\pi_{in}(x), \rho_{ki}\pi_{in}(x), \beta(x)) \\
&= \mu(\pi(x), \pi_{ki}\pi_{im}\pi(x), \rho_{ki}\pi_{im}\pi(x), \beta(x)) \\
&= \mu(\pi(x), \pi_{km}\pi(x), \rho_{km}\pi(x), \beta(x)) \\
&= \pi(x).
\end{aligned}
$$

Consequently, $\pi_{im} \circ \pi'(x) = \pi_{im} \circ \pi(x) = \pi_{in}(x)$ and $\rho_{im} \circ \pi'(x) = \rho_{im} \circ \pi(x) = \rho_{in}(x)$. But if $x \in U \cap \pi'^{-1}(U_{km}) \cap U'_{in}$,

$$
\begin{aligned}
\pi_{km} \circ \pi'(x) &= \pi_{km} \circ \mu(\pi(x), \pi_{kn}(x), \rho_{kn}(x), 1) = \pi_{kn}(x) \\
\rho_{km} \circ \pi'(x) &= \rho_{km} \circ \mu(\pi(x), \pi_{kn}(x), \rho_{kn}(x), 1) = \rho_{kn}(x).
\end{aligned}
$$

So we have a contradiction to the maximality of $\mathcal{J}''$. $\square$

The functions in the preceding Lemma give rise to tubular data on $|\mathfrak{T}|$. For simplicity, we only bother with the unbounded case.

Proposition 4.3.3 *Let $\mathfrak{T}$ be a resolution tower of type UM without boundary. Then $|\mathfrak{T}|$ is a Thom stratified set.*

Proof: Let $W_i = V_i - |\mathcal{A}_i|$ and let $\pi_{ji} : U_{ji} \to W_j$ and $\rho_{ji} : U_{ji} \to [0, \infty)$ satisfy the conclusions of Lemma 4.3.2. Note that the strata of $|\mathfrak{T}|$ are exactly the W_i's. Let $U_i \subset |\mathfrak{T}|$ be $W_i \cup \bigcup_{j>i} U_{ij}$. Define $\pi_i : U_i \to W_i$ by $\pi_i(x) = \pi_{ij}(x)$ if $x \in U_{ij}$ and $\pi_i(x) = x$ if $x \in W_i$. This is well defined since the equivalence relation on $|\mathfrak{T}|$ is generated by $x \sim p_{kj}(x)$ for all $x \in V_{kj}$ and if $x \in U_{ij} \cap V_{kj}$ then $i \le k$ and $\pi_{ij}(x) = \pi_{ik} \circ p_{kj}(x)$ if $i < k$ and $\pi_{ij}(x) = p_{kj}(x)$ if $i = k$. Likewise define $\rho_i : U_i \to [0, \infty)$ by $\rho_i(x) = \rho_{ij}(x)$ if $x \in U_{ij}$ and $\rho_i(x) = 0$ if $x \in W_i$. $\square$

Proposition 4.3.4 *Let S be a connected component of a stratum of $|\mathfrak{T}|$ for a resolution tower $\mathfrak{T}$ of type UM. Then there are compact resolution towers $\mathfrak{T}'$ and $\mathfrak{T}''$ of type UM so that $|\mathfrak{T}'|$ is a link of S, $|\mathfrak{T}''| = \mathfrak{c}|\mathfrak{T}'|$ and $\mathfrak{T}' = \partial\mathfrak{T}''$. Furthermore, if $\mathfrak{T}$ has type R,N,E,S or any combination thereof, so do $\mathfrak{T}'$ and $\mathfrak{T}''$.*

Proof: Let $W_i = V_i - |\mathcal{A}_i|$ and let $\pi_{ji} : U_{ji} \to W_j$ and $\rho_{ji} : U_{ji} \to [0, \infty)$ satisfy the conclusions of Lemma 4.3.2. Let $\pi_i : U_i \to W_i$ and $\rho_i \to [0, \infty)$ be the tubular data for $|\mathfrak{T}|$ constructed as in Proposition 4.3.3. Let k be such that S is a connected component of W_k. Pick any $x \in S$, let $X = \pi_k^{-1}(x)$ and pick a compact neighborhood K of x in X. Let $\delta > 0$ be the minimum of ρ_k on the frontier of K. Pick any ϵ with $0 < \epsilon < \delta$. Let $L = X \cap \rho_k^{-1}(\epsilon)$. Then L is a canonical link of S.

We claim that $L = |\mathfrak{T}'|$ and $\mathfrak{c}L = |\mathfrak{T}''|$ for some $\mathfrak{T}'$ and $\mathfrak{T}''$. In particular, let $\mathfrak{J}' = \{ i \in \mathfrak{J} \mid i > k \}$ and $\mathfrak{J}'' = \{ i \in \mathfrak{J} \mid i \geq k \}$. For any $i > k$ let $V_i' = \pi_{ki}^{-1}(x) \cap \rho_{ki}^{-1}(\epsilon)$ and $\mathcal{A}_i' = \mathcal{A}_i \cap V_i'$. For any $i > j > k$ let $\mathcal{A}_{ji}' = \mathcal{A}_{ji} \cap V_i'$, $V_{ji}' = V_{ji} \cap V_i'$ and $p_{ji}' = p_{ji}|$. For any $i \geq k$ let $V_i'' = \pi_{ki}^{-1}(x) \cap \rho_{ki}^{-1}([0, \epsilon])$ and $\mathcal{A}_i'' = \mathcal{A}_i \cap V_i''$. For any $i > j \geq k$ let $\mathcal{A}_{ji}'' = \mathcal{A}_{ji} \cap V_i''$, $V_{ji}'' = V_{ji} \cap V_i''$ and $p_{ji}'' = p_{ji}|$.

Assertion 4.3.4.1 $\mathfrak{T}'$ *and* $\mathfrak{T}''$ *are resolution towers.*

Proof: We will prove that $\mathfrak{T}''$ is a resolution tower, then note that $\mathfrak{T}' = \partial \mathfrak{T}''$ must be a resolution tower. First of all, if $y \in V_{ji}''$ then $\pi_{kj}p_{ji}(y) = \pi_{ki}(y) = x$ and $\rho_{kj}p_{ji}(y) = \rho_{ki}(y) \in [0, \epsilon]$ so $p_{ji}''(V_{ji}'') \subset V_j''$.

To see I, note that

$$p_{ji}''(V_{ji}'' \cap V_{\ell i}'') = p_{ji}(V_{ji}'' \cap V_{\ell i}'') \subset p_{ji}(V_{ji} \cap V_{\ell i}) \cap V_j'' \subset V_{\ell j} \cap V_j'' = V_{\ell k}''.$$

Condition II, III and IV are now immediate. To see condition V, note that $\partial V_i'' = V_i'$. So if $p_{ji}''(y) \in \partial V_j''$ we have $\rho_{kj}p_{ji}''(y) = \epsilon$ so $\rho_{ki}(y) = \epsilon$ so $y \in \partial V_i''$.

□

Assertion 4.3.4.2 $|\mathfrak{T}'| = L$ *and* $|\mathfrak{T}''| = \mathfrak{c}L$.

Proof: Let $h_i : V_i \to |\mathfrak{T}|$ be the quotient map. Let $L' = X \cap \rho_k^{-1}([0, e])$. By Proposition 4.3.1, $L' = \mathfrak{c}L$. But $V_i' = h_i^{-1}(L)$ and $V_i'' = h_i^{-1}(L')$ and p_{ji}' and p_{ji}'' are the restrictions of p_{ji}. So the maps h_i induce isomorphisms from $|\mathfrak{T}'|$ and $|\mathfrak{T}''|$ to L and L'.

□

Assertion 4.3.4.3 $\mathfrak{T}'$ *and* $\mathfrak{T}''$ *have type UM and also have type R, N, S or E if* $\mathfrak{T}$ *does.*

Proof: We will just do $\mathfrak{T}''$, the result for $\mathfrak{T}'$ follows since $\mathfrak{T}' = \partial \mathfrak{T}''$.

Pick any stratum S of $(V_i'', \mathcal{A}_i'')$ with $S \subset W_{ji}$. To see that $\mathfrak{T}''$ has type U we must show that $p_{ji}''|_S$ is a submersion. Let S' be the stratum of V_i which contains S. We know that $p_{ji}|_{S'}$ is a submersion and $(\pi_{kj}, \rho_{kj}) : U_{kj} - W_{kj} \to W_k \times \mathbf{R}$ is a submersion. Also $\pi_{kj}p_{ji}|_S = \pi_{ki}|_S$ and $\rho_{kj}p_{ji}|_S = \rho_{ki}|_S$. Hence the restriction of p_{ji} to $S = S' \cap V_{ji}''$ is a submersion to V_j''.

Now let us see that $\mathfrak{T}''$ has type M. Pick any point $y \in V_{ji}''$. Pick any chart $\mu : (\mathbf{R}^a, 0) \to (V_k, \pi_{ki}(y))$. By the inverse function theorem, we may pick tico charts $\theta : (\mathbf{R}^c, 0) \to (V_j, p_{ji}(y))$ and $\psi : (\mathbf{R}^b, 0) \to (V_i, y)$ so that $\rho_{kj}\theta(z) =$

$\rho_{ki}(y) + z_{c-a}$ for all z near 0, $\rho_{ki}\psi(z) = \rho_{ki}(y) + z_{b-a}$ for all z near 0, the ℓ-th coordinate of $\mu^{-1}\pi_{kj}\theta(z)$ is $z_{c-\ell+1}$ and the ℓ-th coordinate of $\mu^{-1}\pi_{ki}\psi(z)$ is $z_{b-\ell+1}$. But then $\theta|_{\mathbf{R}^{c-a}}$ and $\psi|_{\mathbf{R}^{b-a}}$ give tico charts for $\mathfrak{T}''$ and show that p''_{ji} is a mico.

In a similar manner, we show that $\mathfrak{T}''$ has type S, N or E if $\mathfrak{T}$ does. It is trivial to show that $\mathfrak{T}''$ has type R if $\mathfrak{T}$ does. $\quad\square$

Note that the above assertions prove the Proposition. $\quad\square$

We find the following result convenient. For various reasons this was proven in several places, see [**G**], [**J**] or [**V**] for a history. We do not require the axiom of the frontier in our stratified sets, but this is not really essential.

Proposition 4.3.5 *Any compact Thom stratified set is homeomorphic to a polyhedron in such a way that each open simplex is a smooth submanifold of a stratum.*

CHAPTER V

ALGEBRAIC STRUCTURES ON
RESOLUTION TOWERS

The main consequence of this chapter is

Theorem 5.0 *If $\mathfrak{T}$ is any compact resolution tower of type SF then the realization $|\mathfrak{T}|$ is homeomorphic to a real algebraic set. In fact $|\mathfrak{T}|$ is isomorphic as a stratified set to the singular stratification of some real algebraic set.*

Proof: By Theorem 4.2.8 and Proposition 4.2.7 we may as well assume $\mathfrak{T}$ is of type $RESF$. But then we may assume $\mathfrak{T}$ is quasialgebraic by Theorem 5.4.1 below. But then Theorems 5.3.1 and 5.3.3 will imply this result. □

In this chapter, all ticos will be regular.

1. Making Tico Maps Algebraic

We start out with a relative version of Theorem 2.8.8, which says that if a map is homotopic to an entire rational function and is already rational on a tico, then it can be made algebraic.

Lemma 5.1.1 *Suppose $(V, \mathcal{A})$ is a compact algebraic tico, W is a nonsingular real Zopen set and $f \colon V \to W$ is a smooth map so that $f|_{|\mathcal{A}|}$ is entire rational and f is homotopic to an entire rational function. Then there are a nonsingular real Zopen set V', a rational diffeomorphism $h \colon V' \to V$ and an entire rational function $g \colon V' \to W$ so that $|\mathcal{A}| \subset V'$, $h|_{|\mathcal{A}|} = $ identity and so that f is ϵ homotopic rel $|\mathcal{A}|$ to gh^{-1}. (More precisely given any neighborhood U of f in the C^∞ topology, there are h, V' and g as above so that gh^{-1} and all maps in the homotopy are in U.*

Proof: There is a smooth function $p \colon V \times S^1 \to W$ so that if a and b are certain distinct points of S^1 then $p(x, a) = f(x)$ for all $x \in V$ and so that $p|_{V \times b}$ is an entire rational function (just take the double of some homotopy of f to an entire rational function). By Theorem 2.8.3 (with $T := U := V := V \times S^1$, $f := p$ and $P := L := V \times b \cup |\mathcal{A}| \times a$) there is a Zclosed subset $Z \subset V \times S^1 \times \mathbf{R}^k$ for some k and an entire rational function $q \colon Z \to W$ and a $Z_0 \subset \text{Nonsing } Z$ so that Z_0 is a union of connected components of Z, $V \times b \times 0 \subset Z_0$, $|\mathcal{A}| \times a \times 0 \subset Z_0$, $q(x,b,0) = p(x,b)$ for all $x \in V$, $q(x,a,0) = f(x)$ for all $x \in |\mathcal{A}|$, $\pi \colon Z_0 \to V \times S^1$ is a diffeomorphism where π is induced by projection $V \times S^1 \times \mathbf{R}^k \to V \times S^1$ and $q|_{Z_0}$ is a close approximation to $p\pi$. By Theorem 2.8.2 there is a nonsingular real algebraic set $V' \subset Z$ so that V' is isotopic to $\pi^{-1}(V \times a)$ by a small isotopy and $|\mathcal{A}| \times a \times 0 \subset V'$. Then projection $V \times S^1 \times \mathbf{R}^k \to V$ induces an entire rational function $h \colon V' \to V$ which is a diffeomorphism since V' is close to $V \times a \times 0$. Also $g = q|_{V'}$ is a close approximation to $p\pi|$, hence it closely approximates fh. So g is ϵ-homotopic to fh rel $|\mathcal{A}| \times a \times 0$. $\square$

If f is a tico map we can strengthen Lemma 5.1.1 to the following proposition, to make the approximation ϵ-tico homotopic to f.

Proposition 5.1.2 *Let $(V, \mathcal{A})$ and $(W, \mathcal{B})$ be compact algebraic ticos and let $f \colon (V, \mathcal{A}) \to (W, \mathcal{B})$ be a tico map so that $f|_{|\mathcal{A}|}$ is an entire rational function and f is homotopic to an entire rational function. Then there is a nonsingular real Zopen set V' and a rational diffeomorphism $h \colon V' \to V$ and an entire rational function $g \colon V' \to W$ so that $|\mathcal{A}| \subset V'$ and so that f and gh^{-1} are ϵ-tico homotopic rel $|\mathcal{A}|$.*

Proof: By replacing $\mathcal{A}$ with the collection of irreducible components of $|\mathcal{A}|$ we may as well assume f has constant exponents by Lemma 3.2.3. For convenience let $\mathcal{A} = \{V_1, \dots, V_m\}$ and $\mathcal{B} = \{W_1, \dots, W_k\}$. Let α_{ji} denote the exponent of V_i in W_j. The proof will be by induction on the quantity $\beta(f) = \sum_{i=1}^m \sum_{j=1}^k \max(0, \alpha_{ji} - 1)$.

First suppose $\beta(f) = 0$. By Lemma 5.1.1 we may find a rational diffeomorphism $h \colon V' \to V$ and an entire rational function $g \colon V' \to W$ so that $|\mathcal{A}| \subset V'$, $h|_{|\mathcal{A}|} = \text{identity}$ and f is ϵ-homotopic to gh^{-1} rel $|\mathcal{A}|$. But since each α_{ji} is 1 or 0, any ϵ-homotopy rel $|\mathcal{A}|$ automatically is a tico homotopy, so this case is done by Lemma 3.2.8.

Now assume $\beta(f) > 0$. After renumbering we may assume that $\alpha_{11} > 1$. In particular we know that $f(V_1) \subset W_1$ by Lemma 3.2.2. Let $f' \colon V \to W \times V$ be $f'(x) = (f(x), x)$. We have the product tico $\mathcal{C} = \mathcal{B} \times V \cup W \times \mathcal{A}$ in $W \times V$. Notice that $f' \colon (V, \mathcal{A}) \to (W \times V, \mathcal{C})$ is a tico map with constant exponents, the exponent of V_i in $W_j \times V$ is α_{ji} and the exponent of V_i in $W \times V_j$ is δ_{ji} the Kronecker delta. Consequently, $\beta(f') = \beta(f)$.

Let $\pi \colon (W'', \mathcal{D}) \to (W \times V, \mathcal{C})$ be the tico blow-up with center $W_1 \times V_1$. Note that this center is fat; in fact it is the intersection of two sheets $W_1 \times V$ and $W \times V_1$. Since f' hits $W_1 \times V_1$ cleanly and $f'^{-1}(W_1 \times V_1)$ has codimension one, we know by Lemmas 2.5.9 and 2.5.6 that there is a unique smooth map $f'' \colon V \to W''$

so that $\pi f'' = f'$. Alternatively, this is true because $f'^* \mathcal{I}^\infty (W_1 \times V_1) = \mathcal{I}^\infty (V_1)$ which is locally principal.

But then Lemma 3.3.5 implies that f'' is a tico map with constant exponents. In particular, let D_i be the strict preimage of $W_i \times V$, let E_i be the strict preimage of $W \times V_i$ and let $P = \pi^{-1} (W_1 \times V_1)$. Then

$$\mathcal{D} = \{\, D_i \mid i = 1, \ldots, k \,\} \cup \{\, E_i \mid i = 1, \ldots, m \,\} \cup \{P\}.$$

By Lemma 3.3.5 the exponent of V_i in D_j is α_{ji} if $i \neq 1$ or $j \neq 1$ and it is $\alpha_{11} - 1$ if $i = j = 1$. The exponent of V_i in E_j is δ_{ij} if $j \neq 1$ and it is 0 if $j = 1$. Finally, the exponent of V_i in P is δ_{i1}. Consequently, $\beta(f'') = \beta(f') - 1 = \beta(f) - 1$ so we will be able to apply induction to f''.

First we show that f'' is homotopic to an entire rational function. By Lemma 5.1.1 we have a homotopy $H: V \times I \to W$ so that $H(x, 0) = f(x)$ for all x and $H|_{V \times 1}$ is entire rational and $H(x, t) = f(x)$ for all $x \in |\mathcal{A}|$. Now $H \times id$ hits $W_1 \times V_1$ cleanly and $(H \times id)^{-1} (W_1 \times V_1) = V_1 \times I$ which has codimension one. Hence $H \times id$ lifts to $H': V \times I \to W''$. But $H'(x, 0) = f''(x)$ for all $x \in V$ by uniqueness of lifting and $H'|_{V \times 1}$ is entire rational, i.e., f'' is homotopic to an entire rational function. Also since $H|_{V_j \times I}$ is entire rational and $V_j \times I \cap H^{-1} (W_1 \times V_1) = (V_j \cap V_1) \times I$ we know that $H'|_{V_j \times I}$ is entire rational for $j \neq 1$, in particular we know that $f''|_{V_j}$ is entire rational for $j \neq 1$. We now show $f''|_{V_1}$ is entire rational. Since the exponent of V_1 in D_1 is $\alpha_{11} - 1 > 0$ we know that $f''(V_1) \subset D_1$. But $\pi|: D_1 \to W_1 \times V$ is a birational isomorphism since the center $W_1 \times V_1$ has codimension one in $W_1 \times V$. Let $\lambda: W_1 \times V \to D_1$ be the inverse of $\pi|$. Then we must have $f''|_{V_1} = \lambda f'|_{V_1}$ so $f''|_{V_1}$ is entire rational.

Now since $\beta(f'') < \beta(f') = \beta(f)$ we can by induction find a rational diffeomorphism $h: V' \to V$ and an entire rational $g': V' \to W''$ so that f'' and $g' h^{-1}$ are ϵ tico homotopic rel $|\mathcal{A}|$. But $f = \pi_1 \pi f''$ where $\pi_1: W \times V \to W$ is projection and π_1 and π are tico maps. So f and $\pi_1 \pi g' h^{-1}$ are ϵ tico homotopic rel $|\mathcal{A}|$. So setting $g = \pi_1 \pi g'$ we are done. $\qquad \square$

Proposition 5.1.3 *Let $(V, \mathcal{A})$ be a compact nonsingular Zopen set with an algebraic tico and let W be a totally algebraic nonsingular Zopen set. Let $f: V \to W$ be a smooth map such that $f|_{|\mathcal{A}|}$ is an entire rational function. Then there is a nonsingular Zopen set V' with $|\mathcal{A}| \subset V'$ and a diffeomorphism $\theta: V' \to V$ with $\theta|_{|\mathcal{A}|} = id$ and an entire rational function $g: V' \to W$ such that g is ϵ-close to $f \circ \theta$ and $g|_{|\mathcal{A}|} = f|_{|\mathcal{A}|}$.*

Proof: By Lemma 5.1.1 we only need find a nonsingular algebraic set V' with $|\mathcal{A}| \subset V'$ and a diffeomorphism $\theta: V' \to V$ with $\theta|_{|\mathcal{A}|} = $ identity so that $f\theta$ is homotopic to an entire rational function. This follows from Theorem 2.8.4, (with $M := V$, $P := |\mathcal{A}|$, $L := |\mathcal{A}|$ and $S := \mathbf{R}^n$). $\qquad \square$

2. Nice Charts on Resolution Towers

The following technical lemma is essentially the generalization of Lemma 3.6.2 to resolution towers.

Lemma 5.2.1 *Let* $\mathfrak{T} = \{V_i, \mathcal{A}_i, p_i\}_\mathfrak{I}$ *be a resolution tower of type* S. *Pick any* $z \in \bigcap_{j \in \mathfrak{I}'} V_{jn}$ *for some indices* n *and* $\mathfrak{I}' \subset \mathfrak{I} - \{n\}$. *Let* $\mathfrak{I}'' = \mathfrak{I}' \cup \{n\}$. *Assume for simplicity of notation that the index* n *is an integer and* $\mathfrak{I}'' = \{0, \ldots, n\}$. *Then there are charts* $\psi_k \colon Z_k \to V_k$ *for* $k \in \mathfrak{I}''$, *so that*

a) Z_k *is an open subset of* $\mathbf{R}^{a_k}$ *with* $a_k = \dim V_k$ *for* $k \in \mathfrak{I}''$.
b) $\psi_k(Z_k)$ *is a neighborhood of* $p_{kn}(z)$ *for all* $k \in \mathfrak{I}'$ *and* $\psi_n(Z_n)$ *is a neighborhood of* z.
c) *There are integers* $a_{k0} < a_{k1} < \cdots < a_{kk} = a_k$, $k \in \mathfrak{I}''$ *so that* $\psi_k^{-1}(V_{jk}) = Z_k \cap \bigcup_{s > a_{kj}}^{a_{kj+1}} \mathbf{R}_s^{a_k}$, *for all* $j < k$.
d) $p_{jk}(\psi_k(Z_k) \cap V_{jk}) \subset \psi_j(Z_j)$ *for all* $j, k \in \mathfrak{I}''$ *with* $j < k$.
e) *For* $j < k$, $t \leq a_k$ *and* $x \in \psi_k^{-1}(V_{jk})$, *let* $\kappa_{jkt}(x)$ *be the* t-*th coordinate of* $\psi_j^{-1} p_{jk} \psi_k(x)$. *Then* $\kappa_{jkt}(x) = x_t$ *for* $t \leq a_{j0}$.
f) *If* $t > a_{j0}$ *then* $\kappa_{jkt}(x) = \prod_{s > a_{j0}}^{a_k} x_s^{\beta_{jkts}}$ *for some integers* β_{jkts}. *Also,* $\beta_{jkts} = 0$ *if* $t \leq a_{ji}$ *and* $s > a_{ki}$ *for some* i.
g) *For any* $i, j, k \in \mathfrak{I}''$ *with* $i < j < k$, $\beta_{ikts} = \beta_{ijts}$ *for* $a_{i0} < s \leq a_{j0}$ *and* $\beta_{ikts} = \sum_{u > a_{j0}}^{a_j} \beta_{ijtu} \beta_{jkus}$ *for* $s > a_{j0}$.

Proof: We reduce to the case of finding ψ_k's satisfying a)-e) and the weaker f') below.

Assertion 5.2.1.1 *Condition g) follows from e) and f).*

Proof: We know that $p_{ik} \psi_k(x) = p_{ij} p_{jk} \psi_k(x)$ for $x \in \psi_k^{-1}(V_{ik} \cap V_{jk})$. Hence for $t > a_{i0}$,

$$\kappa_{ikt}(x) = \prod_{u > a_{i0}}^{a_j} \kappa_{jku}(x)^{\beta_{ijtu}} = \prod_{u > a_{i0}}^{a_{j0}} x_u^{\beta_{ijtu}} \prod_{u > a_{j0}}^{a_j} \prod_{s > a_{j0}}^{a_k} x_s^{\beta_{jkus} \beta_{ijtu}}.$$

Hence g) follows since $\kappa_{ikt}(x)$ is not identically 0 for $x \in \psi_k^{-1}(V_{ik} \cap V_{jk})$.   □

Assertion 5.2.1.2 *It suffices to prove a weaker version of f), namely:*

f') *If* $t > a_{j0}$ *then* $\kappa_{jkt}(x) = \pm \prod_{s > a_{j0}}^{a_k} x_s^{\beta_{jkts}}$ *for some integers* β_{jkts}.

Proof: Suppose we can pick $\psi_k \colon Z_k \to V_k$ for $k = 0, \ldots, n$ satisfying a)-e) and f'). First let us show that $\beta_{jkts} = 0$ if $t \leq a_{ji}$ and $s > a_{ki}$ for some i. Suppose that $\beta_{jkts} \neq 0$ for some $t \leq a_{ji}$ and $s > a_{ki}$. Then

$$p_{jk}(\psi_k(\mathbf{R}_s^{a_k}) \cap V_{jk}) \subset \mathbf{R}_t^{a_j} \subset \bigcup_{\ell < i} V_{\ell j}.$$

But this violates condition III in the definition of resolution tower since $\psi_k(\mathbf{R}_s^{a_k}) \cap V_{jk} - \bigcup_{\ell < i} V_{\ell k} \neq \emptyset$.

We now get rid of the minus signs in f'). What we will do is to modify the ψ_k's by composing them with reflections until f) is satisfied. The first step is to compose each ψ_k with a reflection about $\mathbf{R}_t^{a_k}$ for each $t \le a_{k0}$ such that the t-th coordinate z_t of $\psi_n^{-1}(z)$ is negative. Then a)-e) and f') are still satisfied, but we now have $z_t \ge 0$ for each t. We also know by e) that the t-th coordinate of $\psi_k^{-1} p_{kn}(z)$ is nonnegative if $t \le a_{k0}$.

Suppose that after composing the ψ_k's with reflections about some hyperplanes $\mathbf{R}_t^{a_k}$ with $t > a_{k0}$ that f) holds for all $k < m$. We now compose each ψ_k for $k < m$ with a reflection about $\mathbf{R}_t^{a_k}$ for each $t > a_{k0}$ such that $\kappa_{kmt}(x) = -\prod_{s>a_{k0}}^{a_m} x_s^{\beta_{kmts}}$. With these new ψ_k's we will have $\kappa_{kmt}(x) = \prod_{s>a_{k0}}^{a_m} x_s^{\beta_{kmts}}$. So f) holds for $k = m$, we must only check that f) still holds for $k < m$.

So suppose that $\kappa_{jkt}(x) = -\prod_{u>a_{j0}}^{a_k} x_u^{\beta_{jktu}}$ for some $0 \le j < k < m$ and $t > a_{j0}$. Then for $x \in \psi_m^{-1}(V_{jm} \cap V_{km})$ we have

$$\prod_{s>a_{j0}}^{a_m} x_s^{\beta_{jmts}} = \kappa_{jmt}(x)$$

$$= -\prod_{u>a_{j0}}^{a_k} \kappa_{kmu}(x)^{\beta_{jktu}}$$

$$= -\prod_{u>a_{j0}}^{a_{k0}} x_u^{\beta_{jktu}} \prod_{u>a_{k0}}^{a_k} \prod_{s>a_{k0}}^{a_m} x_s^{\beta_{kmus}\beta_{jktu}}.$$

Since $\psi_m^{-1} p_{mn}(z)$ has s-th coordinates nonnegative for $s \le a_{m0}$ and 0 for $s > a_{m0}$ we may pick an $x \in \psi_m^{-1}(V_{jm} \cap V_{km})$ so that $x_s > 0$ for all $s \le a_{mj}$ and $x_s \ge 0$ for all $s > a_{mj}$. But we showed above that $\beta_{jmts} = 0$ for all $s > a_{mj}$, so the left hand side above is positive and the right hand side is nonpositive, a contradiction. So in fact there are no sign changes and f) still holds for $j < k < m$.

So by induction on m we are done.   □

So let us proceed to find ψ_k's satisfying a)-e) and f'). By induction on n we may assume we have $\psi_k \colon Z_k \to V_k$ for $k = 0, \ldots, n-1$ satisfying a)-e) and f'). By definition of submersive mico we may extend $p_{n-1,n}$ to a submersive tico map $p'_{n-1,n}$ defined on a neighborhood of $V_{n-1,n}$. Then by Lemma 3.6.2 applied to $p'_{n-1,n}$ we may pick some chart $\psi_n \colon Z_n \to V_n$ satisfying a)-e) and also satisfying f') for $j = n-1$ and $k = n$. Suppose that for some $d \ge 0$, f') holds for all $j > d$. We will show that after perturbing ψ_n a bit, f') will hold for all $j \ge d$. Hence by induction we will be done.

So suppose that for some $d \ge 0$, f') holds for all $j > d$. We may suppose $d < n - 1$. Let us look at $\kappa_{dnt}(x)$. If $x \in \psi_n^{-1}(V_{jn} \cap V_{dn})$ for some $j > d$ then

$$\kappa_{dnt}(x) = \kappa_{djt}\left(\psi_j^{-1} p_{jn}\psi_n(x)\right).$$

So if $t \leq a_{d0}$, $\kappa_{dnt}(x) = \kappa_{jnt}(x) = x_t$ since $a_{j0} \geq a_{d0}$. Likewise if $a_{d0} < t$, we have

$$\kappa_{dnt}(x) = \pm \prod_{s>a_{d0}}^{a_j} \kappa_{jns}(x)^{\beta_{djts}} = \pm \prod_{s>a_{d0}}^{a_{j0}} x_s^{\beta_{djts}} \prod_{s>a_{j0}}^{a_j} \prod_{u>a_{j0}}^{a_n} x_u^{\beta_{jnsu}\beta_{djts}}.$$

Define $\beta_{dnts} = \sum_{u>a_{j0}}^{a_j} \beta_{jnus}\beta_{djtu}$ for $a_{j0} < s$ and define $\beta_{dnts} = \beta_{djts}$ for $a_{d0} < s \leq a_{j0}$. Then

$$\kappa_{dnt}(x) = \pm \prod_{s>a_{d0}}^{a_n} x_s^{\beta_{dnts}} \quad \text{for} \quad x \in \psi_n^{-1}(V_{jn} \cap V_{dn}).$$

At first glance, β_{dnts} seems to depend on the choice of j. Let β'_{dnts} be defined using some j' instead of j.

Then for $x \in \psi_n^{-1}(V_{jn} \cap V_{j'n} \cap V_{dn})$,

$$\pm \prod_{s>a_{d0}}^{a_{nd}} x_s^{\beta_{dnts}} = \kappa_{dnt}(x) = \pm \prod_{s>a_{d0}}^{a_{nd}} x_s^{\beta'_{dnts}}.$$

But for a generic point of $\psi_n^{-1}(V_{jn} \cap V_{j'n} \cap V_{dn})$ we know $x_s \neq 0$ for $s < a_{nd}$. So the exponents are equal, so $\beta_{dnts} = \beta'_{dnts}$ for all $t > a_{d0}$ and $s > a_{d0}$.

We can define $\lambda_t \colon \psi_n^{-1}(V_{dn}) \to \mathbf{R}$ by the equation

$$\kappa_{dnt}(x) = \pm \prod_{s>a_{d0}}^{a_n} x_s^{\beta_{dnts}} \lambda_t(x)$$

for $x \in \psi_n^{-1}(V_{dn})$ and for $a_{d0} < t$ where the $\pm$ sign is chosen so that $\lambda_t \circ \psi_n^{-1}(z) > 0$. Then we just showed that $\lambda_t(x) = 1$ for $x \in \psi_n^{-1}(V_{dn} \cap \bigcup_{j>d} V_{jn})$. By Lemma 3.6.1, we know there are γ_{su} for $a_{d0} < s \leq a_n$ and $a_{d0} < u \leq a_d$ so that $\sum_{s>a_{d0}}^{a_n} \beta_{dnts}\gamma_{su} = \delta_{tu}$ the Kronecker delta, for $a_{d0} < t$. For $a_{d0} \geq t$ define $\mu_t \colon Z_n \to \mathbf{R}$ to some extension of κ_{dnt} so that $\mu_t(x) = x_t$ if $x \in \psi_n^{-1}(\bigcup_{j>d} V_{jn})$. Likewise for $s > a_{d0}$, extend λ_s to Z_n so that $\lambda_s = 1$ on $\psi_n^{-1}(\bigcup_{j>d} V_{jn})$. Define $\mu_t(x) = x_t \prod_{s>a_{d0}}^{a_d} \lambda_s(x)^{\gamma_{ts}}$ for $t > a_{d0}$. Define $\mu \colon Z_n \to \mathbf{R}^{a_n}$ by $\mu(x) = (\mu_1(x), \dots, \mu_{a_n}(x))$. Then μ is the identity on $\psi_n^{-1}(\bigcup_{j>d} V_{jn})$.

It is easy to check that the Jacobian matrix of μ is the identity at $\psi_n^{-1}(z)$. Hence $\mu^{-1} \colon Z_n' \to Z_n$ is defined for some neighborhood Z_n' of $\mu\psi_n^{-1}(z)$. But for $x \in \mu\psi_n^{-1}(V_{dn})$ we have

$$\kappa_{dnt}\mu^{-1}(x) = \begin{cases} x_t & \text{for } t \leq a_{d0} \\ \pm \prod_{s>a_{d0}}^{a_n} x_s^{\beta_{dnts}} & \text{for } t > a_{d0} \end{cases}$$

Also $\kappa_{jnt}\mu^{-1}(x) = \kappa_{jnt}(x)$ for all $x \in \mu\psi_n^{-1}(V_{jn})$, $j > d$. So by replacing ψ_n by $\psi_n\mu^{-1}$ we can make f') be satisfied for all $j \geq d$. Therefore we are through by induction. $\square$

Lemma 5.2.2 *Let $\mathfrak{T} = \{V_i, \mathcal{A}_i, p_i\}_{i=0}^{n}$ be a resolution tower of type RS. Let $S \subset |\mathcal{A}_n|$ be a connected fat submanifold of V_n and let k be the largest index so that $S \subset V_{kn}$. Let T be the smallest fat submanifold of V_k so that $p_{kn}(S) \subset T$. Let $G_t \colon (S, S \cap \mathcal{A}_n) \to (T, T \cap \mathcal{A}_k)$, $t \in I$ be an ϵ-tico homotopy rel $|S \cap \mathcal{A}_n|$ with $G_0 = p_{kn}|$. Then there is an ϵ isotopy $H_t \colon V_n \to V_n$, $t \in I$ so that:*

   a) *H_0 is the identity.*
   b) *$H_t(A) = A$ for all $A \in \mathcal{A}_n$.*
   c) *$G_t = p_{kn} H_t|_S$.*
   d) *$p_{in} H_t|_A = p_{in}|_A$ for all sheets $A \in \mathcal{A}_{in}$ with $S \not\subset A$.*

Proof: We will find a smooth vector field $(\omega(x,t), 1)$ on $V_n \times I$ so that:

   i. $\omega(x,t)$ is small.
   ii. $\omega(x,t)$ is tangent to each $A \in \mathcal{A}_n$.
   iii. $dp_{kn}(\omega(x,t)) = \partial G_t / \partial t$ for $x \in S$.
   iv. $dp_{in}(\omega(x,t)) = 0$ for $x \in A$ where A is any sheet in $\mathcal{A}_{in}$ with $S \not\subset A$.

Then integrating this vector field will give us the desired isotopy H_t. Notice that it suffices to find ω locally and piece together with a partition of unity.

So take any $z \in V_n$. If $z \notin S$ we may as well take $\omega = 0$ locally and our conditions will be satisfied. Now suppose $z \in S - |S \cap \mathcal{A}_n|$. Then since $p_{kn}| \colon S - |S \cap \mathcal{A}_n| \to T$ is a submersion we may pick a small family of small vector fields $\omega(x,t)$ on S, tangent to S so that $dp_{kn}\omega(x,t) = \partial G_t(x)/\partial t$. Now just extend $\omega(x,t)$ to a neighborhood of z in V_n so that ω is tangent to each $A \in \mathcal{A}_n$ and all our conditions will be met. (Note that any sheet A with $S \not\subset A$ must have $S \cap A \subset |S \cap \mathcal{A}_n|$, so A is far from z.)

So the remaining case is where $z \in |S \cap \mathcal{A}_n|$. Pick $j_0 < j_1 < \cdots < j_{m-1} < j_m = n$ so that $z = V_{jn}$ if and only if $j = j_i$ for some $i = 0, \ldots, m-1$. By Lemma 5.2.1 we may find charts $\psi_i \colon Z_i \to V_{j_i}$ for $i = 0, \ldots, m$ satisfying the conclusions of Lemma 5.2.1. In particular:

   1) Z_i is an open subset of $\mathbf{R}^{a_i}$
   2) $\psi_m(Z_m)$ is a neighborhood of z
   3) $p_{j_b j_i}(\psi_i(Z_i) \cap V_{j_b j_i}) \subset \psi_b(Z_b)$
   4) If $\kappa_{biq}(x)$ is the qth coordinate of $\psi_b^{-1} p_{j_b j_i} \psi_i(x)$ for $x \in \psi_i^{-1}(V_{j_b j_i})$ then

$$\kappa_{biq}(x) = \begin{cases} x_q & \text{for } q \leq a_{b0} \\ \prod_{s > a_{b0}}^{a_i} x_s^{\beta_{biqs}} & \text{for } q > a_{b0} \end{cases}$$

Let $Q \subset \{1, \ldots, a_m\}$ be the index subset so that $\psi_m^{-1}(S) = \{x \in Z_m \mid x_s = 0 \text{ for all } s \in Q\}$. Let d be such that $k = j_d$. Let $\lambda_q(x,t)$ be the qth coordinate of $\psi_d^{-1} G_t \psi_m(x)$ for $x \in \psi_m^{-1}(S)$.

Suppose $q > a_{d0}$ is such that $\psi_d^{-1}(T) \not\subset \mathbf{R}_q^{a_d}$. Then λ_q is not identically 0 on $\psi_m^{-1}(S) \times I$, so we have $\lambda_q(x,t) = \prod_{s > a_{d0}}^{a_m} x_s^{\beta_{dmqs}} \eta_q(x,t)$ for some smooth η_q. Note that $\eta_q(x,0) = 1$. By repeated applications of Lemma 3.1.1 we know that

$\eta_q\left(x,t\right) \approx 1$, since $\lambda_q\left(x,t\right) \approx \lambda_q\left(x,0\right)$. We may extend λ_q to all of $Z_m \times I$ by ignoring the coordinates x_s with $s \in Q$, i.e., first you project to $\bigcap_{s \in Q} \mathbf{R}_s^{am} \times I$, then apply λ_q. For $q > a_{d0}$ with $\psi_d^{-1}\left(T\right) \subset \mathbf{R}_q^{ad}$ we may set $\eta_q\left(x,t\right) = 1$. Note that $\lambda_q\left(x,t\right) = \kappa_{dmq}\left(x\right)$ for all t if $x \in Z_m \cap \mathbf{R}_s^{am}$ with $s > a_{m0}$ and $s \notin Q$ since G_t is a homotopy rel $|S \cap \mathcal{A}_k|$.

What we need is a family of vector fields $v\left(x,t\right)$ on Z_m, $x \in Z_m$, $t \in I$ so that:

    i. $v\left(x,t\right)$ is small.
    ii. $v\left(x,t\right)$ is tangent to each $\mathbf{R}_s^{am}$ for $s > a_{m0}$.
    iii. $d\kappa_{dmq}\left(v\left(x,t\right)\right) = \partial\lambda_q(x,t)/\partial t$ for all q and $x \in \psi_m^{-1}\left(S\right)$
    iv. $d\kappa_{bmq}\left(v\left(x,t\right)\right) = 0$ for $x \in Z_m \cap \mathbf{R}_s^{am}$, $s \notin Q$ and $a_{mb} < s \leq a_{mb+1}$.

Then letting $\omega\left(x,t\right) = d\psi_m v\left(x,t\right)$ we will be done.

Let $v_s\left(x,t\right)$ denote the sth coordinate of $v\left(x,t\right)$. By Lemma 3.6.1 we may pick γ_{sr}, $a_{d0} < s \leq a_m$, $a_{d0} < r \leq a_d$ so that $\sum_{s > a_{d0}}^{a_m} \beta_{dmqs}\gamma_{sr} = \delta_{qr}$ the Kronecker δ. Define

$$v_s\left(x,t\right) = \begin{cases} x_s \sum_{r > a_{d0}}^{a_d} \gamma_{sr}\partial\eta_r\left(x,t\right)/\partial t & \text{for } s > a_{d0} \\ \partial\lambda_s\left(x,t\right)/\partial t & \text{for } s \leq a_{d0} \end{cases}$$

Clearly $v_s\left(x,t\right)$ is small and it is tangent to each $\mathbf{R}_s^{am}$ for $s > a_{m0}$ since $a_{m0} \geq a_{d0}$.

For $q \leq a_{d0}$ and $x \in \psi_m^{-1}\left(S\right)$,

$$d\kappa_{dmq}\left(v\left(x,t\right)\right) = v_q\left(x,t\right) = \partial\lambda_q(x,t)/\partial t.$$

For $q > a_{d0}$ and $x \in \psi_m^{-1}\left(S\right)$,

$$d\kappa_{dmq}\left(v\left(x,t\right)\right) = \sum_{s > a_{d0}}^{a_m} v_s\left(x,t\right)\partial\kappa_{dmq}\left(x\right)/\partial x_s$$

$$= \sum_{s > a_{d0}}^{a_m} \sum_{r > a_{d0}}^{a_d} x_s\gamma_{sr}\partial\eta_r/\partial t\, \beta_{dmqs}\kappa_{dmq}\left(x\right)/x_s$$

$$= \kappa_{dmq}\left(x\right) \sum_{r > a_{d0}}^{a_d} \left(\partial\eta_r/\partial t \sum_{s > a_{d0}}^{a_m} \beta_{dmqs}\gamma_{sr}\right)$$

$$= \kappa_{dmq}\left(x\right)\partial\eta_q\left(x,t\right)/\partial t = \partial\lambda_q\left(x,t\right)/\partial t.$$

So it remains to show that if we pick $s_0 \notin Q$ with $a_{mb} < s_0 \leq a_{mb+1}$ then $d\kappa_{bmq}v\left(x,t\right) = 0$ for all q and all $x \in Z_m \cap \mathbf{R}_{s_0}^{am}$. So pick such an s_0.

If $b > d$ then we know $\beta_{dmqs_0} = 0$ for all $q > a_{d0}$ by Lemma 5.2.1f. Hence $\eta_q\left(x,t\right) = \lambda_q\left(x,t\right)/\kappa_{dmq}\left(x\right)$ is defined on a dense subset of $Z_m \cap \mathbf{R}_{s_0}^{am}$. But $\lambda_q\left(x,t\right) = \kappa_{dmq}\left(x\right)$ for $x \in Z_m \cap \mathbf{R}_{s_0}^{am}$ so $\eta_q\left(x,t\right) = 1$ on $Z_m \cap \mathbf{R}_{s_0}^{am}$. Hence $\partial\eta_q/\partial t = 0$ so $v\left(x,t\right) = 0$ for $x \in Z_m \cap \mathbf{R}_{s_0}^{am}$.

If $b = d$ we have already seen that $d\kappa_{dmq}(v(x,t)) = \partial\lambda_q(x,t)/\partial t$ but for $x \in \mathbf{R}_{s_0}^{a_m} \cap Z_m$, $\lambda_q(x,t) = \kappa_{dmq}(x)$ is independent of t so $d\kappa_{dmq}(v(x,t)) = 0$ for $x \in \mathbf{R}_{s_0}^{a_m} \cap Z_m$.

Now suppose $b < d$. Then for $q \leq a_{b0}$,

$$d\kappa_{bmq}v(x,t) = v_q(x,t) = \partial\lambda_q(x,t)/\partial t = 0$$

for $x \in \mathbf{R}_{s_0}^{a_m} \cap Z_m$. For $q > a_{b0}$,

$$
\begin{aligned}
d\kappa_{bmq}v(x,t) &= \sum_{s>a_{b0}}^{a_m} v_s(x,t)\,\beta_{bmqs}\kappa_{bmq}(x)/x_s \\
&= \sum_{s>a_{b0}}^{a_{d0}} \partial\lambda_s(x,t)/\partial t\,\beta_{bmqs}\kappa_{bmq}(x)/x_s \\
&\quad + \sum_{s>a_{d0}}^{a_m}\sum_{r>a_{d0}}^{a_d} \gamma_{sr}\partial\eta_r(x,t)/\partial t\,\beta_{bmqs}\kappa_{bmq}(x) \\
&= \sum_{s>a_{b0}}^{a_{d0}} \partial\lambda_s(x,t)/\partial t\,\beta_{bmqs}\kappa_{bmq}(x)/x_s \\
&\quad + \sum_{s>a_{d0}}^{a_m}\sum_{r>a_{d0}}^{a_d}\sum_{u>a_{d0}}^{a_d} \gamma_{sr}\,\partial\eta_r(x,t)/\partial t\,\beta_{bdqu}\beta_{dmus}\kappa_{bmq}(x) \\
&= \sum_{s>a_{b0}}^{a_{d0}} \partial\lambda_s(x,t)/\partial t\,\beta_{bmqs}\kappa_{bmq}(x)/x_s \\
&\quad + \sum_{r>a_{d0}}^{a_d} \partial\eta_r(x,t)/\partial t\,\beta_{bdqr}\kappa_{bmq}(x).
\end{aligned}
$$

But for $x \in Z_m \cap \mathbf{R}_{s_0}^{a_m}$ we know that $\partial\lambda_s(x,t)/\partial t = 0$ so

$$d\kappa_{bmq}v(x,t) = \sum_{r>a_{d0}}^{a_d} \partial\eta_r(x,t)/\partial r\,\beta_{bdqr}\kappa_{bmq}(x).$$

Suppose that $r > a_{d0}$ is such that $\beta_{bdqr} \neq 0$. Then $\beta_{bdqr} > 0$. If $\beta_{dmrs_0} > 0$ then $\beta_{bmqs_0} = \beta_{bdqr}\beta_{dmrs_0} +$ nonnegative stuff so $\beta_{bmqs_0} > 0$. Hence $\kappa_{bmq}(x) = 0$ for all $x \in Z_m \cap \mathbf{R}_{s_0}^{a_m}$. If $\beta_{dmrs_0} = 0$ then $\eta_r(x,t) = \lambda_r(x,t)/\kappa_{dmr}(x)$ and $\kappa_{dmr}(x)$ is nonzero on a dense subset of $Z_m \cap \mathbf{R}_{s_0}^{a_m}$. But $\lambda_r(x,t) = \kappa_{dmr}(x)$ for $x \in Z_m \cap \mathbf{R}_{s_0}^{a_m}$ so $\eta_r(x,t) = 1$ on $Z_m \cap \mathbf{R}_{s_0}^{a_m}$. So for any $r > a_{d0}$ we have $\partial\eta_r(x,t)/\partial t\,\beta_{bdqr}\kappa_{bmq}(x) = 0$ for $x \in Z_m \cap \mathbf{R}_{s_0}^{a_m}$. Hence $d\kappa_{bmq}v(x,t) = 0$ for $x \in Z_m \cap \mathbf{R}_{s_0}^{a_m}$ and we are finished. $\qquad\square$

3. Quasialgebraic Towers are Algebraic

Definition: A resolution tower $\mathfrak{T} = \{V_i, \mathcal{A}_i, p_i\}_{\mathfrak{I}}$ is *quasialgebraic* if each $(V_i, \mathcal{A}_i)$ is an algebraic tico and the following is true. Suppose $S \subset V_{ji}$ is any fat nonsingular Zclosed subset. Let T be the smallest fat nonsingular Zclosed subset of V_j with $p_{ji}(S) \subset T$. Then we require that $p_{ji}|: S \to T$ be homotopic to an entire rational function. Hence, a resolution tower is *algebraic* if it is quasialgebraic and each $p_{ji}: V_{ji} \to V_j$ is an entire rational function.

Theorem 5.3.1 *Any quasialgebraic resolution tower of type RSE is isomorphic to an algebraic resolution tower.*

Proof: Let $\mathfrak{T} = \{V_i, \mathcal{A}_i, p_i\}_{\mathfrak{I}}$ be a quasialgebraic resolution tower of type RSE. What we will show is that there are rational diffeomorphisms $\rho_i: V_i' \to V_i$ and ϵ-automorphisms $h_i: V_i \to V_i$ so that:

1) $h_i(P) = P$ for all sheets $P \in \mathcal{A}_i$ with $i \in \mathfrak{I}$.
2) $p_{ji}': V_{ji}' \to V_j'$ is an entire rational function where $V_{ji}' = \rho_i^{-1}(V_{ji})$ and $p_{ji}' = \rho_j^{-1}h_j^{-1}p_{ji}h_i\rho_i|_{V_{ji}'}$.

Then the diffeomorphisms $h_i\rho_i: V_i' \to V_i$, $i \in \mathfrak{I}$ give an isomorphism from an algebraic resolution tower to our quasialgebraic resolution tower.

Pick a maximal index $n \in \mathfrak{I}$. By induction we may assume that there are rational diffeomorphisms $\rho_i: V_i' \to V_i$, $i \neq n$ and ϵ-automorphisms $h_i: V_i \to V_i$ so that 1) and 2) above hold for $i \neq n$. We will now induct on the number of strata in the subset K defined below.

Suppose that there are a Zclosed subset K of V_n, a rational diffeomorphism $\rho_n: V_n' \to V_n$ and an ϵ-automorphism $h_n: V_n \to V_n$ so that:

i) $h_n(P) = P$ for each sheet $P \in \mathcal{A}_n$.
ii) $p_{jn}'|_{V_{jn}' \cap \rho_n^{-1}(K)}$ is an entire rational function for each $j < n$ where $V_{jn}' = \rho_n^{-1}(V_{jn})$ and $p_{jn}' = \rho_j^{-1}h_j^{-1}p_{jn}h_n\rho_n|_{V_{jn}'}$.
iii) K is a union of strata.

If $K = V_n$ we are done, so suppose that $K \neq V_n$.

Assertion 5.3.1.1 *There is a fat nonsingular irreducible Zclosed subset S of V_n so that $S \cap K = |S \cap \mathcal{A}_n|$.*

Proof: Let $k \geq 0$ be such that all fat nonsingular Zclosed subsets of dimension $< k$ are contained in K, but some k-dimensional fat nonsingular Zclosed subsets are not. Now let S be any irreducible k-dimensional fat nonsingular Zclosed subset not contained in K. Since $|S \cap \mathcal{A}|$ is a union of fat nonsingular Zclosed subsets of dimension $< k$ we know $|S \cap \mathcal{A}| = S \cap K$. □

So pick an S as in Assertion 5.3.1.1. We will show that there is a rational diffeomorphism $r: V_n'' \to V_n'$ and an ϵ-automorphism $h: V_n \to V_n$ so that:

i') $h(P) = P$ for each sheet $P \in \mathcal{A}_n$.

ii') $p''_{jn}|_{S''}$ is an entire rational function for each j with $S \subset V_{jn}$ where $p''_{jn} = \rho_j^{-1} h_j^{-1} p_{jn} h_n h \rho_n r$ and $S'' = r^{-1} \rho_n^{-1} (S)$.

iii') $p_{in} h_n h (x) = p_{in} h_n (x)$ for all $i < n$ and $x \in K \cap V_{in}$.

Then by Lemma 3.3.10 and induction on the number of strata of K we will be done, since we may replace ρ_n by $\rho_n r$ and replace h_n by $h_n h$ and replace K by $K \cup S$.

So let us try to find this r. Pick the largest j so that $S \subset V_{jn}$ and let T be the smallest fat nonsingular Zclosed subset of V_j so that $\rho_{jn} (S) \subset T$. Note that by Lemma 3.3.11 we know that $\rho_{jn}|: (S, S \cap \mathcal{A}_n) \to (T, T \cap \mathcal{A}_j)$ is a tico map. First we shall find a rational diffeomorphism $r' : S' \to \rho_n^{-1} (S)$ so that $p'_{jn} r' : S' \to \rho_j^{-1} (T)$ is homotopic to an entire rational function. This is easy; just take the pullback of the maps $q \rho_n$ and ρ_j where $q : S \to T$ is an entire rational function homotopic to $p_{jn}| : S \to T$.

$$
\begin{array}{ccc}
S' & \xrightarrow{q'} & \rho_j^{-1} (T) \\
r' \downarrow & & \downarrow \rho_j \\
\rho_n^{-1} (S) & \xrightarrow{\rho_n} S \xrightarrow{q} & T
\end{array}
$$

Notice that since h_j and h_n are ϵ-automorphisms they are isotopic to the identity so q is also homotopic to $h_j^{-1} p_{jn} h_n$. Hence the entire rational function q' is homotopic to $\rho_j^{-1} h_j^{-1} p_{jn} h_n \rho_n r' = p'_{jn} r'$.

Let $K'' = (\rho_n r')^{-1} (K) \cap S'$. Since $p'_{jn} r'|_{K''}$ is entire rational we know by Proposition 5.1.2 that there is a rational diffeomorphism $r'' : S'' \to S'$ and an entire rational function $q'' : S'' \to \rho_j^{-1} (T)$ so that $K'' \subset S''$, the restriction $r''|_{K''}$ is the identity and so that q'' and $p'_{jn} r' r''$ are ϵ tico homotopic rel K''. By Theorem 2.8.7 we may find a rational diffeomorphism $r : V''_n \to V'_n$ so that $S'' \subset V''_n$ and $r|_{S''}$ is $r' r''$.

Let $V''_{in} = r^{-1} (V'_{in})$ and $K' = r^{-1} \rho_n^{-1} (K)$. By Lemma 5.2.2 there is an ϵ-isotopy $H_t : V''_n \to V''_n$ so that H_0 is the identity, $H_t r^{-1} \rho_n^{-1} (P) = r^{-1} \rho_n^{-1} (P)$ for all $P \in \mathcal{A}_n$, $p'_{jn} r H_1|_{S''} = q''$ and $p'_{in} r H_t (x) = p'_{in} r (x)$ for all $x \in V''_{in} \cap K'$. Now define an ϵ-automorphism $h : V_n \to V_n$ by $h = \rho_n r H_1 r^{-1} \rho_n^{-1}$. Define $p''_{in} : V''_{in} \to V'_i$ by

$$
p''_{in} = \rho_i^{-1} h_i^{-1} p_{in} h_n h \rho_n r|_{V''_{in}} = p'_{in} r H_1|_{V''_{in}}.
$$

Then $p''_{in}|_{K' \cap V''_{in}} = p'_{in} r|_{K' \cap V''_{in}}$ which is entire rational.

If $S \subset V_{in}$ then we must have $i \leq j$. If $i = j$ we have already seen that $p''_{in}|_{S''}$ is entire rational. If $i < j$ then $p''_{in}|_{S''} = p'_{ij} p''_{jn}|_{S''}$. So $p''_{in}|_{S''}$ is the composition of entire rational functions and so is entire rational. So by Lemma 3.3.10, $p''_{in}|_{r^{-1} \rho_n^{-1} ((K \cup S) \cap V_{in})}$ is entire rational for all i, and we are done. $\square$

Theorem 5.3.2 *If $\mathfrak{T}$ is any compact algebraic resolution tower then $|\mathfrak{T}|$ is homeomorphic to a real algebraic set Y. In fact if $\mathfrak{T}$ is such that $\dim V_i < \dim V_j$ for all $i < j$, then $|\mathfrak{T}|$ is isomorphic as a stratified set to the singular stratification of Y.*

Proof: Let $\mathfrak{T} = \{V_i, \mathcal{A}_i, p_i\}_{\mathfrak{J}}$. If $\dim V_i < \dim V_j$ for all $i < j$, we may assume by Lemma 4.1.2 that $\mathfrak{T}$ is well indexed with bias 0. Pick a maximal $n \in \mathfrak{J}$. Let $\mathfrak{J}' = \mathfrak{J} - \{n\}$ and let $\mathfrak{T}' = \mathfrak{T}[\mathfrak{J}']$ be the truncation of $\mathfrak{T}$, $\mathfrak{T}' = \{V_i, \mathcal{A}_i, p_i\}_{\mathfrak{J}'}$. We may assume by induction that there is a real algebraic set W and entire rational functions $q_i \colon V_i \to W$, $i \in \mathfrak{J}'$ and a homeomorphism $h \colon W \to |\mathfrak{T}'|$ so that for each i, the map $hq_i \colon V_i \to |\mathfrak{T}'|$ is just the quotient map $V_i \subset \bigcup_{j \in \mathfrak{J}'} V_i \to |\mathfrak{T}'|$ and so each $q_i(V_i)$ is an algebraic set. Furthermore, in case $\dim V_i < \dim V_j$ for all $i < j$ we also assume by induction that $q_i(V_i) \supset \operatorname{Sing} q_{i+1}(V_{i+1})$, $i = 0, \dots, n-2$ and $\operatorname{Sing} q_0(V_0) = \emptyset$.

Suppose $x \in V_{in} \cap V_{jn}$ with $j < i < n$. Then $p_{ji}p_{in}(x) = p_{jn}(x)$ so $p_{in}(x)$ and $p_{jn}(x)$ are equivalent in $|\mathfrak{T}'|$. So $hq_jp_{jn}(x) = hq_ip_{in}(x)$ so $q_jp_{jn}(x) = q_ip_{in}(x)$. Hence if $W \subset \mathbf{R}^m$, by repeated applications of Lemma 3.3.10 we may find an entire rational function $r \colon V_n \to \mathbf{R}^m$ so that $r|_{|\mathcal{A}_{in}|} = q_ip_{in}$. Pick some polynomial $s \colon V_n \to \mathbf{R}$ so that $s^{-1}(0) = |\mathcal{A}_n|$. Let $p \colon V_n \to \mathbf{R}^m \times \mathbf{R}$ be defined by $p(x) = (r(x), s(x))$. By Proposition 2.6.1 there are an algebraic set Y, entire rational functions $f \colon V_n \to Y$ and $g \colon W \to Y$ and a homeomorphism $h' \colon Y \to X$ where X is the quotient space $V_n \cup W/\sim$ and $\sim$ is the equivalence relation generated by $x \sim y$ for $x \in V_n$ and $y \in W$ with $p(x) = (y, 0)$. Furthermore $g \colon W \to g(W)$ is a birational isomorphism and if $\dim V_n > \dim W$ then $g(W) \supset \operatorname{Sing} Y$. In particular, each $gq_i(V_i)$ is an algebraic set, and if $\dim V_i < \dim V_j$, all $i < j$, then $gq_i(V_i) \supset \operatorname{Sing} gq_{i+1}(V_{i+1})$ for $i = 0, \dots, n-1$ and $\operatorname{Sing} gq_0(V_0) = \emptyset$.

Define a map $\varphi \colon \bigcup_{i \in \mathfrak{J}} V_i \to Y$ as follows. Let $\varphi|_{V_n} = f$ and let $\varphi|_{V_i} = gq_i$ for $i \neq n$. First of all φ is onto. To see this, pick any $y \in Y$. Then either $y = f(x) = \varphi(x)$ for some $x \in V_n$ or $y = g(w)$ for some $w \in W$. If $y = g(w)$ then $w = q_i(x)$ for some $x \in V_i$ for some i, so $y = \varphi(x)$. It is easy to see that if $\varphi(x) = \varphi(x')$ for $x \in V_i$, $x' \in V_j$, $j \leq i$, then either $x = x'$, $p_{ji}(x) = x'$ or $p_{ki}(x) = p_{kj}(x')$ for some $k < j$. Hence we know that $|\mathfrak{T}|$ is homeomorphic to Y. But now if $\dim V_i < \dim V_j$ for all $i < j$ then Proposition 2.2.16 implies that after perhaps modifying Y a bit we may assume that $\varphi(V_i) = \operatorname{Sing}(\varphi(V_{i+1}))$ for $i = 0, \dots, n-1$ and $\operatorname{Sing} \varphi(V_0) = \emptyset$. So we are done. $\square$

Now Lemma 4.1.3 gives the following:

Corollary 5.3.3 *Let $\mathfrak{T}$ be any compact algebraic resolution tower of type U. Then $|\mathfrak{T}|$ is isomorphic to the singular stratification of some real algebraic set.*

4. RF Towers are Quasialgebraic

In this section we will show that any topological resolution tower of type RF is quasialgebraic.

Theorem 5.4.1 *Let $\mathfrak{T}$ be a topological resolution tower of type RF. Then $\mathfrak{T}$ is isomorphic to a quasialgebraic resolution tower.*

Proof: First we make a few temporary definitions. Let $\mathcal{A}$ be a tico in V. Let $\sigma(\mathcal{A})$ be the set of connected fat submanifolds of V. We will call $B \subset \sigma(\mathcal{A})$ *admissible* if for every $S \in B$ and $T \in \sigma(\mathcal{A})$ then $S \supset T$ implies $T \in B$, in other words if $S \in B$ then any connected fat submanifold of S is also in B.

Let $|B|$ denote $\bigcup_{T \in B} T$. We call $B \subset \sigma(\mathcal{A})$ *algebraic* if V is a nonsingular real Zopen set and each $S \in B$ is a nonsingular Zclosed subset of V. We call $B \subset \sigma(\mathcal{A})$ *totally algebraic* if B is algebraic and furthermore V and each $S \in B$ has totally algebraic homology.

We now start the proof. For convenience, change the indexing set so that $\mathfrak{T} = \{V_i, \mathcal{A}_i, p_i\}_{i=0}^{n}$. Recall that for each $k < n$, $\mathfrak{T}[k]$ denotes the truncation $\mathfrak{T}[k] = \{V_i, \mathcal{A}_i, p_i\}_{i=0}^{k}$. Since each V_i is full we may as well assume each V_i is a nonsingular totally algebraic set by Corollary 2.8.11. This will in fact imply that each connected component of each V_i is irreducible and totally algebraic, but in any case by applying Corollary 2.8.11 to each component separately we can arrange this.

We say a resolution tower $\mathfrak{T}$ satisfies $C(k, B)$ if:

1) The truncation $\mathfrak{T}[k-1]$ is a quasialgebraic resolution tower. Furthermore, $\sigma(\mathcal{A}_i)$ is totally algebraic for all $i < k$.
2) $B \subset \sigma(\mathcal{A}_k)$ is admissible and totally algebraic.
3) For each $i < k$ and fat submanifold $T \subset V_i$, the map $p_{ik}|\colon |B| \cap p_{ik}^{-1}(T) \to T$ is homotopic to an entire rational function.
4) For each $S \in \sigma(\mathcal{A}_k) - B$ there is a nonsingular totally algebraic set W_S and a diffeomorphism $\lambda_S \colon S \to W_S$ so that
   a) If S is a component of V_k then $W_S = S$ and $\lambda_S = $ identity.
   b) $\lambda_S|_{|B| \cap S}$ is an entire rational function.

Assertion 5.4.1.1 *If the resolution tower $\mathfrak{T}$ satisfies $C(k, \sigma(\mathcal{A}_k))$ then it satisfies $C(k+1, \emptyset)$.*

Proof: Conditions 1), 2) and 3) are immediate. So we must only show that for each $S \in \sigma(\mathcal{A}_{k+1})$ there is a nonsingular totally algebraic set W_S diffeomorphic to S. This follows from Corollary 2.8.11.   □

Assertion 5.4.1.2 *If $\mathfrak{T}$ satisfies $C(k, B)$ with $B \neq \sigma(\mathcal{A}_k)$ then $\mathfrak{T}$ is isomorphic to some $\mathfrak{T}''$ satisfying $C(k, B'')$ for some B'' bigger than B.*

Proof: Pick some $S \in \sigma(\mathcal{A}_k) - B$ of smallest dimension. Then if $U \in \sigma(\mathcal{A}_k)$ and $U \subset S$ we must have either $U = S$ or $U \in B$. Hence $B \cup \{S\}$ is admissible.

Suppose first that $\dim S = \dim V_k$, i.e., S is a component of V_k. Then $|B| \supset |\mathcal{A}_k|$ so $\mathfrak{T}$ satisfies $C(k, \sigma(\mathcal{A}_k))$ and we are done.

Now suppose that $\dim S < \dim V_k$, so $S \subset |\mathcal{A}_k|$. Let

$$
\begin{aligned}
\mathcal{D} &= \{T \in \sigma(\mathcal{A}_i) \mid i < k \text{ and } S \subset p_{ik}^{-1}(T)\} \text{ and} \\
\mathcal{C} &= \{U \in \sigma(\mathcal{A}_k) \mid S \subset U\}.
\end{aligned}
$$

By 3), for each $T \in \sigma(\mathcal{A}_i)$ with $i < k$, $p_{ik}|: p_{ik}^{-1}(T) \to T$ is homotopic to a map f_T so that f_T restricted to $|B| \cap p_{ik}^{-1}(T)$ is an entire rational function r_T. Consider the embedding $h: S \to \prod_{T \in \mathcal{D}} T \times \prod_{U \in \mathcal{C}} W_U$ given by $\pi_T \circ h = f_T|_S$ and $\pi_U \circ h = \lambda_U|_S$ where π_T and π_U are the projections to T and W_U respectively. Notice $|S \cap \mathcal{A}_k| = S \cap |B|$ and $h|_{|S \cap \mathcal{A}_k|}$ is an entire rational function. Let $\mathcal{A}'_k$ denote $S \cap \mathcal{A}_k$. By Proposition 5.1.3 there is a nonsingular algebraic set S' with $|\mathcal{A}'_k| \subset S'$ and a diffeomorphism $\theta: S' \to S$ with $\theta|_{|\mathcal{A}'_k|} = id$ and an entire rational function $g: S' \to \prod_{T \in \mathcal{D}} T \times \prod_{U \in \mathcal{C}} W_U$ such that $g|_{|\mathcal{A}'_k|} = h|_{|\mathcal{A}'_k|}$ and g is a ϵ-close to $h \circ \theta$. In particular, if V' is the component of V_k containing S, then $\pi_{V'} \circ g$ is an embedding. By Theorem 2.8.7 there is a nonsingular algebraic set $\overline{V}_k$ and a rational diffeomorphism $\mu: \overline{V}_k \to V_k$ so that $\tau = \mu^{-1} \circ \pi_{V'} \circ g$ is a birational isomorphism of S' to a nonsingular $S'' \subset \overline{V}_k$. Since $h \circ \theta$ is ϵ-close to g and $\pi_S \circ h \circ \theta$ is a diffeomorphism we know that $\pi_S \circ g$ is a rational diffeomorphism. Hence S' is totally algebraic by Lemma 2.7.2. So S'' is totally algebraic also. The embeddings $S' \to V_k$ given by $x \mapsto \theta(x)$ and $x \mapsto \pi_{V'} \circ g$ are ϵ-close and they agree on $|\mathcal{A}'_k|$ since $\theta(x) = \pi_{V'} \circ h \circ \theta(x)$. Hence these embeddings are isotopic rel $|\mathcal{A}'_k|$. We may cover this isotopy with an isotopy $\psi_t: V_k \to V_k$, $t \in I$ rel $|B|$ so that $\psi_0 = $ identity and $\psi_1 \circ \pi_{V'} \circ g = \theta(x)$ for all $x \in S'$.

Now define a resolution tower $\mathfrak{T}'' = \{V_i'', \mathcal{A}_i'', p_i''\}_{i=0}^n$ which is isomorphic to $\mathfrak{T}$ by replacing V_k'' with $\overline{V}_k$. To be precise, let

    i. $V_k'' = \overline{V}_k$ and $V_i'' = V_i$ if $i \neq k$.
    ii. $\mathcal{A}_k'' = (\psi_1 \circ \mu)^* \mathcal{A}_k$ and $\mathcal{A}_i'' = \mathcal{A}_i$ if $i \neq k$.
    iii. $p_{jk}'' = p_{jk} \circ \psi_1 \circ \mu$, $p_{ki}'' = (\psi_1 \circ \mu)^{-1} \circ p_{ki}$ and $p_{ji}'' = p_{ji}$ for $i \neq k \neq j$.

The diffeomorphism $\psi_1 \circ \mu: \overline{V}_k \to V_k$ gives an identification of $\sigma(\mathcal{A}_k)$ with $\sigma(\mathcal{A}_k'')$. Let $B' = \{\mu^{-1}(U) \mid U \in B\} \subset \sigma(\mathcal{A}_k'')$. Let $B'' = B' \cup \{S''\} \subset \sigma(\mathcal{A}_k'')$.

We claim $\mathfrak{T}''$ satisfies $C(k, B'')$. Condition 1) is immediate. Condition 2) follows from Lemma 2.7.2. To see condition 3), pick any fat submanifold $T \subset V_i$ with $i < k$. First suppose that $T \notin \mathcal{D}$. Then

$$
|B''| \cap {p''}_{ik}^{-1}(T) = \mu^{-1}\left(|B| \cap p_{ik}^{-1}(T)\right)
$$

so the restriction of p_{ik}'' is $p_{ik} \circ \mu|$ which is homotopic to the entire rational function $r_T \circ \mu|$. Now suppose that $T \in \mathcal{D}$, so $p_{ik}(S) \subset T$. We denote

$$
K = |B''| \cap {p''}_{ik}^{-1}(T) = S'' \cup \mu^{-1}\left(|B| \cap p_{ik}^{-1}(T)\right).
$$

Recall that $p''_{ik}|_K = p_{ik} \circ \psi_1 \circ \mu|$, so our first step is to homotop $p''_{ik}|_K$ to $f_T \circ \psi_1 \circ \mu|_K$. On $\mu^{-1}\left(|B| \cap p_{ik}^{-1}(T)\right)$ we know that $f_T \circ \psi_1 \circ \mu$ is just the entire rational function $r_T \circ \mu|$. Hence by Lemma 3.3.10 we only need to homotop $f_T \circ \psi_1 \circ \mu|_{S''}$ rel $S'' \cap \mu^{-1}(|B|)$ to an entire rational function and we will be done showing 3). But

$$f_T \circ \psi_1 \circ \mu|_{S''} = f_T \circ \psi_1 \circ \pi_{V'} \circ g \circ \tau^{-1} = f_T \circ \theta \circ \tau^{-1}$$

which is ϵ-close to the entire rational function $\pi_T \circ g \circ \tau^{-1}$ and agrees with it on $S'' \cap \mu^{-1}(|B|)$. Hence they are homotopic rel $S'' \cap \mu^{-1}(|B|)$ and 3) is verified.

So we must only show condition 4). So pick any $U \in \sigma(\mathcal{A}''_k) - B''$. If $\dim U = \dim V_k$ we let $W_U = U$ and $\lambda_U =$ identity. Otherwise, let $U' = \psi_1 \mu(U) \in \sigma(\mathcal{A}_k)$ and let $W_U = W_{U'}$. If $S \not\subset U'$ we let $\lambda_U = \lambda_{U'} \psi_1 \mu$. If $S \subset U'$ then $U' \in \mathcal{C}$, so the embedding $S' \xrightarrow{\theta} S \hookrightarrow U'$ is close to $\lambda_{U'}^{-1} \pi_{U'} g$. Hence there is an ϵ-isotopy $\psi_t^U : U' \to U'$ rel $|B| \cap U'$ so that $\psi_0^U =$ identity and $\psi_1^U \circ \theta = \lambda_{U'}^{-1} \pi_{U'} g$. Define $\lambda_U = \lambda_{U'} \psi_1^U \psi_1 \mu$. We must now only check condition 4b). First we will do the case when $S \not\subset U'$. Then for $x \in |B'| \cap U$,

$$\lambda_U(x) = \lambda_{U'} \psi_1 \mu(x) = \lambda_{U'} \mu(x) = \lambda_{U'}|_{|B| \cap U'} \circ \mu(x)$$

which is an entire rational function. But since $\dim(S'' \cap U) < \dim S''$ we know that $S'' \cap U \subset |B'|$, so $|B'| \cap U = |B''| \cap U$ so this case is proven. Now suppose $S \subset U'$. For $x \in |B'| \cap U$,

$$\lambda_U(x) = \lambda_{U'} \psi_1^U \psi_1 \mu(x) = \lambda_{U'} \mu(x)$$

which is an entire rational function. For $x \in S'' \cap U = S''$,

$$\lambda_U(x) = \lambda_{U'} \psi_1^U \psi_1 \mu(x) = \lambda_{U'} \psi_1^U \theta \tau^{-1}(x) = \pi_{U'} g \tau^{-1}(x)$$

which is also an entire rational function. Hence by Lemma 3.3.10 $\lambda_U|_{|B''| \cap U}$ is entire rational. So $C(k, B'')$ is satisfied.   □

We have already shown we may as well assume $\mathfrak{T}$ satisfies $C(0, \emptyset)$. By induction and Assertion 5.4.1.1 and 5.4.1.2, we find that $\mathfrak{T}$ is isomorphic to a tower satisfying $C(n+1, \emptyset)$, i.e., a quasialgebraic resolution tower.   □

Corollary 5.4.2 *If $(V, \mathcal{A})$ is a closed smooth manifold with a regular tico such that $(V, \mathcal{A})$ is full then there exists a nonsingular algebraic set with an algebraic tico $(V', \mathcal{A}')$ which is totally algebraic such that there is a diffeomorphism $\varphi: V \to V'$ with $\varphi^* \mathcal{A}' = \mathcal{A}$. In fact, each fat submanifold S of V' has $H_*(S, \mathbf{Z}/2\mathbf{Z})$ generated by nonsingular algebraic subsets.*

CHAPTER VI

RESOLUTION TOWER STRUCTURES ON ALGEBRAIC SETS

In this chapter we show that any compact real algebraic set is homeomorphic to the realization of a (real algebraic) resolution tower of type $REFUN$. This homeomorphism is an isomorphism of stratified sets for some algebraic stratification of the algebraic set. In combination with the results in Chapter V it brings us much closer to classifying real algebraic sets in all dimensions since we showed there that any resolution tower of type $REFS$ is isomorphic to an algebraic resolution tower. Thus the only obstruction to our classifying all real algebraic sets is the difference between type UN and type S. We presume that there is an as yet unproven resolution of singularities theorem for rational functions which bridges this difference. In Section 4 of Chapter VII we prove this in low dimensions.

We should mention that although we only explicitly talk about the real algebraic case, the proof will work for complex varieties also. In particular, the following theorem is true.

Theorem 6.0 *Any complex algebraic set is homeomorphic to the realization of a (complex algebraic) resolution tower of type UNE. This homeomorphism is an isomorphism of complex stratified sets for some algebraic stratification of the algebraic set.*

The definition of a complex resolution tower is exactly analogous to the definition of a resolution tower, except that one uses complex manifolds instead of smooth manifolds. We define complex ticos and tico maps below in section 2.

Exercise: Prove Theorem 6.0 while reading through this chapter. This is a good exercise since the proof is the same as the real case except that it is easier since some steps are unnecessary. In particular, all references to type F are ignored and of course one need not resort to complexifying.   ◇

1. Uzunblowups and Fullness

In Chapter II we introduced the notion of an uzunblowup, which is a topological multiblowup which relaxes the idea of an algebraic multiblowup by allowing rational diffeomorphisms. In a similar way, we wish to extend the notion of a tico multiblowup to an algebraic setting.

Definition: We say that $\pi\colon (X',\mathcal{A}') \to (X,\mathcal{A})$ is a *tico uzunblowup* if $\pi = \pi_1 \circ \cdots \circ \pi_n$, $\pi_i\colon (X_i\mathcal{A}_i) \to (X_{i-1},\mathcal{A}_{i-1})$, $(X,\mathcal{A}) = (X_0,\mathcal{A}_0)$, $(X',\mathcal{A}') = (X_n,\mathcal{A}_n)$ and each π_i is either a tico blowup with nonsingular center or a rational diffeomorphism. In case π_i is a rational diffeomorphism then $\mathcal{A}_i = \{\,\pi_i^{-1}(A) \mid A \in \mathcal{A}_{i-1}\,\} = \pi_i^*(\mathcal{A}_{i-1})$.

Thus any algebraic tico multiblowup is a tico uzunblowup. Although any tico uzunblowup is a tico multiblowup, it might not be an algebraic tico multiblowup.

Definitions: We say that a tico uzunblowup $\pi_1 \circ \cdots \circ \pi_n$ as above is *full* if each blowup π_i is full. We say that it is *unobtrusive* if each π_i is unobtrusive. We say that it is *unskewed* if each π_i is unskewed.

The following lemma shows that an uzunblowup of a clean submanifold will induce an uzunblowup of the whole. This induced uzunblowup may not be unique. However, one can show that it will be unique up to rational diffeomorphism.

Lemma 6.1.1 *Let* $(V,\mathcal{A})$ *be an algebraic tico and let* $W \subset V$ *be a compact nonsingular Zclosed subset of V which intersects $\mathcal{A}$ cleanly. Let*

$$(W_n,\mathcal{B}_n) \xrightarrow{\pi_n} (W_{n-1},\mathcal{B}_{n-1}) \to \cdots \xrightarrow{\pi_1} (W_0,\mathcal{B}_0) = (W, W \cap \mathcal{A})$$

be any tico uzunblowup. Then there is a tico uzunblowup

$$(V_n,\mathcal{A}_n) \xrightarrow{\rho_n} (V_{n-1},\mathcal{A}_{n-1}) \to \cdots \xrightarrow{\rho_n} (V_0,\mathcal{A}_0) = (V,\mathcal{A})$$

with the same centers. In other words, if $W_i' \subset V_i$ *is the strict transform of W then there are birational isomorphisms* $h_i\colon W_i \to W_i'$ *so that*

a) W_i' *intersects* $\mathcal{A}_i$ *cleanly.*
b) $h_i^*(W_i' \cap \mathcal{A}_i) = \mathcal{B}_i$.
c) $\rho_i h_i = h_{i-1}\pi_i$.
d) $h_0 = $ *identity.*
e) ρ_i *is a rational diffeomorphism if* π_i *is a rational diffeomorphism.*
f) *If* π_i *is the blowup with center* $L_i \subset W_{i-1}$ *then* ρ_i *is the blowup with center* $h_{i-1}(L_i)$.

Furthermore, if each π_i *is an algebraic tico blowup (i.e., we have an algebraic tico multiblowup) this lemma remains true even if W is not compact.*

Proof: We may as well assume $n = 1$. If π_1 is a rational diffeomorphism, then by Proposition 2.8.7, there is a rational diffeomorphism $\rho_1\colon V_1 \to V$ so that $W_1 \subset V_1$ and $\rho_1| = \pi_1$. If π_1 is a blowup with center L, just let $V_1 = B(V,L)$

and $\rho_1 = \pi(V, L)$. Then L intersects $\mathcal{A}$ cleanly since L intersects $W \cap \mathcal{A}$ cleanly. Hence ρ_1 is a tico blowup so the results follow. $\square$

Proposition 6.1.2 *Let $(X, \mathcal{A})$ be an algebraic tico with X compact. Then there is a full algebraic tico uzunblowup $\pi\colon (Y, \mathcal{B}) \to (X, \mathcal{A})$ so that $(Y, \mathcal{B})$ is full. Furthermore, all the centers of π are irreducible and have dimension less than $\dim X$.*

Proof: We first find a nondegenerate full tico uzunblowup $\rho\colon (X', \mathcal{A}') \to (X, \mathcal{A})$ with irreducible centers so that X' is full. The proof that we can do this is identical to the proof of Theorem 2.9.5 except that we require all uzunblowups we construct to be full tico uzunblowups.

So we may as well assume that X is full. Let

$$\mathcal{C} = \{ A \in \mathcal{A} \mid (A, A \cap \mathcal{A}) \text{ is not a full tico} \}.$$

If $\mathcal{C}$ is empty we are done, so pick $C \in \mathcal{C}$. By induction on dimension, we know that there is a nondegenerate full tico uzunblowup with irreducible centers $\pi\colon (C', \mathcal{E}) \to (C, C \cap \mathcal{A})$ so that $(C', \mathcal{E})$ is full. By Lemma 6.1.1, π induces a nondegenerate full tico uzunblowup $\pi'\colon (X'', \mathcal{A}'') \to (X, \mathcal{A})$ with the same centers so that $C' \subset X''$, $\mathcal{E} = C' \cap \mathcal{A}''$ and $\pi' \mid = \pi$. By Lemma 3.4.1 we know that if

$$\mathcal{C}' = \{ A \in \mathcal{A}'' \mid (A, A \cap \mathcal{A}'') \text{ is not a full tico} \}$$

then $\mathcal{C}'$ is contained in the strict transform of $\mathcal{C} - \{C\}$. So $\mathcal{C}'$ has fewer members than $\mathcal{C}$ so after a finite number of steps we are done. $\square$

Proposition 6.1.3 *Let $(X, \mathcal{A})$ be a full algebraic tico with X compact. Let $\pi\colon (Y, \mathcal{B}) \to (X, \mathcal{A})$ be an algebraic tico uzunblowup with centers lying over some $W \subset X$. Then there is a full algebraic tico uzunblowup $\pi'\colon (Z, \mathcal{C}) \to (Y, \mathcal{B})$ with all centers irreducible and lying over $\pi^{-1}(W)$ so that $(Z, \mathcal{C})$ is full.*

Proof: We will prove this by induction on the dimension of X and the number of $B \in \mathcal{B}$ so that the tico $(B, \mathcal{B} \cap B)$ is not full. To start off, notice that if $(B, \mathcal{B} \cap B)$ is full for all $B \in \mathcal{B}$, then Y is full by Lemma 3.4.2. Hence $(Y, \mathcal{B})$ is full.

Now assume $(B, \mathcal{B} \cap B)$ is not full for some $B \in \mathcal{B}$.

Assertion 6.1.3.1 *There is a full tico uzunblowup $\pi''\colon (B', \mathcal{B}') \to (B, \mathcal{B} \cap B)$ with irreducible centers lying over $\pi^{-1}(W) \cap B$ so that $(B', \mathcal{B}')$ is full.*

Proof: First suppose that B is the strict transform of some $A \in \mathcal{A}$. Then $\pi\mid\colon (B, \mathcal{B} \cap B) \to (A, \mathcal{A} \cap A)$ is an algebraic tico uzunblowup. The assertion then follows from this Proposition by induction on $\dim X$.

Now suppose that B is not the strict transform of some $A \in \mathcal{A}$. Then $\pi(B) \subset W$. By Proposition 6.1.2 there is a full algebraic tico uzunblowup $\pi''\colon (B', \mathcal{B}') \to$

$(B, \mathcal{B} \cap B)$ with irreducible centers so that $(B', \mathcal{B}')$ is full. Since $B \subset \pi^{-1}(W)$ we know that all centers lie over $\pi^{-1}(W) \cap B = B$. □

So take a full uzunblowup π'' as in Assertion 6.1.3.1. By Lemma 6.1.1 the uzunblowup π'' induces a full tico uzunblowup $\pi''' \colon (Y', \mathcal{B}'') \to (Y, \mathcal{B})$ with irreducible centers lying over $\pi^{-1}(W) \cap B$ with $B' \subset Y'$ the strict transform of B. By Lemma 3.4.1 we know that $\mathcal{B}''$ has fewer sheets B'' so that $(B'', B'' \cap \mathcal{B}'')$ is not full. So by induction we are done. □

2. Complex Ticos and Complexifications

Definitions: We may define complex ticos in a way similar to our definition of ticos. A *complex tico* $\mathcal{A}$ in a complex manifold M is a collection of immersed complex codimension 1 proper submanifolds of M in general position. A complex tico is *regular* if all its sheets are embedded. A complex tico is *algebraic* if it is regular, if M is a nonsingular complex Zopen set and each sheet of $\mathcal{A}$ is a nonsingular complex Zclosed subset of M.

Definition: Let $(V, \mathcal{A})$ and $(W, \mathcal{B})$ be complex ticos. Then a complex analytic function $f \colon V \to W$ is a *complex tico map* if $f^{-1}(|\mathcal{B}|) \subset |\mathcal{A}|$.

Let us see why this gives the local expression analogous to real tico maps. Let $\mathbf{C}_i^n$ denote $\{(z_1, \dots, z_n) \in \mathbf{C}^n \mid z_i = 0\}$. Pick $x \in V$ and pick analytic coordinates $\psi \colon (U, 0) \to (V, x)$ and $\theta \colon (U', 0) \to (W, f(x))$ with U and U' neighborhoods of 0 in $\mathbf{C}^m$ and $\mathbf{C}^n$ so that $\psi^{-1}(|\mathcal{A}|) = \bigcup_{i=1}^{a} \mathbf{C}_i^m \cap U$, $\theta^{-1}(|\mathcal{B}|) = \bigcup_{i=1}^{b} \mathbf{C}_i^n \cap U'$, and $f\psi(U) \subset \theta(U')$. Let $\lambda_i \colon U \to \mathbf{C}$ be the i-th coordinate of $\theta^{-1}f\psi$. Since $f^{-1}(|\mathcal{B}|) \subset |\mathcal{A}|$ we know that $\lambda_i^{-1}(0) \subset U \cap \bigcup_{j=1}^{a} \mathbf{C}_j^m$ for $i \le b$. Hence by Lemma 6.2.1 below we know that $\lambda_i(z) = \prod_{j=1}^{a} z_j^{a_{ij}} \varphi_i(z)$ for some analytic φ_i with $\varphi_i(0) \ne 0$. So we get the same local expression as in the real case.

Lemma 6.2.1 *Let $f \colon U \to \mathbf{C}$ be a complex analytic map where U is some ball around the origin in $\mathbf{C}^m$. Suppose $f^{-1}(0) \subset \bigcup_{i=1}^{k} \mathbf{C}_i^m$. Then $f(z) = \prod_{i=1}^{k} z_i^{\alpha_i} g(z)$ for some integers α_i and some analytic g with $g(0) \ne 0$.*

Proof: Pick maximal α_i's so that $f(z) = \prod_{i=0}^{k} z_i^{\alpha_i} g(z)$ for some analytic g. If $g(0) \ne 0$ we are done. So suppose $g(0) = 0$. Notice $g^{-1}(0) \subset f^{-1}(0) \subset \bigcup_{i=1}^{k} \mathbf{C}_i^m$. But every irreducible component of $g^{-1}(0)$ must have codimension one. Since each $\mathbf{C}_i^m$ is irreducible and $g^{-1}(0) \subset \bigcup_{i=1}^{k} \mathbf{C}_i^m$, this means that each irreducible component K of $g^{-1}(0)$ must be $\mathbf{C}_i^m \cap U$ for some $i \le k$. But then $g(z) = z_i h(z)$ for the analytic $h(z) = \int_0^1 \partial g / \partial z_i (z_1, \dots, tz_i, \dots, \dots, z_n) \, dt$ which contradicts the maximality of the α_i's. □

Lemma 6.2.2 *Let* $f\colon V \to W$ *be an analytic function between real analytic manifolds. Let* $\mathcal{A}$ *and* $\mathcal{B}$ *be analytic ticos in* V *and* W *respectively (i.e., they are regular ticos and each sheet is an analytic submanifold). Suppose that complexifications exist and for some complexifications,* $f_{\mathbf{C}}\colon (V_{\mathbf{C}}, \mathcal{A}_{\mathbf{C}}) \to (W_{\mathbf{C}}, \mathcal{B}_{\mathbf{C}})$ *is a complex tico map. Then* $f\colon (V, \mathcal{A}) \to (W, \mathcal{B})$ *is a tico map.*

Proof: It suffices to prove this locally. So we may assume that V and W are open sets around the origin in $\mathbf{R}^n$ and $\mathbf{R}^m$ respectively, $\mathcal{A}$ is $\{\mathbf{R}_i^n \cap V\}_{i=1}^k$ and $\mathcal{B}$ is $\{\mathbf{R}_i^m \cap W\}_{i=1}^p$. But then $\mathcal{A}_{\mathbf{C}}$ is $\{\mathbf{C}_i^n \cap V_{\mathbf{C}}\}_{i=1}^k$ and $\mathcal{B}_{\mathbf{C}}$ is $\{\mathbf{C}_i^m \cap W_{\mathbf{C}}\}_{i=1}^p$. Let f_i denote the i-th coordinate of f. Then $f_{i_{\mathbf{C}}}^{-1}(0) \subset \bigcup \mathbf{C}_j^n \cap V_{\mathbf{C}}$ if $i \leq p$ so Lemma 6.2.1 implies that $f_{i\mathbf{C}}(z) = \prod z_j^{\alpha_j} g(z)$ for some integers α_j and analytic g with $g(0) \neq 0$. Plugging in z real gives us the desired result. $\square$

At some point in this chapter we will need to use Hironaka's resolution of singularities (Theorem 2.5.11) To understand his paper we must have some understanding of schemes. What follows is a dictionary which will allow the reader unfamiliar with schemes to translate the language in [**H**] to the more accessible language of complex algebraic sets. For the purposes of this chapter, the reader may consider an affine algebraic scheme X over $\mathbf{R}$ to be a complex Zopen set X which is invariant under complex conjugation. Let σ denote the complex conjugation involution. The scheme is irreducible if $X = X' \cup \sigma(X')$ for some irreducible complex Zopen set X' and it is nonsingular if X is a nonsingular complex Zopen set. A (reduced) subscheme Y of X is a complex Zclosed subset Y of X which is also invariant under complex conjugation. A morphism $f\colon X \to Y$ between schemes is an equivariant rational function, in other words a rational function whose coefficients are real (perhaps not defined on all of X).

If $D \subset X$ is a nonsingular subscheme of a scheme X then the monoidal transformation of X with center D is just the blowup of X with center D. A subscheme E of a nonsingular scheme X has only normal crossings if and only if $E = |\mathcal{A}|$ for some complex algebraic tico $\mathcal{A}$ in X so that each sheet of $\mathcal{A}$ is invariant under complex conjugation, i.e., each sheet is a nonsingular subscheme. If E has only normal crossings and $D \subset X$ is a nonsingular subscheme then E has only normal crossings with D if and only if D intersects the above $\mathcal{A}$ cleanly.

Notice that if an algebraic scheme over $\mathbf{R}$ is nonsingular and irreducible and if X is the complex variety associated to it, then either X has no real points or X is irreducible. This is because $X = X' \cup \sigma(X')$ with X' irreducible. But the intersection of two different irreducible components of an algebraic set is always contained in the singularity set by Proposition 2.2.6. Hence either $X' \cap \sigma(X') = X'$ or $X' \cap \sigma(X') = \emptyset$. But the real points of X are contained in the fixed point set of σ which is contained in $X' \cap \sigma(X')$. So if X has real points we must have $X' = \sigma(X') = X$ so X is irreducible.

If $V \subset \mathbf{R}^n$ is a real algebraic set we may associate to V an algebraic scheme over $\mathbf{R}$, namely $\mathrm{Spec}\,(\mathbf{R}[x_1,\dots,x_n]/\mathfrak{I}\,(V))$. The complex variety we identify with this scheme turns out to be $V_{\mathbf{C}}$, the complexification of V.

For example, let $X = \{(x,y) \in \mathbf{C}^2 \mid x^2 + y^2 + 1 = 0\}$. Then X is a non-singular irreducible complex algebraic set invariant under complex conjugation. It represents a nonsingular scheme over $\mathbf{R}$ with no real points. The scheme is $\mathrm{Spec}\,\big(\mathbf{R}[x,y]/\langle x^2 + y^2 + 1\rangle\big)$.

For another example, let $X = \{\,(x,y) \in \mathbf{C}^2 \mid \big(x^2 + y^2\big)^2 + 1 = 0\,\}$. Then X is a nonsingular complex algebraic set invariant under complex conjugation, but it is reducible. Its two components are

$$X_\pm = \{(x,y) \in \mathbf{C}^2 \mid x^2 + y^2 \pm \sqrt{-1} = 0\}.$$

Notice $\sigma\,(X_+) = X_-$ and $\sigma\,(X_-) = X_+$. It represents an irreducible nonsingular scheme over $\mathbf{R}$ with no real points, namely $\mathrm{Spec}\,\Big(\mathbf{R}[x,y]/\langle(x^2+y^2)^2 + 1\rangle\Big)$.

For a nonsingular scheme with real points, let $X = \{(x,y) \in \mathbf{C}^2 \mid x^2+y^2-1 = 0\}$. Then X is a nonsingular irreducible complex algebraic set invariant under complex conjugation. The scheme is $\mathrm{Spec}\,\big(\mathbf{R}[x,y]/\langle x^2 + y^2 - 1\rangle\big)$.

Lemma 6.2.3 *Let* $(X,\mathcal{A})$ *be a complex algebraic tico and let* $W \subset X$ *be a complex Zclosed subset. Suppose that* W, X *and each* $S \in \mathcal{A}$ *are all invariant under complex conjugation. Then there is a complex algebraic tico multiblowup* $\pi\colon (X',\mathcal{A}') \to (X,\mathcal{A})$ *so that the strict transform* W' *of* W *is nonsingular and* W' *intersects* $\mathcal{A}'$ *cleanly. Furthermore, all the centers of the multiblowup are invariant under complex conjugation, have irreducible real points and lie over the algebraic set* Z *where* $Z = Z' \cup \mathrm{Sing}\,W$ *and*

$$Z' = \{\,y \in \mathrm{Nonsing}\,W \mid W \text{ does not intersect } \mathcal{A} \text{ cleanly near } y\,\}.$$

Proof: This is just an interpretation of Hironaka's resolution theorem in [**H**]. First of all we may as well assume that X is either irreducible or the disjoint union of an irreducible algebraic set and its complex conjugate. Now think of X as being a nonsingular irreducible algebraic scheme over $\mathbf{R}$. Think of W, Z and the sheets of $\mathcal{A}$ as being reduced subschemes. Then in the terminology of [**H**], $(((|\mathcal{A}|;\,;W)\,,W - Z)$ is a resolution datum of type $\mathcal{R}_I^{N,n}$ with open restriction on X, where $N = \dim_{\mathbf{C}} X$ and $n = \dim_{\mathbf{C}} W$. By Theorem $I_2^{N,n}$ of [**H**] there is a finite succession of monoidal transformations $f = \{f_i\colon X_{i+1} \to X_i\}$, $0 \le i < m$ and $X_0 = X$ which is permissible for $(((|\mathcal{A}|;\,;W)\,,W - Z)$ such that $f^*\,(((|\mathcal{A}|;\,;W))$ is resolved everywhere. Let B_i be the center of f_i and let $\mathcal{R}_i = (((E_i;\,;W_i)\,,U_i)$ for $0 \le i \le m$ be the resolution data on X_i which is the transform of $(((|\mathcal{A}|;\,;W)\,,W - Z)$. Since f is permissible we know that each B_i is irreducible and nonsingular, E_i has only normal crossings with B_i and $B_i \supset W_i - U_i$. By definition of the transform in [**H**] we know that $E_0 = |\mathcal{A}|$, $W_0 = W$, $U_0 = W - Z$, $E_{i+1} = f_i^{-1}\,(E_i \cup B_i) = f_i^{-1}\,(B_i)\cup$ the strict transform of E_i. Also W_{i+1} is the strict transform of W_i and $U_{i+1} = f_i^{-1}\,(U_i)$. Since $f^*\,(((|\mathcal{A}|;\,;W)) = (E_m;\,;W_m)$

is resolved everywhere we know that W_m is nonsingular and E_m has only normal crossings with W_m. Since $U_{i+1} = f_i^{-1}(U_i)$ for all i we know that each center B_i lies over Z.

We claim that f actually gives a complex tico multiblowup and if $\mathcal{A}_i$ is the tico in X_i which is the total transform of $\mathcal{A}$, then $|\mathcal{A}_i| = E_i$. This is because $|\mathcal{A}_{i+1}| = f_i^{-1}(|\mathcal{A}_i| \cup B_i) = f_i^{-1}(E_i \cup B_i) = E_{i+1}$ if $|\mathcal{A}_i| = E_i$ and f_i is a tico blowup since E_i has only normal crossings with B_i, hence B_i intersects $\mathcal{A}_i$ cleanly. $\qquad\square$

Lemma 6.2.4 *Let $\pi\colon (X', \mathcal{A}') \to (X_{\mathbf{C}}, \mathcal{A}_{\mathbf{C}})$ be a complex algebraic tico multiblowup of a complexification of a real algebraic tico $(X, \mathcal{A})$. Suppose that the centers are all invariant under complex conjugation. Then there is a real algebraic tico multiblowup $\rho\colon (X'', \mathcal{A}'') \to (X, \mathcal{A})$ so that $(X'', \mathcal{A}'')$ is the real points of $(X', \mathcal{A}')$ and the centers of ρ are the real points of the centers of π.*

Proof: Depending on ones point of view, this result is either tautological or easy. Our official point of view for this book is that real algebraic sets are all affine. This simplifies things and keeps the results on a concrete, down to earth level. However, as we pointed out in Chapter II, we cannot do the same with complex algebraic sets because the blowup of an affine complex algebraic set is not affine. However, if we replace all the blowups in the multiblowup π with the semiblowups of Chapter II we get this lemma. $\qquad\square$

From Lemma 6.2.3 we can derive the following consequence for real algebraic sets.

Lemma 6.2.5 *Let $(X, \mathcal{A})$ be an algebraic tico and suppose $W \subset X$ is a Zclosed subset. Then there is an algebraic tico multiblowup $\pi\colon (Y, \mathcal{B}) \to (X, \mathcal{A})$ whose centers are irreducible and lie over W so that $\pi^{-1}(W) = |\mathcal{B}'|$ for some $\mathcal{B}' \subset \mathcal{B}$.*

Proof: First consider the real case. Let $(X_{\mathbf{C}}, \mathcal{A}_{\mathbf{C}})$ be a complexification of $(X, \mathcal{A})$. Let $W_{\mathbf{C}} \subset X_{\mathbf{C}}$ be the complexification of W. By Lemma 6.2.3 there is an equivariant complex algebraic tico multiblowup $\pi'\colon (X', \mathcal{A}') \to (X_{\mathbf{C}}, \mathcal{A}_{\mathbf{C}})$ so that the centers all lie over $W_{\mathbf{C}}$ and have irreducible real points and so that the strict transform W' of $W_{\mathbf{C}}$ is nonsingular and intersects $\mathcal{A}'$ cleanly. By Lemma 6.2.4 this induces a real tico multiblowup $\pi''\colon (X'', \mathcal{A}'') \to (X, \mathcal{A})$ with all centers irreducible and lying over W so that the strict transform W'' of W is nonsingular and intersects $\mathcal{A}''$ cleanly. Note that $\pi^{-1}(W) = W'' \cup |\mathcal{B}''|$ for some $\mathcal{B}'' \subset \mathcal{A}''$ since $\pi^{-1}(W)$ contains the strict transform of the inverse image of all centers of the multiblowup. In particular, $\mathcal{B}'' = \mathcal{A}'' - \{$ strict transforms of $A \in \mathcal{A}\}$. Let $(Y, \mathcal{B}) \to (X'', \mathcal{A}'')$ be the tico multiblowup which blows up the irreducible components of W'' one by one. Then the composition $(Y, \mathcal{B}) \to (X'', \mathcal{A}'') \to (X, \mathcal{A})$ satisfies the conclusions of our lemma.

The complex version is done similarly only of course it is not necessary to complexify.   □

Lemma 6.2.6 *Suppose* $f: V \to W$ *is an entire rational function between two nonsingular real algebraic sets and suppose* $\mathcal{A}$ *and* $\mathcal{B}$ *are algebraic ticos in* V *and* W *respectively. Then there is an algebraic tico multiblowup* $\pi: V' \to V$ *with irreducible centers so that if* $\mathcal{A}'$ *is the total transform of* $\mathcal{A}$ *then* $f\pi: (V', \mathcal{A}') \to (W, \mathcal{B})$ *is a tico map.*

Furthermore, if $Z \subset V$ *is a closed subset so that the restriction*

$$f|: (V - Z, (V - Z) \cap \mathcal{A}) \to (W, \mathcal{B})$$

is a tico map then we may assume the centers of π *all lie over* Z.

Proof: Let $(V_{\mathbf{C}}, \mathcal{A}_{\mathbf{C}})$ be a complexification of $(V, \mathcal{A})$ and let $(W_{\mathbf{C}}, \mathcal{B}_{\mathbf{C}})$ be a complexification of $(W, \mathcal{B})$. Let $f_{\mathbf{C}}$ be the complexification of f. Then after perhaps shrinking $V_{\mathbf{C}}$ a bit we may assume that $f_{\mathbf{C}}: V_{\mathbf{C}} \to W_{\mathbf{C}}$ is defined on all of $V_{\mathbf{C}}$. Let $Y \subset V_{\mathbf{C}}$ be the Zariski closure of $f_{\mathbf{C}}^{-1}(|\mathcal{B}_{\mathbf{C}}|) - |\mathcal{A}_{\mathbf{C}}|$.

Assertion 6.2.6.1 f *is locally a tico map at* $y \in V$ *if and only if* $y \in V - Y$. *In particular,* Z *contains the real points of* Y.

Proof: Notice that

$$f_{\mathbf{C}}|: (V_{\mathbf{C}} - Y, (V_{\mathbf{C}} - Y) \cap \mathcal{A}_{\mathbf{C}}) \to (W_{\mathbf{C}}, \mathcal{B}_{\mathbf{C}})$$

is a complex tico map since

$$f_{\mathbf{C}}^{-1}(|\mathcal{B}_{\mathbf{C}}|) \cap (V_{\mathbf{C}} - Y) \subset |\mathcal{A}_{\mathbf{C}}|.$$

Hence by Lemma 6.2.2,

$$f|: (V - Y, (V - Y) \cap \mathcal{A}) \to (W, \mathcal{B})$$

is a real tico map. So one direction is proven. Now suppose f is locally a tico map at $y \in V$. Pick integers a and b and pick polynomials $p_i: V \to \mathbf{R}$, $i = 1, \ldots, a$ and $q_i: W \to \mathbf{R}$, $i = 1, \ldots, b$ so that $|\mathcal{A}| \cap U = \bigcup_{i=1}^{a} p_i^{-1}(0) \cap U$ and $|\mathcal{B}| \cap U' = \bigcup_{i=1}^{b} q_i^{-1}(0) \cap U'$ for some neighborhood U of y in V and U' of $f(y)$ in W and so that $(p_1, \ldots, p_a) = V \to \mathbf{R}^a$ has rank a at y and $(q_1, \ldots, q_b): W \to \mathbf{R}^b$ has rank b at $f(y)$. After shrinking U a bit, we know from the fact that $f|_U$ is a tico map that $q_i f(x) = \prod_{j=1}^{a} p_j(x)^{\alpha_{ij}} r_i(x)$ for some nonzero smooth functions $r_i: U \to \mathbf{R} - 0$. By repeated applications of Lemma 2.2.11 we see that r_i is an entire rational function, and U may be taken to be Zariski open. Notice that $q_{i\mathbf{C}} f_{\mathbf{C}}(x) = \prod_{j=1}^{a} p_{j\mathbf{C}}(x)^{\alpha_{ij}} r_{i\mathbf{C}}(x)$ and $r_{i\mathbf{C}}(y) \neq 0$. Hence

$$U'' \cap f_{\mathbf{C}}^{-1}(|\mathcal{B}_{\mathbf{C}}|) \subset U'' \cap \bigcup_{j=1}^{a} p_{j\mathbf{C}}^{-1}(0) \subset U'' \cap |\mathcal{A}_{\mathbf{C}}|$$

for some Zariski open neighborhood U'' of y in $V_{\mathbf{C}}$. So $Y \subset V_{\mathbf{C}} - U''$ so $y \in V - Y$.

□

By Lemma 6.2.3 there is an equivariant complex algebraic tico multiblowup $\pi' \colon (V'', \mathcal{A}'') \to (V_{\mathbf{C}}, \mathcal{A}_{\mathbf{C}})$ so that all centers lie over Y and have irreducible real points, the strict transform Y'' of Y is nonsingular and Y'' intersects $\mathcal{A}''$ cleanly. After further blowups with centers the irreducible components of Y'' we may as well assume that Y'' is empty. Let $\pi \colon (V', \mathcal{A}') \to (V, \mathcal{A})$ be the corresponding real algebraic tico multiblowup obtained from Lemma 6.2.4. Its centers lie over the real points of Y, hence over Z. Also V'' is a complexification of V', $\mathcal{A}''$ minus the complexification $\mathcal{A}'_{\mathbf{C}}$ of $\mathcal{A}'$ is a collection of sheets with no real points and π' is the complexification of π. Notice that $\pi'^{-1} f_{\mathbf{C}}^{-1} (|\mathcal{B}_{\mathbf{C}}|) \subset \pi'^{-1} (Y \cup |\mathcal{A}_{\mathbf{C}}|) = |\mathcal{A}''|$. Hence $f_{\mathbf{C}}\pi' \colon (V'', \mathcal{A}'') \to (W_{\mathbf{C}}, \mathcal{B}_{\mathbf{C}})$ is a complex tico map, so $f\pi \colon (V', \mathcal{A}') \to (W, \mathcal{B})$ is a tico map by Lemma 6.2.2.   □

Proposition 6.2.7 *Let $f \colon X \to Y$ be an entire rational function between real Zopen sets X and Y. Suppose X is nonsingular and $\mathcal{A}$ is an algebraic tico in X. Let $\pi \colon Y' \to Y$ be an uzunblowup with centers lying over $Z \subset Y$.*

Then there is a tico uzunblowup $\mu \colon X' \to X$ and an entire rational function $g \colon X' \to Y'$ so that $\pi g = f\mu$ and all the centers of μ are irreducible and lie over $f^{-1}(Z)$. Furthermore, if Y' is nonsingular and $\mathcal{D}'$ is an algebraic tico in Y' with $\pi(|\mathcal{D}'|) \subset Z$ we may guarantee that g is a tico map.

Proof: We will prove only the first part. The second part, that g can be taken to be a tico map, will then follow from Lemma 6.2.6.

It suffices to consider the case where π is a single rational diffeomorphism or blowup. If π is a rational diffeomorphism then Y must be nonsingular so we may just take the pullback.

$$
\begin{array}{ccc}
X' & \xrightarrow{g} & Y' \\
\mu \downarrow & & \downarrow \pi \\
X & \xrightarrow{f} & Y
\end{array}
$$

More explicitly, $X' = \{ (x, y) \in X \times Y' \mid \pi(y) = f(x) \}$, $g(x, y) = y$ and $\mu(x, y) = x$. Then μ is a rational diffeomorphism, $\mu^{-1}(x) = (x, \pi^{-1}f(x))$.

If π is a blowup, this Lemma follows from Chapter 0, Section 5 of [**H**] using the same translation between schemes and real algebraic sets we used in Lemma 6.2.3. It is amusing to note that we can also prove this using Proposition 3.3.9 and resolution of singularities. Suppose the center of π is L and $\mathfrak{I}(L) = \langle h_1, \dots, h_n \rangle$. Think of X as a scheme over $\mathbf{R}$ or equivalently, take its complexification and do everything equivariantly. By a Sard's theorem argument, there is a linear transformation $A \colon \mathbf{R}^n \to \mathbf{R}^n$ and an algebraic tico $\mathcal{C}$ in $X - f^{-1}(L)$ so that $|\mathcal{A}| - f^{-1}(L) \subset |\mathcal{C}|$ and so $f' \colon (X - f^{-1}(L), \mathcal{C}) \to (\mathbf{R}^n, \mathcal{E})$ is a tico map where $f' =$

$A(h_1 \circ f, \ldots, h_n \circ f)$ and $\mathcal{E} = \{\mathbf{R}_i^n\}_{i=1}^n$. By Lemma 6.2.5, we may assume after doing a tico multiblowup of the tico $(X, \mathcal{A})$ that $f^{-1}(L)$ is a tico. After further multiblowups with centers lying over $f^{-1}(L)$, we may assume that $\mathcal{C}$ extends to a tico on X. But now Proposition 3.3.9 says that after further multiblowups, the map f lifts. A more careful argument could show that in fact the centers of blowups not only lie over $f^{-1}(L)$, but lie over the set of points at which $f^*(\mathcal{I}(L))$ is not locally principal. Of course the the proof of resolution of singularities in [**H**] is entwined with the proof that such lifts exist, so strictly speaking this proof is circular. But there are other proofs of resolution of singularities, c.f., [**BM**]. $\square$

Proposition 6.2.8 *Let X and Y be real Zopen sets. Suppose X is nonsingular and $\mathcal{A}$ is an algebraic tico in X. Let $f \colon |\mathcal{B}| \to Y$ be an entire rational function for some $\mathcal{B} \subset \mathcal{A}$. Suppose $\lambda \colon Y' \to Y$ is an uzunblowup with centers lying over some $Z \subset Y$ and suppose $W \subset X$ is a Zclosed subset containing $f^{-1}(Z)$. Then there is a tico uzunblowup $\pi \colon (X', \mathcal{A}') \to (X, \mathcal{A})$ and an entire rational function $h \colon |\mathcal{B}'| \to Y'$ where $\mathcal{B}'$ is the strict transform of $\mathcal{B}$ so that*

    a) *The centers of π are all irreducible and lie over W.*
    b) *$\lambda h = f\pi|$.*
    c) *$\pi^{-1}(W) = |\mathcal{E}|$ for some algebraic tico $\mathcal{E} \subset \mathcal{A}'$.*
    d) *If Y' is nonsingular and $\mathcal{D}$ is an algebraic tico in Y' with $\lambda(|\mathcal{D}|) \subset Z$ then h is a mico.*
    e) *If $(X, \mathcal{A})$ is full, then $(X', \mathcal{A}')$ is full also.*

Proof: Suppose that for some $\mathcal{C} \subset \mathcal{B}$ (possibly empty) there is an entire rational function $g \colon |\mathcal{C}| \to Y'$ so that $\lambda g = f|$. Furthermore, if the conditions of d) are met, then suppose $g| \colon (T, T \cap (\mathcal{A} - \mathcal{B})) \to (Y', \mathcal{D})$ is a tico map for all $T \in \mathcal{C}$. If $\mathcal{C} \neq \mathcal{B}$, pick any $S \in \mathcal{B} - \mathcal{C}$. By Proposition 6.2.7 we know there is an algebraic tico uzunblowup $\mu \colon (S', \mathcal{E}') \to (S, S \cap \mathcal{A})$ and an entire rational function $r \colon S' \to Y'$ so that the centers of μ are irreducible and lie over $S \cap W$, so that $\lambda r = f\mu$ and so that if the conditions of d) hold, $r \colon (S', \mathcal{E}') \to (Y', \mathcal{D})$ is a tico map. Let $\mathcal{E}'' \subset \mathcal{E}'$ be the strict transform of $S \cap \mathcal{B}$. For reasons which will become apparent later, we would like to have $\mu(L') \not\subset W$ for each irreducible component L' of each sheet $L \in \mathcal{E}''$. This is easy to arrange after further blowups since if $\mu(L') \subset W$ we may just blow up S' with center L'. Since L' has codimension one, this blowup map is just the identity. However, the strict transform of L is $L - L'$ so we have improved matters. So we may as well assume that $\mu(L') \not\subset W$ for each irreducible component of L' of each sheet $L \in \mathcal{E}''$. The reader may well be puzzled as to why we bother to blow up L' since the blowup map is just the identity. The reason is that we will eventually extend the blowup to all of X, then this blowup will become meaningful.

Assertion 6.2.8.1 *Let K be the strict transform of $S \cap |\mathcal{C}|$. Then $r|_K = g\mu|_K$.*

Proof: Pick any $x \in K$. Then $x \in L$ for some strict transform L of a sheet of $S \cap \mathcal{C}$. Let L' be the irreducible component of L which contains x. Since $\mu(L') \not\subset W$ we know $\dim\left(\mu^{-1}(W) \cap L'\right) < \dim L'$ by Lemma 2.2.9. So we may take $x' \in L' - \mu^{-1}(W)$ with x' close to x. But $f\mu(x') \notin Z$, hence $\lambda^{-1}f\mu(x')$ has just one point so $r(x') = g\mu(x')$ since $\lambda r(x') = f\mu(x') = \lambda g\mu(x')$. So $r(x) = g\mu(x)$ by continuity.   □

By Lemma 6.1.1 we have an extension $\mu\colon (X'', \mathcal{A}'') \to (X, \mathcal{A})$ of μ to a tico uzunblowup with the same centers so $S' \subset X''$ and $\mathcal{E}' = S' \cap \mathcal{A}''$. Note then that if $\mathcal{C}' \subset \mathcal{A}''$ is the strict transform of $\mathcal{C}$ then $S' \cap |\mathcal{C}'| = K$. By Lemma 3.3.11 there is an entire rational function $g'\colon S' \cup |\mathcal{C}'| \to Y'$ so that $g'|_{S'} = r$ and $g'|_{|\mathcal{C}'|} = g\mu|$. But then $\lambda g' = f\mu$.

Assertion 6.2.8.2 *If the conditions of d) are satisfied then for $T' = S'$ and for all $T' \in \mathcal{C}'$, $g'\colon (T', T' \cap (\mathcal{A}'' - \mathcal{B}'')) \to (Y', \mathcal{D})$ is a tico map where $\mathcal{B}''$ is the strict transform of $\mathcal{B}$.*

Proof: Take $T' \in \mathcal{C}'$ which is the strict transform of some $T \in \mathcal{C}$. We know that the restriction $g|\colon (T, T \cap (\mathcal{A} - \mathcal{B})) \to (Y', \mathcal{D})$ is a tico map. But then $g'|\colon (T', T' \cap (\mathcal{A}'' - \mathcal{B}'')) \to (Y', \mathcal{D})$ is a tico map by Lemmas 3.2.4 and 3.3.4. So now consider $T' = S'$. We know that $g'| = r\colon (S', S' \cap \mathcal{A}'') \to (Y', \mathcal{D})$ is a tico map since $\mathcal{E}' = S' \cap \mathcal{A}''$. Take any irreducible component L' of any sheet $L \in \mathcal{E}'' = S' \cap \mathcal{B}''$. If $r(L') \subset |\mathcal{D}|$ then $f\mu(L') = \lambda r(L') \subset \lambda(|\mathcal{D}|) \subset Z$ so $\mu(L') \subset f^{-1}(Z) \subset W$, a contradiction. So $r(L') \not\subset |\mathcal{D}|$, so $\dim\left(L' \cap r^{-1}(|\mathcal{D}|)\right) < \dim L$ by Lemma 2.2.9. Hence $r^{-1}(|\mathcal{D}|) \subset |\mathcal{E}' - \mathcal{E}''|$ so $r\colon (S', \mathcal{E}' - \mathcal{E}'') \to (Y', \mathcal{D})$ is a tico map by Lemma 3.2.5. But $\mathcal{E}' - \mathcal{E}'' = S' \cap (\mathcal{A}'' - \mathcal{B}'')$.   □

So after replacing $(X, \mathcal{A})$ by $(X'', \mathcal{A}'')$, f by $f\mu$, W by $\mu^{-1}(W)$, $\mathcal{B}$ by $\mathcal{B}''$, g by g' and $\mathcal{C}$ by $\mathcal{C}' \cup \{S'\}$ we see we have enlarged $\mathcal{C}$. Hence we may as well assume that $\mathcal{C} = \mathcal{B}$. By Lemma 6.2.5 we may as well assume that W is the realization of an algebraic tico $\mathcal{E} \subset \mathcal{A}$. By Proposition 6.1.3, if our original $(X, \mathcal{A})$ was full we can blow up further so that $(X', \mathcal{A}')$ will be full. So we have proven a), b), c) and e). But d) follows from Lemma 3.7.1.   □

3. Extending Algebraic Resolution Towers

Definition: An *uzunblowup* of an algebraic resolution tower $\mathfrak{T} = \{V_i, \mathcal{A}_i, p_i\}_\mathcal{J}$ is an algebraic resolution tower $\mathfrak{T}' = \{V_i', \mathcal{A}_i', p_i'\}_\mathcal{J}$ and an unobtrusive, unskewed tico uzunblowup $\pi_i\colon (V_i', \mathcal{A}_i') \to (V_i, \mathcal{A}_i)$ for each $i \in \mathcal{J}$ so that the uzunblowups π_i are compatible with the maps in p_i and p_i'. More precisely, $\mathcal{A}_{ji}'$ is the tower transform of $\mathcal{A}_{ji}$ and $\pi_j p_{ji}'(x) = p_{ji}\pi_i(x)$ for all $x \in |\mathcal{A}_{ji}'|$.

In other words, by forgetting the algebraic structure the uzunblowups π_i become multiblowups and we ask that $\{\pi_i\}$ be a multiblowup of $\mathfrak{T}$.

We first show that an uzunblowup of part of a resolution tower can be extended to the whole tower.

Proposition 6.3.1 *Let* $\mathfrak{T}$ *be an algebraic resolution tower and let* $\{\pi_i\}_{\mathfrak{I}'} \colon \mathfrak{T}' \to$ $\mathfrak{T}[\mathfrak{I}']$ *be an uzunblowup of the truncation* $\mathfrak{T}[\mathfrak{I}']$ *for some lower index subset* $\mathfrak{I}' \subset \mathfrak{I}$. *Then there is an uzunblowup* $\{\pi_i'\}_{\mathfrak{I}} \colon \mathfrak{T}'' \to \mathfrak{T}$ *so that* $\mathfrak{T}''[\mathfrak{I}'] = \mathfrak{T}'$ *and* $\pi_i' = \pi_i$ *for all* $i \in \mathfrak{I}'$. *Furthermore, if the centers of each* π_i *lie over some* $Z_i \subset V_i$ *for all* $i \in \mathfrak{I}'$ *then the centers of each* π_i' *lie over* Z_i' *where* $Z_i' = Z_i \cap |\mathcal{A}_i|$ *for all* $i \in \mathfrak{I}'$ *and* $Z_i' = \bigcup_{j<i} p_{ji}^{-1}(Z_j')$ *if* $i \in \mathfrak{I} - \mathfrak{I}'$. *Additionally, if* $\mathfrak{T}$ *and* $\mathfrak{T}'$ *are both type F, we may may require that* $\mathfrak{T}''$ *have type F also.*

Proof: Let $\mathfrak{T} = \{V_i, \mathcal{A}_i, p_i\}_{\mathfrak{I}}$ and $\mathfrak{T}' = \{V_i', \mathcal{A}_i', p_i'\}_{\mathfrak{I}'}$. It suffices to prove the case $\mathfrak{I}' = \mathfrak{I} - \{n\}$ for some maximal index $n \in \mathfrak{I}$. By repeated applications of Proposition 6.2.8 (with $\mathcal{B} := \mathcal{A}_{jn}$, $Y := V_j$, $f := p_{jn}$ and $Y' := V_j'$ for $j < n$), there is a tico uzunblowup $\pi_n \colon (V_n', \mathcal{A}_n') \to (V_n, \mathcal{A}_n)$ with unobtrusive unskewed centers so that if $\mathcal{A}_{jn}'$ is the tower transform of $\mathcal{A}_{jn}$ then there are entire rational functions $p_{jn}' \colon |\mathcal{A}_{jn}'| \to V_j'$ so that $\pi_j p_{jn}' = p_{jn} \pi_n|$, $0 \le j < n$ and so that the centers lie over Z_n'. By Proposition 6.2.8e or 6.1.3 we may also assume that $(V_n', \mathcal{A}_n')$ is full if $(V_n, \mathcal{A}_n)$ is full. Now the Proposition follows from Lemma 4.2.3, setting $V_i'' = V_i'$, $\mathcal{A}_{ji}'' = \mathcal{A}_{ji}'$ and $p_{ji}'' = p_{ji}'$ for $j < i \le n$. $\qquad\square$

Lemma 6.3.2 *Let* $\mathfrak{T} = \{V_i, \mathcal{A}_i, p_i\}_{\mathfrak{I}}$ *be a compact algebraic resolution tower and suppose that for each* $i \in \mathfrak{I}$ *we have a subtico* $\mathcal{B}_i \subset \mathcal{A}_i$ *and an entire rational function* $q_i \colon |\mathcal{B}_i| \to Y$ *to some algebraic set* Y. *Suppose also that the maps* q_i *are compatible with each other. In other words,* $q_j p_{ji}(x) = q_i(x)$ *for all* $x \in |\mathcal{B}_i| \cap p_{ji}^{-1}(|\mathcal{B}_j|)$ *if* $j < i$.

$$
\begin{array}{ccccc}
 & & |\mathcal{B}_i| & & \\
 & \nearrow & & \searrow^{q_i} & \\
p_{ji}^{-1}(|\mathcal{B}_j|) \cap |\mathcal{B}_i| & & & & Y \\
 & \searrow^{p_{ji}} & & \nearrow^{q_j} & \\
 & & |B_j| & &
\end{array}
$$

Let $Z \subset Y$ *and* $Z_i \subset |\mathcal{A}_i|$ *be algebraic sets so that* $Z_i \supset q_i^{-1}(Z)$ *for all* $i \in \mathfrak{I}$ *and* $Z_i \supset p_{ji}^{-1}(Z_j)$ *for all* $j < i$. *Suppose* $\lambda \colon Y' \to Y$ *is an uzunblowup with centers lying over* Z. *Then there is an uzunblowup* $\{\pi_i\}_{\mathfrak{I}} \colon \mathfrak{T}' \to \mathfrak{T}$ *so that:*

a) *The centers of* π_i *all lie over* Z_i.

b) *Let* $\mathcal{B}_i'$ *be the strict transform* $\mathcal{B}_i$. *Then there are entire rational functions* $h_i \colon |\mathcal{B}_i'| \to Y'$ *so that* $\lambda h_i = q_i \pi_i|$.

c) *Each* $\pi_i^{-1}(Z_i)$ *is the realization of an algebraic tico* $\mathcal{E}_i$.

d) *If* Y' *is nonsingular and* $\mathcal{D}$ *is an algebraic tico in* Y' *with* $\lambda(|\mathcal{D}|) \supset Z$ *then each* h_i *is a mico.*

e) *If* $\mathfrak{T}$ *has type F then* $\mathfrak{T}'$ *also has type F.*

$$
\begin{array}{ccccccc}
|\mathcal{E}_i| & \hookrightarrow & V_i' & \hookleftarrow & |\mathcal{B}_i'| & \xrightarrow{h_i} & Y' \\
\downarrow & & \downarrow \pi_i & & \downarrow & & \downarrow \lambda \\
Z_i & \hookrightarrow & V_i & \hookleftarrow & |\mathcal{B}_i| & \xrightarrow{q_i} & Y \\
\cup & & & & \cup & & \cup \\
q_i^{-1}(Z) & & = & & q_i^{-1}(Z) & \longrightarrow & Z
\end{array}
$$

Proof: Let n be a maximal index in $\mathfrak{J}$ and let $\mathfrak{J}^* = \mathfrak{J} - \{n\}$. By induction on the size of $\mathfrak{J}$, there is an uzunblowup $\{\pi_i^*\}_{\mathfrak{J}^*} \colon \mathfrak{T}^* \to \mathfrak{T}[\mathfrak{J}^*]$ satisfying a)-e). By Proposition 6.3.1 we may extend this to an uzunblowup $\{\pi_i''\}_{\mathfrak{J}} \colon \mathfrak{T}'' \to \mathfrak{T}$ so that $\mathfrak{T}''[\mathfrak{J}^*] = \mathfrak{T}^*$, $\pi_i'' = \pi_i^*$ for all $i \in \mathfrak{J}^*$ and the centers of π_n'' lie over $\bigcup_{i<n} p_{in}^{-1}(Z_i) \subset Z_n$. Let $\mathcal{B}_n''$ and $\mathcal{A}_n''$ be the strict transforms of $\mathcal{B}_n$ and $\mathcal{A}_n$ over π_n''. By Proposition 6.2.8 there is a tico uzunblowup $\pi_n' \colon (V_n', \mathcal{A}_n') \to (V_n'', \mathcal{A}_n'')$ and an entire rational function $h_n \colon |\mathcal{B}_n'| \to Y'$ where $\mathcal{B}_n'$ is the strict transform of $\mathcal{B}_n''$ so that a)-e) hold. Let $(V_i', \mathcal{A}_i') = (V_i'', \mathcal{A}_i'')$ and $\pi_i' = $ the identity for $i \neq n$. Then by Lemma 4.2.3, $\{\pi_i'\}_{\mathfrak{J}} \colon \mathfrak{T}' \to \mathfrak{T}''$ is an uzunblowup, so setting $\pi_i = \pi_i'' \circ \pi_i'$ we are done. $\square$

The following does two things, it lifts a rational tower morphism to a blowup and makes it a tico tower morphism as well.

Proposition 6.3.3 *Let* $\pi \colon \mathfrak{T}' \to \mathfrak{T}$ *be an uzunblowup of algebraic resolution towers and let* $f \colon \mathfrak{T}'' \to \mathfrak{T}$ *be a rational tower morphism of algebraic resolution towers. Then there is an uzunblowup* $\mu \colon \mathfrak{T}''' \to \mathfrak{T}''$ *and a rational tico tower morphism* $g \colon \mathfrak{T}''' \to \mathfrak{T}'$ *so that the following commutes.*

$$
\begin{array}{ccc}
\mathfrak{T}''' & \xrightarrow{g} & \mathfrak{T}' \\
\downarrow \mu & & \downarrow \pi \\
\mathfrak{T}'' & \xrightarrow{f} & \mathfrak{T}
\end{array}
$$

Proof: Let $\mathfrak{T} = \{V_i, \mathcal{A}_i, p_i\}_{\mathfrak{J}}$, let $\mathfrak{T}' = \{V_i', \mathcal{A}_i', p_i'\}_{\mathfrak{J}}$ and let $\mathfrak{T}'' = \{V_i'', \mathcal{A}_i'', p_i''\}_{\mathfrak{J}''}$. Let $n \in \mathfrak{J}$ be a maximal index. Let $\alpha \colon \mathfrak{J}'' \to \mathfrak{J}$ be the index map for f and let $\mathfrak{J} = \mathfrak{J} - \{n\}$ and $\mathfrak{J}'' = \alpha^{-1}(\mathfrak{J})$. By induction we may assume there is an uzunblowup $\mu' \colon \mathfrak{T}^* \to \mathfrak{T}''[\mathfrak{J}'']$ and a rational tico tower morphism $g' \colon \mathfrak{T}^* \to \mathfrak{T}'[\mathfrak{J}]$ so that $\pi g' = f \mu'$. By Proposition 6.3.1 we may assume there is an uzunblowup $\mu'' \colon \mathfrak{T}^{**} \to \mathfrak{T}''$ so that $\mathfrak{T}^* = \mathfrak{T}^{**}[\mathfrak{J}'']$ and $\mu' = \mu''|_{\mathfrak{T}^*}$. Let $\mathfrak{T}^{**} = \{V_i^{**}, \mathcal{A}_i^{**}, p_i^{**}\}_{\mathfrak{J}''}$.

Let $\pi_n \colon V_n' \to V_n$ be the blowup given by π. Then the centers of π_n lie over $|\mathcal{A}_n|$ since π_n is unobtrusive.

If $\alpha(i) = n$, let $f_i \mu_i'' \colon V_i^{**} \to V_n$ be the map given by $f\mu'' \colon \mathfrak{T}^{**} \to \mathfrak{T}$. Then by Proposition 6.2.8, there is a tico uzunblowup $\lambda_i \colon (V_i''', \mathcal{A}_i''') \to (V_i^{**}, \mathcal{A}_i^{**})$ and an entire rational tico map $g_i \colon (V_i''', \mathcal{A}_i''') \to (V_n', \mathcal{A}_n')$ so that $\pi_n g_i = f_i \mu_i'' \lambda_i$ and all the centers of λ_i are irreducible and lie over $(f_i \mu_i'')^{-1}(|\mathcal{A}_n|)$. But $f_i^{-1}(|\mathcal{A}_n|) = |\mathcal{A}_i^*|$ since f is a tower morphism and $\mu''^{-1}_i(|\mathcal{A}_i^*|) = |\mathcal{A}_i^{**}|$ by Lemma 4.2.1. So $(f_i \mu_i'')^{-1}(|\mathcal{A}_n|) = |\mathcal{A}_i^{**}|$ and so the centers of λ_i lie over $|\mathcal{A}_i^{**}|$, hence the centers are unobtrusive. But the centers are also unskewed by Lemma 3.3.3. Also, since α is order preserving, if $\alpha(i) = n$ then i is maximal in $\mathcal{J}''$. Hence the λ_i induce an uzunblowup $\{\lambda_i\}_{\mathcal{J}''} \colon \mathfrak{T}''' \to \mathfrak{T}^{**}$ where $\mathfrak{T}'''[\mathcal{J}''] = \mathfrak{T}^{**}[\mathcal{J}''] = \mathfrak{T}^*$ and $\lambda_i = \text{identity}$ for $\alpha(i) \neq n$. Define $g \colon \mathfrak{T}''' \to \mathfrak{T}'$ by $g|_{\mathfrak{T}^*} = g'$ and let $g_i \colon V_i''' \to V_n'$ be as above if $\alpha(i) = n$. Then g is a tico tower morphism by Lemma 4.2.4. So letting $\mu = \{\lambda_i\}_{\mathcal{J}''} \circ \mu''$ we are done. $\square$

4. Resolution Towers for Algebraic Sets

Definition: A *partial resolution* of an algebraic set X is a nested collection of algebraic subsets $X_m \subset X_{m+1} \subset \cdots \subset X_n = X$, proper entire rational functions $q_i \colon V_i \to X_i$ for $i = m+1, \dots, n$ and a resolution tower $\{V_i, \mathcal{A}_i, p_i\}_{i=m}^n$ so that:

1) $\dim X_{i-1} < \dim X_i$ for $m+1 \le i \le n$.

2) $X_{i-1} \supset \operatorname{Sing} X_i$ for $m+1 \le i \le n$.

3) $q_i^{-1}(X_j) = \bigcup_{k \le j} V_{ki}$ for $m \le j < i \le n$.

4) $q_j p_{ji} = q_i|_{V_{ji}}$ for $m+1 \le j < i \le n$.

5) $q_i| \colon V_i - |\mathcal{A}_i| \to X_i - X_{i-1}$ is a diffeomorphism for $m+1 \le i \le n$.

We call $n - m$ the *height* of the partial resolution.

Notice that the maps q_i induce a homeomorphism from the realization of the tower minus V_m to $X - X_m$. The stratification of the realization corresponds to the stratification of $X - X_m$ with strata $X_i - X_{i-1}$. This indicates the source of the term partial resolution since it resolves $X - X_m$.

Any real algebraic set has a partial resolution, for example, with $n = m$, $X = X_m$ and $V_m = $ a point.

Proposition 6.4.1 *Suppose* $\mathfrak{T} = \{V_i, \mathcal{A}_i, p_i\}_{i=m}^n$, $q_i \colon V_i \to X_i$, $i > m$ *is a partial resolution of* X *with height* $n - m$ *and* X *is compact. Then there is a partial resolution of* X *with height* $n - m + 1$.

Furthermore, if $\mathfrak{T}$ *is of type* F, M *or* U *(or any combination), we may suppose the new partial resolution is of type* F, M *or* U *also.*

Proof: We may assume, by collapsing V_m to a point, that $\dim V_m = 0$ and $\mathcal{A}_m = \emptyset$. It is easy to see that $\mathfrak{T}$ still remains a resolution tower and it is still of type F, M or U if it was originally so.

By [**H**] and Theorem 2.9.5 there is an uzunblowup $g' \colon X_m' \to X_m$ so that all of its centers have dimension less than $\dim X_m$ and X_m' is nonsingular and full. Let A be the union of the projection of the centers of g' to X_m. Then A is a semialgebraic set of dimension less than $\dim X_m$. Also $A \supset \operatorname{Sing} X_m$ since X_m' is nonsingular. Let A_i be the union of the critical values of $q_i| \colon S - q_i^{-1}(\operatorname{Sing} X_m) \to X_m - \operatorname{Sing} X_m$ for all strata S of the tico stratification of V_i with $S \subset V_{mi}$. Then each A_i is a semialgebraic set and $\dim A_i < \dim X_m$ by Sard's theorem.

Let X_{m-1} be the Zariski closure of $A \cup \bigcup_{i=m+1}^{n} A_i$. Then $\dim X_{m-1} < \dim X_m$ and $X_{m-1} \supset \operatorname{Sing} X_m$.

By Lemma 6.2.5 and 6.1.3 (applied with A empty) there is an uzunblowup $g'' \colon X''_m \to X'_m$ with centers lying over $g'^{-1}(X_{m-1})$ so that $g''^{-1}g'^{-1}(X_{m-1}) = |A''_m|$ where (X''_m, A''_m) is a full algebraic tico. Let $g \colon X''_m \to X_m$ be $g'g''$.

Notice that

$$p_{ji}^{-1} q_j^{-1}(X_{m-1}) = V_{ji} \cap q_i^{-1}(X_{m-1}) \subset q_i^{-1}(X_{m-1}).$$

Hence by Lemma 6.3.2 (with $Z_i := q_i^{-1}(X_{m-1})$, $Z := X_{m-1}$, $Y := X_m$, $Y' := X''_m$, $B_i := A_{mi}$ and $q_i := q_i|_{V_{mi}}$) there is an uzunblowup $\{\pi_i\}_{i=m}^{n} \colon \mathfrak{T}' \to \mathfrak{T}$ with $\mathfrak{T}' = \{V'_i, A'_i, p'_i\}_{i=m}^{n}$ so that the centers of each π_i lie over $q_i^{-1}(X_{m-1})$, $\pi_i^{-1} q_i^{-1}(X_{m-1}) = |\mathcal{E}_i|$ for some $\mathcal{E}_i \subset A'_i$ and there are entire rational micos $h_i \colon |B'_i| \to X''_m$ so that $g h_i = q_i \pi_i$ for $i = m+1, \dots, n$ where B'_i is the strict transform of A_{mi} for $i = m+1, \dots, n$. Furthermore, $\mathfrak{T}'$ is of type F if $\mathfrak{T}$ is of type F.

$$
\begin{array}{ccccccc}
|\mathcal{E}_i| & \hookrightarrow & V'_i & \hookleftarrow & |B'_i| & \xrightarrow{\;h_i\;} & X''_m = V''_m \\
& & \downarrow{\scriptstyle\pi_i} & & \downarrow & & \downarrow{\scriptstyle g} \\
\downarrow & & V_i & \hookleftarrow & |A_{mi}| & \xrightarrow{\;q_i|\;} & X_m \\
& & & & \cup & & \cup \\
q_i^{-1}(X_{m-1}) & = & q_i^{-1}(X_{m-1}) & & & \longrightarrow & X_{m-1}
\end{array}
$$

Let $V''_i = V'_i$ for $i = m+1, \dots, n$, let $V''_m = X''_m$ and $V''_{m-1} = $ a point. Let $A''_i = A'_i$ for $m+1 \le i \le n$, let $A''_{ji} = A'_{ji}$ for $m+1 \le j < i \le n$ and let $A''_{m-1i} = \mathcal{E}_i$. Let $A''_{mi} = A'_{mi} - \mathcal{E}_i$ for $m+1 \le i \le n$. (Recall that by convention, A'_{ji} is the tower transform of A_{ji}). We have already defined $A''_{m-1,m} = A''_m$ above, recall $|A''_m| = g^{-1}(X_{m-1})$.

Notice $B'_i \supset A''_{mi}$. Define $p''_{ji} \colon |A''_{ji}| \to V''_j$ as follows. If $j \ge m+1$ let $p''_{ji} = p'_{ji}$. If $j = m$, $p''_{mi} = h_i|$. If $j = m-1$, p''_{m-1i} is the unique map to $V''_{m-1} = $ a point. Let $p''_i = \{p''_{ji}\}_{i=m-1}^{n}$. Define $q''_i \colon V''_i \to X_i$ for $i = m, \dots, n$ by $q''_i = q_i \pi_i$ if $i \ge m+1$ and $q''_m = g$. Then we claim that $\mathfrak{T}'' = \{V''_i, A''_i, p''_i\}_{i=m-1}^{n}$, $q''_i \colon V''_i \to X_i$, $i = m, \dots, n$ is a partial resolution of X.

First we must show that $\mathfrak{T}''$ is a resolution tower. Let us prove condition I.

Assertion 6.4.1.1 $p''_{ji}\left(V''_{ji} \cap V''_{ki}\right) \subset V''_{kj}$ for $m-1 \le k < j < i \le n$.

Proof: Since $\mathfrak{T}'$ is a resolution tower, we know this for $k > m$.

If $k = m-1$ and $j > m$ then

$$q_j \pi_j p''_{ji}\left(V''_{ji} \cap V''_{m-1,i}\right) = q_j p_{ji} \pi_i\left(V''_{ji} \cap V''_{m-1,i}\right)$$

$$\subset q_j p_{ji}\left(V_{ji} \cap q_i^{-1}(X_{m-1})\right) = q_i\left(V_{ji} \cap q_i^{-1}(X_{m-1})\right) \subset X_{m-1}.$$

So $p''_{ji}\left(V''_{ji} \cap V''_{m-1,i}\right) \subset \pi_j^{-1} q_j^{-1}(X_{m-1}) = V''_{m-1,j}$.

If $k = m - 1$ and $j = m$ then $g p''_{mi} = q_i \pi_i|$ so

$$g p''_{mi} \left(V''_{mi} \cap V''_{m-1,i} \right) = q_i \pi_i \left(V''_{mi} \cap V''_{m-1,i} \right) \subset q_i \pi_i \left(V''_{m-1,i} \right) \subset X_{m-1}.$$

So $p''_{mi} \left(V''_{mi} \cap V''_{m-1,i} \right) \subset g^{-1} \left(X_{m-1} \right) = V''_{m-1,m}.$
If $k = m$ then

$$q_j \pi_j p''_{ji} \left(V''_{ji} \cap V''_{mi} - V''_{m-1i} \right) = q_j p_{ji} \pi_i \left(V''_{ji} \cap V''_{mi} - V''_{m-1i} \right)$$

$$\subset q_j p_{ji} \left(V_{ji} \cap V_{mi} - q_i^{-1} \left(X_{m-1} \right) \right) = q_i \left(V_{ji} \cap V_{mi} - q_i^{-1} \left(X_{m-1} \right) \right)$$

$$\subset q_i \left(V_{ji} \cap V_{mi} \right) - X_{m-1} \subset X_m - X_{m-1}.$$

Hence $p''_{ji} \left(V''_{ji} \cap V''_{mi} - V''_{m-1,i} \right) \cap V''_{m-1j}$ is empty. But $V''_{mi} \subset V'_{mi}$ so

$$p''_{ji} \left(V''_{ji} \cap V''_{mi} \right) \subset p'_{ji} \left(V'_{ji} \cap V'_{mi} \right) \subset V'_{mj} = V''_{mj} \cup V''_{m-1j}$$

So by continuity, $p''_{ji} \left(V''_{ji} \cap V''_{mi} \right) \subset V''_{mj}.$ □

Let us now prove condition III.

Assertion 6.4.1.2 $p''^{-1}_{ji} \left(\bigcup_{k<\ell} V''_{kj} \right) = V''_{ji} \cap \left(\bigcup_{k<\ell} V''_{ki} \right)$ if $m \le \ell \le j < i \le n$

Proof: If $\ell > m$ then

$$p''^{-1}_{ji} \left(\bigcup_{k<\ell} V''_{kj} \right) = p'^{-1}_{ji} \left(\bigcup_{k<\ell} V'_{kj} \right) = V'_{ji} \cap \left(\bigcup_{k<\ell} V'_{ki} \right) = V''_{ji} \cap \left(\bigcup_{k<\ell} V''_{ki} \right).$$

If $j > \ell = m$ then

$$p''^{-1}_{ji} \left(\bigcup_{k<m} V''_{kj} \right) = p''^{-1}_{ji} \pi_j^{-1} q_j^{-1} \left(X_{m-1} \right) = V''_{ji} \cap \pi_i^{-1} p_{ji}^{-1} q_j^{-1} \left(X_{m-1} \right)$$

$$= V''_{ji} \cap \pi_i^{-1} q_i^{-1} \left(X_{m-1} \right) = V''_{ji} \cap \bigcup_{k<m} V''_{ki}.$$

If $j = \ell = m$ then

$$p''^{-1}_{mi} \left(\bigcup_{k<m} V''_{km} \right) = V''_{mi} \cap h_i^{-1} g^{-1} \left(X_{m-1} \right)$$

$$= V''_{mi} \cap \pi_i^{-1} q_i^{-1} \left(X_{m-1} \right) = V''_{mi} \cap \bigcup_{k<m} V''_{ki}.$$

□

Assertion 6.4.1.3 $p''_{kj} p''_{ji} (x) = p''_{ki} (x)$ for all $x \in V''_{ji} \cap V''_{ki}$, $m - 1 \le k < j < i \le n$.

Proof: This is trivial if $k \neq m$ so assume $k = m$. Suppose $x \in V''_{ji} \cap V''_{mi} - V''_{m-1i}$. Then $p''_{mj}p''_{ji}(x) = h_j p'_{ji}(x)$ and $p''_{ki}(x) = h_i(x)$. But

$$gh_j p'_{ji}(x) = q_j \pi_j p'_{ji}(x) = q_j p_{ji} \pi_i(x) = q_i \pi_i(x) = gh_i(x).$$

Also

$$gh_i(x) = q_i \pi_i(x) \in q_i \left(V_{ji} \cap V_{mi} - q_i^{-1}(X_{m-1}) \right) \subset X_m - X_{m-1}$$

and g is one to one on $X''_m - g^{-1}(X_{m-1})$ so $h_j p'_{ji}(x) = h_i(x)$. Hence $p''_{mj}p''_{ji}| = p''_{mi}|$ on $V''_{ji} \cap V''_{mi}$ by continuity.   □

So $\mathfrak{T}''$ is a resolution tower. To see that it gives a partial resolution, we have already shown that $X_{m-1} \supset \operatorname{Sing} X_m$ and $\dim X_{m-1} < \dim X_m$. But

$$q''_i{}^{-1}(X_j) = \pi_i^{-1}q_i^{-1}(X_j) = \pi_i^{-1}(\bigcup_{k \leq j} V_{ki}) = \bigcup_{k \leq j} V''_{ki} \text{ if } j \geq m.$$

$$q''_i{}^{-1}(X_{m-1}) = \pi_i^{-1}q_i^{-1}(X_{m-1}) = V''_{m-1i} \text{ if } j = m-1 \text{ and } m < i.$$

$$q''_m{}^{-1}(X_{m-1}) = g^{-1}(X_{m-1}) = V''_{m-1m} \text{ if } j = m-1 \text{ and } i = m.$$

It is trivial that $q''_j p''_{ji} = q''_i|$ for $m \leq j < i \leq n$. So we have a partial resolution of X since:

Assertion 6.4.1.4 *For $i \geq m$, $q''_i|: V''_i - |\mathcal{A}''_i| \to X_i - X_{i-1}$ is a diffeomorphism.*

Proof: For $i \geq m+1$, $q''_i|$ is the composition of $\pi_i|: V''_i - |\mathcal{A}''_i| \to V_i - |\mathcal{A}_i|$ (which is a diffeomorphism since the centers of π_i lie over $q_i^{-1}(X_{m-1}) \subset |\mathcal{A}_i|$) and $q_i|: V_i - |\mathcal{A}_i| \to X_i - X_{i-1}$ which is a diffeomorphism by assumption. Hence $q''_i|$ is a diffeomorphism. Likewise $q''_m|$ is the map $g|: X''_m - g^{-1}(X_{m-1}) \to X_m - X_{m-1}$ which is a diffeomorphism since the centers of g lie over X_{m-1}.   □

Assertion 6.4.1.5 *If $\mathfrak{T}$ is of type M, then $\mathfrak{T}''$ if of type M also.*

Proof: Proposition 4.2.7 implies that $\mathfrak{T}'$ has type M. So we know that $p''_{ji}: V''_i \to V''_{ji}$ is a mico for $j \geq m+1$ since it is just $p'_{ji}: V'_{ji} \to V'_j$. But $p''_{mi}: V''_{mi} \to V''_m$ is a mico since it is a restriction of the mico $h_i: |\mathcal{B}'_i| \to X''_m$. So $\mathfrak{T}''$ is of type M.   □

Assertion 6.4.1.6 *If $\mathfrak{T}$ is of type F, then $\mathfrak{T}''$ if of type F also.*

Proof: If $\mathfrak{T}$ is of type F, we know that $\mathfrak{T}'$ is of type F, hence $(V''_i, \mathcal{A}''_i)$ is full for all $i > m+1$. But $(V''_m, \mathcal{A}''_m) = (X''_m, \mathcal{A}''_m)$ which is full. Also V''_{m-1} is a point which is full. Hence $\mathfrak{T}''$ is of type F.   □

Assertion 6.4.1.7 *If $\mathfrak{T}$ is of type U, then $\mathfrak{T}''$ if of type U also.*

Proof: We claim that $\mathfrak{T}'$ has type U also. To see this, note that the centers of π_i lie over $q_i^{-1}(X_{m-1})$ and $q_i^{-1}(X_{m-1}) \subset q_i^{-1}(X_m) = V_{mi}$. Hence for $j > m$ any stratum S of $V'_{ji} - \bigcup_{k<j} V'_{ki}$ is mapped diffeomorphically by π_i to a stratum of $V_{ji} - \bigcup_{k<j} V_{ki}$ which is submersed by p_{ji} to $V_j - |\mathcal{A}_j|$ which is mapped

diffeomorphically to $V_j' - |\mathcal{A}_j'|$ by π_j^{-1}. Since $p_{ji}'|_S = \pi_j^{-1} p_{ji} \pi_i|_S$ this means that p_{ji}' submerses S. But if T is any stratum of V_{mi}', $p_{mi}'\colon T \to V_m'$ is certainly a submersion since $\dim V_m' = 0$.

So to show that $\mathfrak{T}''$ has type U we only need to show that $p_{mi}''|\colon V_{mi}'' - V_{m-1i}'' \to V_m'' - V_{m-1m}'' = X_m'' - g^{-1}(X_{m-1})$ is a submersion on each stratum. But $g\colon X_m'' - g^{-1}(X_{m+1}) \to X_m - X_{m-1}$ is a diffeomorphism so we only need to show that $gp_{mi}''| = gh_i|\colon V_{mi}'' - V_{m-1i}'' \to X_m - X_{m-1}$ submerses each stratum. But $gh_i| = q_i \pi_i|$ and $\pi_i|\colon V_{mi}'' - V_{m-1i}'' \to V_{mi} - q_i^{-1}(X_{m-1})$ is a diffeomorphism on each stratum so that result follows from the fact that $q_i\colon V_{mi} - q_i^{-1}(X_{m-1}) \to X_m - X_{m-1}$ submerses each stratum. $\qquad\square$

Theorem 6.4.2 *Let X be a real algebraic set. Then there is an (algebraic) resolution tower $\mathfrak{T}$ of type UNE so that $|\mathfrak{T}|$ is homeomorphic to X, in fact the stratification of $|\mathfrak{T}|$ corresponds to an algebraic stratification of X, i.e., one for which all skeleta are Zclosed subsets. Furthermore, $\mathfrak{T}$ can be type F if X is compact.*

Proof: By Proposition 4.2.7 and Theorem 4.2.8 it suffices to find $\mathfrak{T}$ of type FUM with $|\mathfrak{T}| = X$ since we may then find a multiblowup $\mathfrak{T}'$ of $\mathfrak{T}$ so that $\mathfrak{T}'$ has type $FUNE$. But $|\mathfrak{T}| = |\mathfrak{T}'|$ since all the multiblowups are unobtrusive.

If X is noncompact we may find by Lemma 2.6.2 an algebraic set Y homeomorphic to the one point compactification of X. Suppose we can find a resolution tower $\mathfrak{T} = \{V_i, \mathcal{A}_i, p_i\}_\mathfrak{I}$ of type UNE and a homeomorphism $h\colon |\mathfrak{T}| \to Y$. Let $x \in Y$ be such that $Y - x$ is homeomorphic to X. Let $g_i\colon V_i \to |\mathfrak{T}|$ be the canonical quotient map. Let $V_i' = V_i - g_i^{-1} h^{-1}(x)$, let $\mathcal{A}_i' = V_i' \cap \mathcal{A}_i$ and let p_i' be the collection of the restrictions of p_{ji} to $V_{ji} \cap V_i'$. Then $\mathfrak{T}' = \{V_i', \mathcal{A}_i', p_i'\}_\mathfrak{I}$ is a resolution tower of type UNE and $|\mathfrak{T}'| = |\mathfrak{T}| - h^{-1}(x) = X$.

Actually with a bit of work, one can get $\mathfrak{T}'$ to be type F also, but it does not seem worth it in the noncompact case.

So it suffices to consider the case where X is compact. Pick any $n > \dim X$. Now X has a trivial partial resolution of height 0, $\mathfrak{T}'' = \{V_n'', \emptyset, \emptyset\}$ with $V_n'' = $ a point. Notice $\mathfrak{T}''$ has type FUM. But then repeated applications of Proposition 6.4.1 show that X must have a partial resolution $\mathfrak{T}' = \{V_i', \mathcal{A}_i', p_i'\}_{i=0}^n$, $X_0 \subset X_1 \subset \cdots \subset X_n = X$, $q_i\colon V_i' \to X_i$, $i = 1, \ldots, n$ where $\mathfrak{T}'$ has type FUM. But since $n > \dim X$ and $\dim X_0 \leq \dim X - n$ we must have X_0 empty. But then $q_i^{-1}(X_0) = |\mathcal{A}_{0i}'| = V_{0i}'$ is empty for all i. Let $V_i = V_i'$ for $i > 0$. Let $\mathcal{A}_i = \mathcal{A}_i'$ and $p_i = p_i'$. Then $\mathfrak{T} = \{V_i, \mathcal{A}_i, p_i\}_{i=1}^n$ is a resolution tower of type FUM and the maps q_i induce a homeomorphism from $|\mathfrak{T}|$ to X. $\qquad\square$

CHAPTER VII

THE CHARACTERIZATION OF THREE DIMENSIONAL ALGEBRAIC SETS

In this chapter we give a simple method for determining whether a 3 - dimensional stratified set is homeomorphic to a real algebraic set. We also give a method for determining whether 3 - dimensional stratified sets are isomorphic to the singular stratification of a real algebraic set.

In contrast to the rest of this book, we will be interested in the case where resolution towers have boundary.

Definitions: We will call a resolution tower *closed* if it is compact without boundary. Let $\mathfrak{TR}_m$ and $\mathcal{AR}_m$ be the closed m-dimensional towers in $\mathfrak{TR}_{UM}$ and $\mathcal{AR}_{UM}$ respectively. A $\mathfrak{TR}$ *structure* $\mathfrak{T}$ on a stratified set X is a $\mathfrak{T} \in \mathfrak{TR}_{UM}$ such that X is isomorphic to the realization $|\mathfrak{T}|$. We call a stratified set a $\mathfrak{TR}$ *space* if it has a $\mathfrak{TR}$ structure.

Definitions: Define an *Euler space* to be a topological space such that each point has a neighborhood homeomorphic to the cone on a compact set with even Euler characteristic. We define an *Euler polyhedron* to be a polyhedron X so that X and each skeleton of X is an Euler space. An *Euler stratified set* is a locally conelike stratified set X so that X and all skeleta of X are Euler spaces. An *Euler Thom stratified set* is a Thom stratified set X so that X and all skeleta of X are Euler spaces.

Exercise: Show that a polyhedron or locally conelike stratified set is an Euler space if the link of each simplex or stratum has even Euler characteristic. ◇

Exercise: Show that a $\mathfrak{TR}$ space is an Euler Thom stratified set. ◇

It is well known that algebraic sets are triangulable stratified sets [**L2**]. In fact any algebraic stratification is an Euler stratified set [**Su1**]. Furthermore, any algebraic stratification can be refined to satisfy the Whitney conditions and it will then be a Euler Thom stratified set, [**Wa**], [**GWPL**].

Real algebraic sets of dimension ≤ 2 are easily classified by these structures [**AK4**], [**AK6**] and [**BD2**] in fact up to homeomorphism:

$$\left\{ \begin{array}{c} \text{Compact real algebraic} \\ \text{sets of dimension} \leq 2 \end{array} \right\} \equiv \left\{ \begin{array}{c} \text{Compact } \mathcal{TR} \text{ spaces} \\ \text{of dimension} \leq 2 \end{array} \right\}$$

$$\equiv \left\{ \begin{array}{c} \text{Compact triangulable Euler} \\ \text{spaces of dimension} \leq 2 \end{array} \right\}$$

The goal of this chapter is to generalize this theorem to dimension 3. However this does not generalize directly. This is because dimension 3 is the first dimension where there are exotic Euler spaces; that is there are 3-dimensional Euler spaces which cannot be homeomorphic to real algebraic sets. So we encounter new topological invariants of algebraic sets in this dimension. To be more specific let X be a triangulable 3-dimensional Euler space and X_0 be the set of vertices of some triangulation of X. Then for each $x \in X_0$ there are four invariants $\theta_i(x) \in \mathbf{Z}/2\mathbf{Z}$ so that up to homeomorphism ([**AK17**]):

$$\left\{ \begin{array}{c} \text{Compact real algebraic} \\ \text{sets of dimension 3} \end{array} \right\} \equiv \left\{ \begin{array}{c} \text{Compact } \mathcal{TR} \text{ spaces} \\ \text{of dimension 3} \end{array} \right\}$$

$$\equiv \left\{ \begin{array}{c} \text{Compact 3 dimensional triangulable Euler spaces } X \\ \text{so that } \theta_i(x) = 0 \text{ for all } x \in X_0 \text{ and } i = 0, 1, 2, 3 \end{array} \right\}$$

In the notation of section 1 below, $\theta_i(x)$ is the Euler characteristic of the link of x in $\mathcal{Z}_i(X)$.

These obstructions can be obtained by other means, see [**CK**]. This paper [**CK**] also has other interesting results, for example that if $V \subset X$ are Zopen sets and V is irreducible, then there is a proper Zclosed subset $Y \subset V$ so that the mod 4 Euler characteristic of the link of $V - Y$ in X is constant.

Exercise: Prove this result using resolution towers. Hint: use Propositions 2.3.2 and 4.3.4 and Theorem 6.4.2.   ⋄

The argument in Chapter VI can be used to show that if X is an analytic set of dimension < 4 and $x \in X$ then there is a neighborhood U of x in X which has a $\mathcal{TR}$ structure. Hence the local conditions $\theta_i = 0$ are met, so we have the following:

Corollary 7.0 *Every compact real analytic set of dimension less than four is homeomorphic to a real algebraic set.*

The next question to ask is whether we can get a classification of the singular stratifications of algebraic sets up to stratified set isomorphism, not just homeomorphism. Here we must be careful, since the process of finding a resolution tower for an algebraic set involved refining the stratification. If we do not allow such refinement, we will give an example which shows that at the moment this is not a reasonable problem in general dimensions.

Take any polynomial $p\colon \mathbf{R}^k \to \mathbf{R}$ and define a polynomial map $\alpha_p\colon \mathbf{R}^2 \times \mathbf{R}^k \to \mathbf{R}$ by $\alpha_p(x, y, z) = xy(x - y)(x - p(z)y)$. Let

$$V_p = \{\, (u, v, x, y, z) \in \mathbf{R}^4 \times \mathbf{R}^k \mid u^2 + v^4 - v^2 \alpha_p(x, y, z) = 0 \,\}.$$

Then $\operatorname{Sing} V_p = W = \{\, (u, v, x, y, z) \in \mathbf{R}^4 \times \mathbf{R}^k \mid u = v = 0 \,\}$. Since $\operatorname{Sing} W = \emptyset$, the singular stratification of V_p has two strata, $V_p - W$ and W.

Note that if $\alpha_p(x_0, y_0, z_0) > 0$ then $V_p \cap \{(u, v, x_0, y_0, z_0)\}$ is a figure 8 and if $\alpha_p(x_0, y_0, z_0) \leq 0$ it is a point. Thus the set

$$Z_p = \{\, (u, v, x, y, z) \in \mathbf{R}^4 \times \mathbf{R}^k \mid u = v = 0,\ \alpha_p(x, y, z) = 0 \,\}$$

is topologically distinguished as the frontier in W of $W \cap \operatorname{Cl}(\operatorname{Nonsing} V_p)$ (unless p is the constant 0 or 1, we will assume it is not).

We now show that the local description of the singular stratification of V_p is very complicated, and in fact to classify all such involves classifying up to diffeomorphism the local behavior of all polynomials p.

Suppose $h\colon U \to U'$ is an isomorphism from an open neighborhood U of $0 \in V_p$ to an open subset U' of V_q for some polynomials p and q. Let $Y = \{(0, 0, 0, 0, z)\}$. Now since h is an isomorphism of stratified sets, we know that $h(U \cap W) = U' \cap W$ and h restricts to a diffeomorphism of $U \cap W$ to $U' \cap W$. Since Z_p is topologically determined, we also know that $h(Z_p \cap U) = Z_q \cap U'$. But points of Y are topologically distinguished as the points of intersection of the four hypersurfaces $x = 0$, $y = 0$, $x = y$ and $x = p(z)y$ which make up Z_p. Hence $h(Y \cap U) = Y \cap U'$. But now the tangent spaces of the four hypersurfaces have a cross ratio which must be preserved by h. This cross ratio is $p(z)$. Consequently, if we define the germ $h'\colon \mathbf{R}^k \to \mathbf{R}^k$ by $h(0, 0, 0, 0, z) = (0, 0, 0, 0, h'(z))$ then we know that $qh'(z)$ is either $p(z)$, $1 - p(z)$, $1/p(z)$, $1/(1 - p(z))$, $p(z)/(p(z) - 1)$ or $1 - 1/p(z)$. As a result, the germs of p and q must be equivalent after a smooth change of coordinates. Consequently, classifying singular stratifications of algebraic sets requires as a minimum classifying germs of polynomials.

Notice that if we refine the stratification of V_p so that $Z_p - Y$ is a stratum then all these problems disappear since $h|_W$ need not be a diffeomorphism so the cross ratio need not be preserved.

1. Obstructions

In this section we identify the obstructions for a compact three dimensional polyhedron to be homeomorphic to a real algebraic set. These obstructions are topological invariants, local and easy to compute.

Theorem 7.1.1 *Let X be a compact 3-dimensional triangulable topological space. Then there are closed one dimensional subspaces $\mathcal{Z}_i(X)$ of X for $i = 0, 1, 2, 3$ so that:*

a) X is homeomorphic to a real algebraic set if and only if X and each $\mathcal{Z}_i(X)$ for $i = 0, 1, 2, 3$ are Euler spaces.

b) The subsets $\mathcal{Z}_i(X)$ do not depend on the triangulation of X.

c) The subsets $\mathcal{Z}_i(X)$ are locally defined, i.e., the germ of $\mathcal{Z}_i(X)$ at a point x is determined by the germ of X at x.

Furthermore, if we give X any particular Euler Thom stratification and X is homeomorphic to a real algebraic set then a refinement of the given stratification is isomorphic to the singular stratification of a real algebraic set.

Theorem 7.1.2 *Let X be a compact three-dimensional Euler Thom stratified set. Then there are closed one-dimensional subspaces $\mathcal{Z}_i(X)$ of X for $i = 4, 5, \ldots, 11$ so that:*

a) *If all normal bundles of strata of X are trivial, then X is isomorphic to the singular stratification of a real algebraic set if and only if $\mathcal{Z}_i(X)$, $i = 0, \ldots, 11$ are all Euler spaces. (Here $\mathcal{Z}_i(X)$ for $i = 0, 1, 2, 3$ are as in Theorem 7.1.1).*

b) *The sets $\mathcal{Z}_i(X)$ may depend on the stratification of X.*

c) *The sets $\mathcal{Z}_i(X)$ are locally defined, i.e., the germ of $\mathcal{Z}_i(X)$ at a point x is determined by the germ of the stratified set X at x.*

One implication of these results is that the obstructions to an Euler Thom stratified set X being algebraic are all concentrated on the 0-skeleton X_0. At each point of X_0 we assign 12 elements of $\mathbf{Z}/2\mathbf{Z}$. If the first four are 0 at all points of X_0 then X is homeomorphic (in fact PL isomorphic) to a real algebraic set. If all 12 are zero at each point of X_0 and the normal bundles of all strata are trivial then X is isomorphic to the singular stratification of a real algebraic set. It is possible to say much more than Theorem 7.1.2 does about when a stratified set is isomorphic to a real algebraic set, however it does not seem worthwhile to do so. As the example at the end of the previous section showed, a classification of algebraic stratifications of real algebraic sets would be exceedingly complicated so to have any hope of understanding them at all, one should be allowed to refine the stratification. But then you end up in the situation of Theorem 7.1.1 anyway.

We will delay until section 3 the proofs of Theorems 7.1.1 and 7.1.2. We will define the $\mathcal{Z}_i$'s in this section, but first we define some intermediate subspaces.

Definitions: Let X be a locally conelike stratified set. For every component U of the codimension one stratum of X we have an integer $\kappa(U)$ which is the number of points in the link of U. For $i = 0, \ldots, 7$ let $\mathcal{K}_i(X)$ be the union of the components U so that $\kappa(U) \equiv i$ mod 8. For every component W of the codimension two stratum of X and any $i = 0, \ldots, 7$ let $\alpha_i(W)$ be the number of vertices v in the link of W so that $\kappa(v) \equiv i$ mod 8. (Notice v is codimension one in the link so $\kappa(v)$ is defined.) Another way of thinking of $\alpha_i(W)$ is that it is the unsigned multiplicity of W in the boundary of $\mathcal{K}_i(X)$. Now for $a, b, c, d = 0, 1$

define $\mathcal{E}_{abcd}(X)$ to be the closure of the union of all the components W of the codimension two stratum of X so that

$$
\begin{aligned}
\alpha_0(W) &\equiv a \bmod 2, \\
\alpha_6(W) &\equiv b \bmod 2, \\
\alpha_0(W) + \alpha_4(W) &\equiv 2c \bmod 4, \\
\alpha_2(W) + \alpha_6(W) &\equiv 2d \bmod 4.
\end{aligned}
$$

We shall define the subspaces $\mathcal{Z}_i(X)$ as follows.

1) $\mathcal{Z}_0(X) = \mathcal{E}_{1110}(X) \cup \mathcal{E}_{1111}(X)$

2) For $i = 1, 2, 3$, we set $\mathcal{Z}_i(X) = \mathcal{E}_{ab00}(X) \cup \mathcal{E}_{ab01}(X)$, where ab is the binary representation of i with leading zeros, i.e.,

$$
\begin{aligned}
\mathcal{Z}_1(X) &= \mathcal{E}_{0100}(X) \cup \mathcal{E}_{0101}(X) \\
\mathcal{Z}_2(X) &= \mathcal{E}_{1000}(X) \cup \mathcal{E}_{1001}(X) \\
\mathcal{Z}_3(X) &= \mathcal{E}_{1100}(X) \cup \mathcal{E}_{1101}(X)
\end{aligned}
$$

3) For $i = 4, \ldots, 11$, we set $\mathcal{Z}_i(X) = \mathcal{E}_{abc1}(X)$ where abc is the binary representation of $i - 4$ with leading zeroes. So

$$
\begin{aligned}
\mathcal{Z}_4(X) &= \mathcal{E}_{0001}(X) \\
\mathcal{Z}_5(X) &= \mathcal{E}_{0011}(X) \\
\mathcal{Z}_6(X) &= \mathcal{E}_{0101}(X) \\
\mathcal{Z}_7(X) &= \mathcal{E}_{0111}(X) \\
\mathcal{Z}_8(X) &= \mathcal{E}_{1001}(X) \\
\mathcal{Z}_9(X) &= \mathcal{E}_{1011}(X) \\
\mathcal{Z}_{10}(X) &= \mathcal{E}_{1101}(X) \\
\mathcal{Z}_{11}(X) &= \mathcal{E}_{1111}(X)
\end{aligned}
$$

Definition: We call these $\mathcal{Z}_i(X)$'s the *characteristic subsets* of X.

Let us look at some examples. Let Y_0 be the two-dimensional space of Figure VII.1.1 obtained by taking the suspension of a three leaf rose, joining

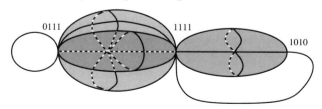

FIGURE VII.1.1. Y_0

it to the suspension of the union of a point and a figure eight at one suspension point and joining a circle to the other. We refine the stratification by putting a one stratum on the suspension of the three leaf rose, thus making it an Euler stratified set.

The numbers $0111, 1111, 0010$ and 1010 give the mod 2 values of α_0, α_6, $(\alpha_0 + \alpha_4)/2$ and $(\alpha_2 + \alpha_6)/2$ for the indicated vertices. Suppose Z_0 is the suspension of Y_0. Then $\mathcal{Z}_0(Z_0)$ is the suspension of the point marked 1111 and all the other $\mathcal{Z}_i(Z_0)$'s are empty, $i = 1, 2, 3$. So Z_0 is not homeomorphic to a real algebraic set since $\mathcal{Z}_0(X)$ is an interval and so it is not an Euler space.

The next example is Y_1. It is obtained by taking the suspension of a three leaf rose and attaching a figure eight. Again we refine the stratification by putting a one stratum on the suspension of the three leaf rose, making it an Euler stratified set. If Z_1 is the suspension of Y_1 then $\mathcal{Z}_1(Z_1)$ is the suspension of the point marked 0101 and all the other $\mathcal{Z}_i(Z_1)$'s are empty.

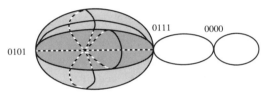

FIGURE VII.1.2. Y_1

We let Y_2 be the space obtained by taking the suspension of the figure eight and joining it to an interval and the suspension of three points. If Z_2 is the suspension of Y_2 then $\mathcal{Z}_2(Z_2)$ is the suspension of the point marked 1000. All other $\mathcal{Z}_i(Z_2)$'s are empty.

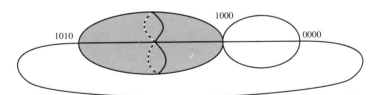

FIGURE VII.1.3. Y_2

The final example is Y_3, obtained from Y_0 by attaching two lines. Again we refine the stratification, making it an Euler stratified set. Let Z_3 be the suspension of Y_3. Then $\mathcal{Z}_3(Z_3)$ is the suspension of the point marked 1101, and $\mathcal{Z}_0(Z_3)$, $\mathcal{Z}_1(Z_3)$ and $\mathcal{Z}_2(Z_3)$ are empty.

The above examples show that there is no redundancy in the obstructions $\mathcal{Z}_i$, $i = 0, 1, 2, 3$. The spaces Z_i have the property that $\mathcal{Z}_i(Z_i)$ is not an Euler space but $\mathcal{Z}_j(Z_i)$ is empty for $j \neq i$. In fact $Y_4 = \mathbf{RP}^2$ and Y_i, $i = 0, 1, 2, 3$ form a basis for two-dimensional stratification independent bordism of resolvable

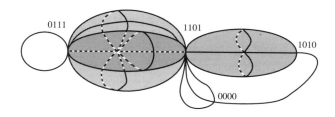

FIGURE VII.1.4. Y_3

stratified sets. In other words, given any compact polyhedron X of dimension ≤ 2 there is a resolution tower with boundary $\mathfrak{T}$ so that the realization of $\partial\mathfrak{T}$ is homeomorphic to the disjoint union of X and some of the Y_i's, $i = 0, \dots, 4$. It is also easy to determine which of the Y_i's will appear. If X has odd Euler characteristic then $Y_4 = \mathbf{RP}^2$ will appear and if $\mathcal{Z}_i(X)$ has an odd number of points then Y_i will appear, $i = 0, 1, 2, 3$.

Although the stratifications of Y_i and Z_i were not presented as Euler stratified sets, we could refine them to be such and the sets $\mathcal{Z}_i(Z_j)$ would be unchanged for $i = 0, 1, 2, 3$.

Let us do another example. Let Y be the disjoint union of $\mathbf{RP}^2$ and a point. Let Z be the suspension of Y. Then $\mathcal{Z}_i(Z)$ is empty, $i = 0, 1, 2, 3$ so Z is homeomorphic to a real algebraic set. We can make Z into an Euler space by letting the 0 stratum of Y be a point p in $\mathbf{RP}^2$ union the isolated point q of Y, letting the 1 stratum be empty and letting the 2 stratum be $\mathbf{RP}^2 - p$. Suspend this to get an Euler stratification of Z. Then $\mathcal{E}_{abcd}(Z)$ is empty unless $abcd = 0000$ when $\mathcal{E}_{0000}(Z)$ is the suspension of $\{p, q\}$. So $\mathcal{Z}_i(Z)$ is empty for $i = 0, \dots, 11$ so this stratification of Z is isomorphic to the singular stratification of some algebraic set.

Exercise: Using Lemma 7.1.6 below, show that for any Euler stratification of the suspension $Z = \sum(\mathbf{RP}^2 \cup q)$, then each $\mathcal{E}_{abcd}(Z)$ is an Euler space. Thus if all normal bundles are trivial, Theorem 7.1.2 says this stratification can be made algebraic. $\diamond$

We will delay the proofs of Theorems 7.1.1 and 7.1.2 until section 3. Meanwhile we will prove some useful Lemmas.

Lemma 7.1.3 *Let X be an Euler stratified set and let L be a link of some stratum Y. Then:*

a) *L is an Euler stratified set.*
b) *Each skeleton of L has even Euler characteristic.*
c) *The 1-stratum of L has an even number of components homeomorphic to $\mathbf{R}$.*

Furthermore, if X is an Euler polyhedron then:

d) *L is an Euler polyhedron.*

e) L *has an even number of* i-*simplices for each* i.

Proof: The links of L are links of corresponding strata of X so they have even Euler characteristic so a) and d) are true. The skeleta of L are links of Y in the skeleta of X so they all have even Euler characteristics so b) is true. If X is a polyhedron then

$$\chi\left(L^{i}\right) - \chi\left(L^{i-1}\right) = \sum_{j=0}^{i} (-1)^{j} m_{j} - \sum_{j=0}^{i-1} (-1)^{j} m_{j} = (-1)^{i} m_{i}$$

where m_{i} is the number of i-simplices and L^{j} is the j-skeleton of L. But $\chi\left(L^{i}\right)$ and $\chi\left(L^{i-1}\right)$ are both even so m_{i} is even, so e) is true.

To prove c) note that the components of the 1-stratum of L which are not homeomorphic to $\mathbf{R}$ are circles and so do not contribute to $\chi\left(L^{1}\right)$. So the number of components of the one stratum of L homeomorphic to $\mathbf{R}$ is $\chi\left(L^{0}\right) - \chi\left(L^{1}\right)$ which is even. $\qquad\square$

Lemma 7.1.4 *Let* L *be a link of a codimension two stratum* W *of an Euler stratified set* X. *Then the number* e *of components of the 1-stratum of* L *homeomorphic to* $\mathbf{R}$ *is congruent to* $\alpha_{2}\left(W\right) + 2\alpha_{4}\left(W\right) + 3\alpha_{6}\left(W\right) \mod 4$.

Proof: Each such component has two ends. So if L^{0} is the 0-skeleton of L we know that $2e = \sum_{v \in L^{0}} \kappa\left(v\right)$. By Lemma 7.1.3a, L is an Euler stratified set so $\alpha_{i}\left(W\right) \equiv 0$ for i odd. So $\mod 8$ we have

$$2e = \sum_{v \in L^{0}} \kappa\left(v\right) \equiv \sum_{i=0}^{7} \sum_{v \in \mathcal{K}_{i}(L)} i = \sum_{i=0}^{7} i\alpha_{i}\left(W\right) = \sum_{i=0}^{3} 2i\alpha_{2i}\left(W\right).$$

The lemma follows. $\qquad\square$

Lemma 7.1.5 *If* X *is an Euler stratified set then every component* W *of the codimension two stratum of* X *is contained in a unique* $\mathcal{E}_{abcd}\left(X\right)$.

Proof: It is immediate from the definition that W is contained in at most one $\mathcal{E}_{abcd}\left(X\right)$. So we must only show that $\alpha_{0}\left(W\right) + \alpha_{4}\left(W\right)$ and $\alpha_{2}\left(W\right) + \alpha_{6}\left(W\right)$ are both even. Let L be the link of W. Let e be the number of components of the 1-stratum of L which are homeomorphic to $\mathbf{R}$. Then by Lemma 7.1.3c we know e is even. By Lemma 7.1.4 we know $e \equiv \alpha_{2}\left(W\right) + 2\alpha_{4}\left(W\right) + 3\alpha_{6}\left(W\right) \mod 4$ so $0 \equiv \alpha_{2}\left(W\right) + \alpha_{6}\left(W\right) \mod 2$. But $\sum_{i=0}^{3} \alpha_{2i}\left(W\right)$ is the number of points in L^{0}. This is even so $\alpha_{0}\left(W\right) + \alpha_{4}\left(W\right)$ is even also. $\qquad\square$

Lemma 7.1.6 *Let L be the link of a codimension three stratum of an Euler stratified set. For $a, b, c, d = 0, 1$ let n_{abcd} be the number of points in $\mathcal{E}_{abcd}(L)$. Then*

  a) $\sum_{c=0}^{1} \sum_{d=0}^{1} (n_{11cd} + n_{abcd})$ *is even for any $a, b = 0, 1$.*
  b) $\sum_{a=0}^{1} \sum_{b=0}^{1} (n_{ab00} + n_{ab11})$ *is even.*

Proof: Let e_i be the number of components of $\mathcal{K}_i(L)$ which are homeomorphic to $\mathbf{R}$. We have $2e_i = \sum_{w \in L^0} \alpha_i(w)$ so

$$
\begin{aligned}
e_0 + e_4 &= \frac{1}{2} \sum_{w \in L^0} \alpha_0(w) + \alpha_4(w) \\
&\equiv \sum_{a,b,d=0,1} \sum_{w \in \mathcal{E}_{ab1d}(L)} 1 \bmod 2 \\
&= \sum_{a,b,d=0,1} n_{ab1d}.
\end{aligned}
$$

Likewise $e_2 + e_6 = \sum_{a,b,c=0,1} n_{abc1}$. But then mod 2 we have

$$
0 \equiv \chi(L^1) = \sum_{a,b,c,d} n_{abcd} - \sum_{i=0}^{3} e_{2i} \equiv \sum_{a,b=0,1} n_{ab00} - n_{ab11}
$$

and b) follows.

Now let us prove a). We have

$$
0 \equiv 2e_0 = \sum_{v \in L^0} \alpha_0(v) \equiv \sum_{b,c,d=0,1} n_{1bcd} \bmod 2.
$$

Likewise $0 \equiv 2e_6 \equiv \sum_{a,c,d=0,1} n_{a1cd} \bmod 2$. Hence a) is true except possibly in the case $a = b = 0$. But $\sum_{a,b,c,d} n_{abcd}$ is even since it is the number of points in the zero stratum of L by Lemma 7.1.5. So

$$
\sum_{c,d} (n_{11cd} - n_{00cd}) = \sum_{b,c,d} n_{1bcd} + \sum_{a,c,d} n_{a1cd} - \sum_{a,b,c,d} n_{abcd}
$$

is even. So a) is true. $\square$

Proposition 7.1.7 *The characteristic subsets $\mathcal{Z}_i(X)$ for $i = 0, 1, 2, 3$ are topological invariants of a three dimensional locally conelike stratifiable set X. Furthermore the germ of $\mathcal{Z}_i(X)$ at a point x depends only on the germ of X at x.*

Proof: First we show that $\mathrm{Cl}(\mathcal{K}_i(X))$ is independent of the stratification if $i \neq 2$. Pick a stratification of X. Let X^* be the intrinsic stratification of X, the coarsest topological stratification you can put on X, see **[K]**, **[Si]**. In particular, the i skeleton of X^* consists of points x so that there is a y in the i skeleton of X and a homeomorphism $h \colon (U, x) \to (V, y)$ where U and V are neighborhoods of

x and y in X. This X^* is a topological invariant of X. Note that for each i, the i skeleton of X^* is a union of components of strata of X. Since $\dim X \leq 3$, X^* is necessarily a smooth stratification.

Every component U of the 2-stratum of X is either a submanifold of the 3-stratum of X^* (so $\kappa(U) = 2$) or it is an open subset of some component U' of the two stratum of X^* (so $\kappa(U) = \kappa(U')$). But the closure of any component of the two stratum of X^* is a union of closures of some components of the two stratum of X. Hence

$$\mathrm{Cl}\left(\mathcal{K}_i\left(X\right)\right) = \mathrm{Cl}\left(\mathcal{K}_i\left(X^*\right)\right) \quad \text{if } i \neq 2.$$

Now take a component W of the one stratum of X. Suppose that $W \subset \mathcal{Z}_i\left(X\right)$ for some $i = 0, 1, 2, 3$. Then either

1) W is in the three stratum of X^* or
2) W is in some component U' of the two stratum of X^* or
3) W is in the one stratum of X^*.

In case 1), $\alpha_i\left(W\right) = 0$ if $i \neq 2$ since W is not in $\mathrm{Cl}\left(\mathcal{K}_i\left(X\right)\right) = \mathrm{Cl}\left(\mathcal{K}_i\left(X^*\right)\right)$. So W is in some $\mathcal{E}_{000d}\left(X\right)$. Consequently W is not in $\mathcal{Z}_i\left(X\right)$ for $i = 0, 1, 2, 3$, a contradiction.

In case 2), there are two vertices in the link of W which correspond to the link of W in U'. All the rest of the vertices have $\kappa = 2$, so $\alpha_j\left(W\right) = 0$ if $j \neq 2$ and $j \not\equiv \kappa\left(U'\right) \bmod 8$ and $\alpha_j\left(W\right) = 2$ if $2 \neq j \equiv \kappa\left(U'\right) \bmod 8$. Again α_0 and α_6 are even so W is not in $\mathcal{Z}_i\left(X\right)$ for $i = 0, 1, 2, 3$.

So case 3) must hold, W is in a component W' of the one stratum of X^*. Then $\alpha_i\left(W\right) = \alpha_i\left(W'\right)$ for $i \neq 2$.

Consequently $\mathcal{Z}_0\left(X\right)$ is the closure of the union of all components W' of the 1-skeleton of X^* so that $\alpha_0\left(W'\right) \equiv 1 \bmod 2$, $\alpha_6\left(W'\right) \equiv 1 \bmod 2$ and $\alpha_0(W') + \alpha_4(W') \equiv 2 \bmod 4$. Likewise $\mathcal{Z}_i\left(X\right)$ is the closure of the union of all components W' of the 1-skeleton of X^* so that $\alpha_0\left(W'\right) \equiv a \bmod 2$, $\alpha_6\left(W'\right) \equiv b \bmod 2$ and $\alpha_0\left(W'\right) + \alpha_4\left(W'\right) \equiv 0 \bmod 4$ where $i = 1, 2, 3$ and $ab = 01, 10, 11$ respectively. Since this depends only on the intrinsic stratification X^* which is a topological invariant, we know $\mathcal{Z}_i\left(X\right)$ is a topological invariant for $i = 0, 1, 2, 3$. Clearly the above description of $\mathcal{Z}_i\left(X\right)$ is purely local. $\qquad \square$

Proposition 7.1.8 *Let X be a compact three dimensional Euler polyhedron so that the characteristic subsets $\mathcal{Z}_i\left(X\right)$ for $i = 0, 1, 2, 3$ are all Euler spaces. Let X' be the first barycentric subdivision of X. Then X' is an Euler polyhedron and $\mathcal{Z}_i(X')$ is an Euler space for $i = 0, \dots, 11$.*

Proof: Let $X_i' = \mathcal{Z}_i\left(X'\right)$ for $i = 0, \dots, 11$ be the characteristic subsets in X'.

Assertion 7.1.8.1 X'_5, X'_{10} and X'_{11} are empty. $X'_i = \mathcal{E}_{abc0}(X) \cup \mathcal{E}_{abc1}(X)$ for $i = 6, 7, 8, 9$ and $abc = 010, 011, 100, 101$ respectively. X'_4 is the union of the 1-simplices of X' which go from barycenters of 3-simplices of X to barycenters of 2-simplices or vertices of X. So $\mathcal{E}_{abcd}(X') = \mathcal{E}_{abc0}(X) \cup \mathcal{E}_{abc1}(X)$ if $d \equiv a + b \bmod 2$ and either a or b is nonzero. Likewise $\mathcal{E}_{abcd}(X')$ is empty if $d \not\equiv a + b \bmod 2$ and either a, b or c is nonzero.

Proof: First let us look at 1-simplices of X' which lie in open 3-simplices of X. By examining the barycentric subdivision of the tetrahedron we see that those going from the barycenter of the 3-simplex to the barycenter of a 2-simple or a vertex have $\alpha_i = 0$ for $i \neq 2$ and $\alpha_2 = 6$. Hence they are in $\mathcal{E}_{0001}(X') = X'_4$. Those going to a barycenter of a 1-simplex have $\alpha_i = 0$ for $i \neq 2$ and $\alpha_2 = 4$. Hence they are in $\mathcal{E}_{0000}(X')$ so they are in no X'_i.

Now we shall look at those 1-simplices of X' which lie in an open 2-simplex U of X. Let $i = \kappa(U)$. Then each of these 1-simplices is a face of $2 + i$ of the 2-simplices of X'. Of these, 2 are in U and the other i lie inside the i 3-simplices of X which touch U. So if $i \not\equiv 2 \bmod 8$ we have $\alpha_2 = i$ and $\alpha_i = 2$ and we have $\alpha_j = 0$ for $j \neq 2, i$. If $i \equiv 2 \bmod 8$ we have $\alpha_2 = i + 2$ and $\alpha_j = 0$ for $j \neq 2$. So if $i \equiv 0 \bmod 4$ these one simplices lie in $\mathcal{E}_{0010}(X')$ and if $i \equiv 2 \bmod 4$ they lie in $\mathcal{E}_{0000}(X')$. In any case they are in no X'_i.

Finally let us look at 1-simplices W' of X' which lie in an open 1-simplex W of X. The link L' of W' in X' is the first barycentric subdivision of the link L of W in X. Then $\alpha_i(W') = \alpha_i(W)$ if $i \neq 2$ and $\alpha_2(W') = \alpha_2(W) + g$ where g is the number of edges in L. So $\alpha_0(W) = \alpha_0(W')$, $\alpha_6(W) = \alpha_6(W')$ and $\alpha_0(W) + \alpha_4(W) = \alpha_0(W') + \alpha_4(W')$. But by Lemma 7.1.4,

$$g \equiv \alpha_2(W) + 2\alpha_4(W) + 3\alpha_6(W) \bmod 4.$$

So mod 4 we have

$$\begin{aligned}
\alpha_2(W') + \alpha_6(W') &\equiv \alpha_2(W) + g + \alpha_6(W) \\
&\equiv 2(\alpha_2(W) + \alpha_4(W)) \\
&\equiv 2(\alpha_6(W) + \alpha_0(W))
\end{aligned}$$

Consequently, if $W \subset \mathcal{E}_{abcd}(X)$ then $W' \subset \mathcal{E}_{abcd'}(X')$ where $d' \equiv a + b \bmod 2$.

$\square$

We know by Proposition 7.1.7 that $X'_i = \mathcal{Z}_i(X)$ for $i = 0, 1, 2, 3$ so they are all Euler spaces. We also have $X'_6 = X'_1 = \mathcal{Z}_1(X)$ and $X'_8 = X'_2 = \mathcal{Z}_2(X)$ so these are both Euler spaces. By Lemma 7.1.6a and the fact that X'_0 and X'_3 are Euler spaces, we know that $\bigcup_{c,d} \mathcal{E}_{abcd}(X')$ is an Euler space for each $a, b = 0, 1$. Taking $ab = 01$ and 10, by Assertion 7.1.8.1 we then get $X'_6 \cup X'_7$ and $X'_8 \cup X'_9$ are both Euler spaces. So X'_7 and X'_9 are both Euler spaces. The only one left is X'_4. Take a vertex v of X. Then by Assertion 7.1.8.1 there is one branch of X'_4 coming into v for each 2-simplex in the link of v in X. But this is an even number of branches by Lemma 7.1.3e. Now take a barycenter v of a 2-simplex W of X. There is one branch of X'_4 coming into v for each 3-simplex containing W, i.e., there are $\kappa(W)$ branches but $\kappa(W)$ is even. Finally consider a barycenter v of a 3-simplex of X. There is one branch of X'_4 coming into v for each 0 or 2-simplex in the boundary of the 3-simplex, i.e. there are 8 branches. So X'_4 is an Euler space.   □

Lemma 7.1.9 *Let X be a polyhedron which is also an Euler space. Then the first barycentric subdivision X' of X is an Euler polyhedron.*

Proof: The link of an open i simplex A of X' in X' is the join of the first barycentric subdivision of the boundary of a j-simplex in X with the first barycentric subdivision L' of L. Here A is contained in an open $i + j$ simplex of X whose link in X is L.

The first claim is that all the strata of L' have even Euler characteristic. This is because for $k > 0$ the number of open k simplices in the first barycentric subdivision of the n simplex $\triangle^n$ which are in the interior of $\triangle^n$ is even. Hence L' has an even number of simplices in each dimension > 0. But the mod 2 Euler characteristic of L' is 0 and is also the number of simplices in L'. Hence L' also has an even number of 0-simplices. Since it has an even number of simplices in each dimension, each skeleton of L' has even Euler characteristic. The result now follows from Lemma 7.1.10 below.   □

Lemma 7.1.10 *If K and L are compact polyhedra so that each skeleton of K and of L has even Euler characteristic, then each skeleton of the join $K * L$ has even Euler characteristic.*

Proof: The property that each skeleton have even Euler characteristic is equivalent to there being an even number of simplices of each dimension. The result then follows from the observation that the i-simplices of $K * J$ are either i-simplices of K, i-simplices of J or joins of j simplices of J and $i - j - 1$ simplices of K. There are an even number of simplices of each type. Hence $J * K$ has an even number of simplices in each dimension.   □

2. The Cobordism Groups

In order to prove Theorems 7.1.1 and 7.1.2 it turns out we need to understand when a resolution tower with dimension ≤ 2 bounds. It is possible to define the relevant bordism groups and compute them in these low dimensions, but the groups tend to get complicated. For example, bordism of two dimensional towers is $\mathbf{Z}/2\mathbf{Z}^\infty$. Therefore it is more practical to make some simplifications. As an exercise the reader can try to figure out the more general cobordism groups, the calculations are similar to those given below only there is more stuff to keep track of.

The first simplification we make is to assume all our towers are well indexed. By Proposition 4.1.2 and Lemma 4.1.3 there is no reason not to do so. The next simplification we make is to just calculate when a tower is a weak boundary as defined below, in other words we ignore the given sheet decomposition $\mathcal{A}_{ji}$ of each V_{ji}. (Equivalently we could assume each $\mathcal{A}_{ji}$ has at most one nonempty sheet.)

Definition: We say that $\mathfrak{T} = \{V_i, \mathcal{A}_i, p_i\}_{\mathfrak{I}}$ is a *weak boundary* of the resolution tower $\mathfrak{T}' = \{V_i', \mathcal{A}_i', p_i'\}_{\mathfrak{I}'}$ if $\mathfrak{I} \subset \mathfrak{I}'$, $V_i = \partial V_i'$ for all $i \in \mathfrak{I}$, $\partial V_i' = \emptyset$ if $i \in \mathfrak{I}' - \mathfrak{I}$ and $V_{ji} = \partial V_{ji}'$ and $p_{ji} = p_{ij}'|$ if $j < i$.

By Theorem 4.2.8, there is no reason not to just look at when a tower is a weak boundary. After all, the actual sheet decomposition of V_{ji} is unimportant, what is important is V_{ji} itself. This simplifies things quite a bit, but at this point we would still have infinitely generated two dimensional bordism. The final simplification is to limit the exponents of the p_{ji}'s to be ≤ 2. In particular:

Definition: We say a resolution tower $\mathfrak{T} = \{V_i, \mathcal{A}_i, p_i\}_{\mathfrak{I}}$ of type M has exponents $\leq k$ if for each mico p_{ji}, all the local exponents are $\leq k$.

In these low dimensions anyway, limiting the exponents to be ≤ 2 is no problem since there are only two topological types of monomial maps from $\mathbf{R}$ to $\mathbf{R}$, and these are given by the monomials x and x^2.

Definition: Let $\mathfrak{TR}_n'$ denote the set of closed, n dimensional, well indexed resolution towers of type UM which have all exponents ≤ 2. Assume the bias is 0 also.

If $\mathfrak{T}$ is a compact well indexed resolution tower of type UM, bias 0 and dimension n for $n = 0, 1, 2$ we will define below an additive invariant $\rho_n(\mathfrak{T}) \in \mathbf{Z}/2\mathbf{Z}^{a_n}$ for $a_n = 1, 2, 67$ so that the following is true:

Theorem 7.2.1 *If $\mathfrak{T} \in \mathfrak{TR}_n$ is a weak boundary of a compact resolution tower of type UM for $n = 0, 1, 2$ then $\rho_n(\mathfrak{T}) = 0$. The maps $\rho_0|_{\mathfrak{TR}_0'}$ and $\rho_1|_{\mathfrak{TR}_1'}$ are onto. Furthermore, the following are equivalent:*

a) *$\mathfrak{T} \in \mathfrak{TR}_n'$ is a weak boundary of a compact resolution tower of type UM.*
b) *$\rho_n(\mathfrak{T}) = 0$.*
c) *$\mathfrak{T} \in \mathfrak{TR}_n'$ is a weak boundary of a compact resolution tower $\mathfrak{T}'$ of type S so that $|\mathfrak{T}'|$ is the cone on $|\mathfrak{T}|$ and so that $\mathfrak{T}'$ has all exponents ≤ 2.*

When we say the ρ_n are additive we mean that ρ_n of a disjoint union is the sum of the ρ_n's.

It turns out that $\rho_2|_{\mathfrak{TR}_2'}$ is not onto, its image is a subgroup of index 4 but it is more convenient to use ρ_2 than an invariant mapping onto $\mathbf{Z}/2\mathbf{Z}^{65}$.

The next question is whether you can detect the bordism class of a $\mathfrak{T} \in \mathfrak{TR}_n'$ from its realization $|\mathfrak{T}|$. In other words, can you determine $\rho_n(\mathfrak{T})$ from the stratified set $|\mathfrak{T}|$. If $n = 0$ or 1 you can, but for $n = 2$ the answer is no, you only end up determining roughly one fourth of ρ_2.

More specifically, let $\mathcal{S}_n$ denote the set of locally conelike smooth stratified sets of dimension n. We define $\rho'_n \colon \mathcal{S}_n \to \mathbf{Z}/2\mathbf{Z}^{b_n}$ for $n = 0, 1, 2$ and $b_n = 1, 2, 15$ as follows. Pick $X \in \mathcal{S}_n$ and let X_i denote the i-skeleton of X. Let χ' denote the mod 2 Euler characteristic. We let:

$$
\begin{aligned}
\rho'_0(X) &= \chi'(X) \\
\rho'_1(X) &= (\chi'(X_0), \chi'(X)) \\
\rho'_2(X) &= (\chi'(X_0), \chi'(X_1), \chi'(X), \chi'(\mathcal{Z}_0(X)), \dots, \chi'(\mathcal{Z}_{11}(X)))
\end{aligned}
$$

We have a map $\nu_n \colon \mathcal{TR}_n \to \mathcal{S}_n$ where $\nu_n(\mathfrak{T}) = |\mathfrak{T}|$. then we will prove the following:

Theorem 7.2.2 *For $n = 0$, 1 and 2 there are surjective linear maps*

$$
\rho''_n \colon \mathbf{Z}/2\mathbf{Z}^{a_n} \to \mathbf{Z}/2\mathbf{Z}^{b_n}
$$

so that

$$
\rho''_n \circ \rho_n|_{\mathcal{TR}_n} = \rho'_n \circ \nu.
$$

In particular, ρ''_0 and ρ''_1 are isomorphisms.

$$
\begin{array}{ccc}
\mathcal{TR}_n & \xrightarrow{\rho_n} & \mathbf{Z}/2\mathbf{Z}^{a_n} \\
\downarrow \nu_n & & \downarrow \rho''_n \\
\mathcal{S}_n & \xrightarrow{\rho'_n} & \mathbf{Z}/2\mathbf{Z}^{b_n}
\end{array}
$$

Let us now get more specific and define these maps ρ_n, discover their properties and prove Theorems 7.2.1 and 7.2.2.

Definition: We will use the notation that if K is a finite set then $\sharp K$ will be the number of points in K.

The case $n = 0$ is very easy. In this case, $\mathfrak{T} = \{ V_0, \emptyset, \emptyset \}$ and V_0 is a compact 0-dimensional manifold, i.e., a finite number of points. Then $\rho_0(\mathfrak{T}) = \sharp V_0 \bmod 2$.

Clearly ρ_0 is onto $\mathbf{Z}/2\mathbf{Z}$ (consider V_0 with just one point). Also $\rho_0(\mathfrak{T}) = \rho'_0(|\mathfrak{T}|)$ so we may set $\rho'' =$ the identity and thus prove Theorem 7.2.2 for $n = 0$. If $\mathfrak{T} = \partial \mathfrak{T}'$ for a compact $\mathfrak{T}'$ then $V_0 = \partial V'_0$ where V'_0 is a compact one dimensional manifold. So V_0 has an even number of points so $\rho_0(\mathfrak{T}) = 0$. So to prove Theorem 7.2.1 for $n = 0$ we must only show that if $\mathfrak{T} \in \mathcal{TR}'_0$ and $\rho_0(\mathfrak{T}) = 0$ then $\mathfrak{T}$ is a weak boundary of a compact resolution tower $\mathfrak{T}'$ of type S so that $|\mathfrak{T}'|$ is the cone on $|\mathfrak{T}|$.

So assume $\mathfrak{T} \in \mathcal{TR}'_0$ and $\rho_0(\mathfrak{T}) = 0$. Since $\sharp V_0$ is even we may write $V_0 = \{q_{-11}, q_{11}, q_{-12}, q_{12}, \dots, q_{-1m}, q_{1m}\}$, in other words we may pair up the points of V_0 in some arbitrary way. Let $\mathbf{m} = \{1, \ldots, m\}$. Set $V'_0 = [-1,1] \times \mathbf{m}$, $V'_{-1,0} = 0 \times \mathbf{m}$, $V'_{-1} =$ a point and $p'_{-1,0} \colon V'_{-1,0} \to V'_{-1}$ the only map possible. Then this gives a resolution tower $\mathfrak{T}'$ of type S and $\mathfrak{T} = \partial \mathfrak{T}'$ and $|\mathfrak{T}'| = \mathfrak{c}|\mathfrak{T}|$. So Theorem 7.2.1 is proven for $n = 0$.

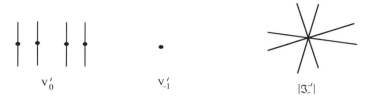

$$V'_0 \qquad\qquad V'_{-1} \qquad\qquad |\mathfrak{T}'|$$

FIGURE VII.2.1. $\mathfrak{T}'$ and its resolution

Let us now look at the case $n = 1$. Take any $\mathfrak{T} \in \mathcal{TR}_1$. If $\mathfrak{T} = \{V_i, \mathcal{A}_i, p_i\}_{i=0}^1$ let c_d for $d = 0, 1$ be the number of points x in V_0 so that $\sharp p_{01}^{-1}(x)$ is congruent to $d \bmod 2$. We then define $\rho_1(\mathfrak{T}) = (c_0, c_1)$.

Suppose that $\mathfrak{T}$ is weakly isomorphic to the boundary of some compact resolution tower $\mathfrak{T}'$. Then we know $V_0 = \partial V'_0$ so $\sharp V_0$ is even, i.e., $c_0 + c_1$ is even. Also we know that $V_{01} = \partial V'_{01}$ so $\sharp V_{01}$ is even. But mod 2 we have $\sharp V_{01} = c_1$. Consequently $\rho_1(\mathfrak{T}) = (0, 0)$.

Now suppose $\mathfrak{T} \in \mathcal{TR}_1$ and $\rho_1(\mathfrak{T}) = 0$. We will show in Theorem 7.5.1 below that $\mathfrak{T}$ is a weak boundary of a compact resolution tower $\mathfrak{T}'$ of type S so that $|\mathfrak{T}'|$ is the cone on $|\mathfrak{T}|$ and all exponents of $\mathfrak{T}'$ are ≤ 2. Thus Theorem 7.2.1 will be proven for $n = 1$.

We can see that ρ_1 is onto by considering ρ_1 of the two following towers:

1) $V_1 =$ the circle S^1, $V_{01} = \emptyset$, $V_0 =$ a point. This has $\rho_1 = (1, 0)$.
2) $V_1 = S^1$, $V_{01} =$ a point in S^1, $V_0 =$ a point. This has $\rho_1 = (0, 1)$.

Finally let us find ρ''_1. The zero skeleton of $|\mathfrak{T}|$ is V_0 and $c_0 + c_1 \equiv \sharp V_0 \bmod 2$. But it is easy to compute that the mod 2 Euler characteristic of $|\mathfrak{T}|$ is c_0. Thus $\rho''_1(c_0, c_1) = (c_0 + c_1, c_0)$ which is clearly an isomorphism.

We now proceed to the case $n = 2$. First we indicate a few useful properties of resolution towers with small depth.

Lemma 7.2.3 *Let $\mathfrak{T} = \{V_i, \mathcal{A}_i, p_i\}_{i=0}^2$ be a well indexed resolution tower of type UM. Let V^2_{02} denote the double point set of V_{02}. Then*

a) *V_{12}, $V_{12} \cap V_{02}$, V^2_{02} and V_{01} are all submanifolds of V_2 or V_1.*
b) *If S and T are two different sheets of $\mathcal{A}_2$ and $S \cap T \neq \emptyset$ then either S or T is in $\mathcal{A}_{02}$.*
c) *Any three different sheets of $\mathcal{A}_2$ have empty intersection. Any two sheets of $\mathcal{A}_1$ have empty intersection.*
d) *The maps $p_{01} \colon V_{01} \to V_0$, $p_{02}| \colon V_{12} \cap V_{02} \to V_0$, $p_{12}| \colon V_{12} \cap V_{02} \to V_{01}$ and $p_{02}| \colon V^2_{02} \to V_0$ are covering projections.*

e) *For each sheet $S \in \mathcal{A}_{02}$, if $g \colon S' \to V_2$ is the immersion associated to S, then $p_{02} \circ g \colon S' \to V_0$ is a proper submersion, hence a locally trivial fibration.*

f) *$\mathfrak{T}$ has type S.*

Proof: Let $\mathfrak{T}$ have bias m. Let S and T be two different sheets of $\mathcal{A}_{12}$. If $S \cap T \neq \emptyset$, then by transversality of the sheets we know that $S \cap T$ has dimension m. But by type U we know that p_{12} submerses a nonempty open subset of $S \cap T$ which is a contradiction since a nonempty m dimensional manifold cannot submerse to an $m + 1$ dimensional manifold. So b) is shown.

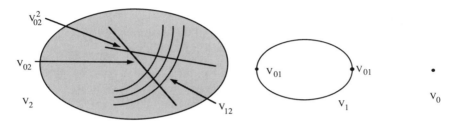

FIGURE VII.2.2. A typical tower in $\mathfrak{T}\mathcal{R}_2$

Likewise, by type U an open dense subset of the double point set of V_{12} submerses to V_1 and so the double point set must be empty. So V_{12} is a manifold. If $V_{12} \cap V_{02}$ is not a manifold, then $V_{12} \cap V_{02}^2$ is nonempty. But $V_{12} \cap V_{02}^2$ has dimension $m - 1$ and p_{02} submerses an open dense subset of it to V_0, which has dimension m. So $V_{12} \cap V_{02}^2$ is empty, so $V_{12} \cap V_{02}$ is a manifold. Likewise, the triple point set of V_{02} and the double point set of V_{01} are empty so V_{02}^2 and V_{01} are manifolds.

To see c), note that we showed above that the double point set of V_{01} and the triple point set of $V_{12} \cup V_{02}$ are empty.

Conclusion d) follows from c) and the fact that the maps are proper submersions (and hence local diffeomorphisms) between manifolds of the same dimension. (To see that $p_{12}| \colon V_{12} \cap V_{02} \to V_{01}$ is a local diffeomorphism, note that its composition with the local diffeomorphism p_{01} is the local diffeomorphism $p_{02}|$.) (Of course the degree of the cover could vary from component to component of the base.)

Conclusion e) is immediate from type U. Finally, to see conclusion f), note that type U immediately implies type S everywhere except $V_{02} \cap V_{12}$. But for any $x \in V_{02} \cap V_{12}$, pick open embeddings $\theta \colon \mathbf{R} \times \mathbf{R}^m \to V_{12}$ and $\psi \colon \mathbf{R} \times \mathbf{R}^m \to V_1$ so that $\theta(0,0) = x$, $\psi(0,0) = p_{12}(x)$, $\theta^{-1}(V_{02}) = 0 \times \mathbf{R}^m$, $\psi^{-1}(V_{01}) = 0 \times \mathbf{R}^m$ and $p_{12}\theta(\mathbf{R} \times \mathbf{R}^m) \subset \psi(\mathbf{R} \times \mathbf{R}^m)$. Then $\psi^{-1}p_{12}\theta(t,x) = (t^d\varphi(t,x), \mu(t,x))$ with $\varphi(0,0) \neq 0$ and furthermore by type U we know $\mu \colon 0 \times \mathbf{R}^m \to \mathbf{R}^m$ is a submersion, i.e., an open embedding. Then after composing ψ with the inverse of the local diffeomorphism $(t, x) \mapsto (t|\varphi(t,x)|^{1/d}, \mu(t,x))$ we see that p_{12} has type S.					□

We now give the definition of ρ_2.

Pick $\mathfrak{T} = \{V_i, \mathcal{A}_i, p_i\}_{i=0}^{2}$, a compact well indexed resolution tower of type UM, bias 0 and dimension 2. We define a map $\zeta \colon V_{02} \cap V_{12} \to \{1, 2\}$ as follows. We know by Lemma 3.7.1 that $p_{12} \colon (V_{12}, V_{12} \cap \mathcal{A}_{02}) \to (V_1, \mathcal{A}_1)$ is a tico map. Pick any $x \in V_{02} \cap V_{12}$. Let K be the sheet of $V_{12} \cap \mathcal{A}_{02}$ containing x and let L be the sheet of $\mathcal{A}_1$ containing $p_{12}(x)$. Then we let $\zeta(x)$ be the mod 2 exponent of K in L for the map p_{12}. In other words, p_{12} looks like $t \mapsto t^{\zeta(x)}$ near x after a continuous change of coordinates.

We now define $\zeta' \colon V_{01} \to \mathbf{Z}/2\mathbf{Z}^2$ as follows. For any $x \in V_{01}$ we let

$$\zeta'(x) = \left(\sharp(p_{12}^{-1}(x) \cap \zeta^{-1}(1)), \sharp(p_{12}^{-1}(x) \cap \zeta^{-1}(2)) \right).$$

In other words, $\zeta'(x) = (\zeta_1'(x), \zeta_2'(x))$ where $\zeta_d'(x)$ is the mod 2 number of points y of $p_{12}^{-1}(x)$ so that $\zeta(y) = d$.

Let $f \colon Q \to V_2$ be the immersion associated to V_{02}. We know that if $x \in V_0$ then $f^{-1}p_{02}^{-1}(x)$ is a union of circles. Let $\xi_0(x) \in \mathbf{Z}/2\mathbf{Z}$ be the mod 2 number of these circles C so that the restriction to C of the normal bundle of the immersion is nontrivial. (The normal bundle of the immersion is the quotient bundle $f^*(TV_2)/TQ$ where $f^*(TV_2)$ is the pullback of the tangent bundle of V_2, so its fibre over any point y is $T_{f(y)}V_2/df(T_yQ)$.) For example, if $\mathcal{A}_{02}$ is regular then $\xi_0(x)$ is the mod 2 number of components S of sheets of $\mathcal{A}_{02}$ so that $p_{02}(S) = \{x\}$ and S has nontrivial normal bundle.

For each point $x \in V_0$ we define $\xi_1(x) \in \mathbf{Z}/2\mathbf{Z}$ to be the mod 2 number of points in $V_{02}^2 \cap p_{02}^{-1}(x)$.

Finally for each point $x \in V_0$ and $a, b \in \mathbf{Z}/2\mathbf{Z}$ we define $\xi_{2ab}(x) \in \mathbf{Z}/2\mathbf{Z}$ to be the mod 2 number of points $y \in p_{01}^{-1}(x)$ so that $\zeta'(y) = (a, b)$. In other words,

$$\xi_{2ab}(x) \equiv \sharp \left(p_{01}^{-1}(x) \cap \zeta'^{-1}(a, b) \right).$$

So we get a map $\xi \colon V_0 \to (\mathbf{Z}/2\mathbf{Z})^6$ given by

$$\xi(x) = (\xi_0(x), \xi_1(x), \xi_{200}(x), \xi_{201}(x), \xi_{210}(x), \xi_{211}(x)).$$

For each component T of V_1 define $\omega(T) \in \mathbf{Z}/2\mathbf{Z}$ to be the mod 2 degree of the proper map $p_{12} \colon V_{12} \cap p_{12}^{-1}(T) \to T$, i.e., the mod 2 number of points in $p_{12}^{-1}(x)$ for any $x \in T - V_{01}$. For $i = 0, 1$, let V_1^i be the union of components T of V_1 so that $\omega(T) = i$.

At last we can define the invariant $\rho_2(\mathfrak{T}) \in \mathbf{Z}/2\mathbf{Z}^{67} = \mathbf{Z}/2\mathbf{Z}^3 \oplus \mathbf{Z}/2\mathbf{Z}^{\mathbf{Z}/2\mathbf{Z}^6}$. For $i = 0, 1$ let $\eta_i(\mathfrak{T})$ be the number of components of $p_{12}^{-1}(V_1^i)$ which have nontrivial normal bundle in V_2. Let $\eta_2(\mathfrak{T})$ be the mod 2 Euler characteristic of V_2. For any

$a, b, c, d, e, f \in \mathbf{Z}/2\mathbf{Z}$ define $\eta_{3abcdef}(\mathfrak{T}) \in \mathbf{Z}/2\mathbf{Z}$ to be $\sharp\xi^{-1}(a, b, c, d, e, f)$ mod 2. We now define

$$\rho_2(\mathfrak{T}) = \Big(\eta_0(\mathfrak{T}), \eta_1(\mathfrak{T}), \eta_2(\mathfrak{T}), \eta_{3000000}(\mathfrak{T}), \dots, \eta_{3111111}(\mathfrak{T}) \Big).$$

We will now show what happens to ρ_2 if you blow up a point in V_{02}. Note that Proposition 7.2.4 below implies that $\rho_2(\mathfrak{T})$ is not determined by $|\mathfrak{T}|$. The notation $\xi_k(\mathfrak{T}, x)$ just means the functions ξ_k defined above, but with respect to the resolution tower $\mathfrak{T}$.

Proposition 7.2.4 Suppose $\mathfrak{T} = \{V_i, \mathcal{A}_i, p_i\}_{i=0}^2 \in \mathcal{TR}_2$. Pick any $x \in V_0$ and $y \in p_{02}^{-1}(x)$. Let $\mathfrak{T}'$ be the tico tower blowup of $\mathfrak{T}$ with center y.
    Then $\eta_2(\mathfrak{T}') = \eta_2(\mathfrak{T})+1$. For any $a, b \in \mathbf{Z}/2\mathbf{Z}$ and $z \in V_0$ we have $\xi_{2ab}(\mathfrak{T}', z) = \xi_{2ab}(\mathfrak{T}, z)$. We also have $\xi_i(\mathfrak{T}', z) = \xi_i(\mathfrak{T}, z)$ for any $z \in V_0 - x$ and $i = 0, 1$. However the other invariants change in the following ways:

   a) If $y \in V_{02}^2$ then
        i) $\eta_0(\mathfrak{T}') = \eta_0(\mathfrak{T})$ and $\eta_1(\mathfrak{T}') = \eta_1(\mathfrak{T})$.
        ii) $\xi_0(\mathfrak{T}', x) = \xi_0(\mathfrak{T}, x) + 1$ and $\xi_1(\mathfrak{T}', x) = \xi_1(\mathfrak{T}, x) + 1$.
   b) If $y \in V_{02} - (V_{12} \cup V_{02}^2)$ then
        i) $\eta_0(\mathfrak{T}') = \eta_0(\mathfrak{T})$ and $\eta_1(\mathfrak{T}') = \eta_1(\mathfrak{T})$.
        ii) $\xi_0(\mathfrak{T}', x) = \xi_0(\mathfrak{T}, x)$ and $\xi_1(\mathfrak{T}', x) = \xi_1(\mathfrak{T}, x) + 1$.
   c) If $y \in V_{02} \cap V_{12}$ and if $d = 0, 1$ is such that $p_{12}(y) \in V_1^d$ then
        i) $\eta_d(\mathfrak{T}') = \eta_d(\mathfrak{T}) + 1$ and $\eta_{1-d}(\mathfrak{T}') = \eta_{1-d}(\mathfrak{T})$.
        ii) $\xi_0(\mathfrak{T}', x) = \xi_0(\mathfrak{T}, x)$ and $\xi_1(\mathfrak{T}', x) = \xi_1(\mathfrak{T}, x) + 1$.

Proof: That $\eta_2(\mathfrak{T}') = \eta_2(\mathfrak{T}) + 1$ follows from the fact that V_2' is the connected sum of V_2 with $\mathbf{RP}^2$. All the other changes follow from the fact that blowing up a point in a curve in a surface gives the normal bundle a half twist, and the fact that the sheet $\pi^{-1}(y)$ in V_{02}' has a nontrivial normal bundle. $\square$

We now give a bundle result which indicates why we are always counting nontrivial bundles over the circle in our invariants.

Lemma 7.2.5 Let $\pi \colon E \to S$ be a line bundle over a compact surface with boundary. Let n be the number of components C of ∂S such that the bundle restricted to C is nontrivial. Then n is even.
    Conversely, if $\rho \colon F \to T$ is a line bundle over a closed 1-manifold and the number of components of T with nontrivial bundle is even, then there is a line bundle $\pi \colon E \to S$ over a compact surface S so that $\partial S = T$, $F = \pi^{-1}(T)$ and $\rho = \pi|_F$.

Proof: This is actually just an easy computation in $\mathfrak{N}_1(BO(1))$, but we give a more elementary proof. Let the bundle π be classified by a map $f \colon S \to BO(1) = \mathbf{RP}^\infty$. Then $f_*([\partial S])$ is null homologous in $H_1(BO(1); \mathbf{Z}/2\mathbf{Z}) = \mathbf{Z}/2\mathbf{Z}$. But the nontrivial bundle over the circle represents the generator of $H_1(BO(1); \mathbf{Z}/2\mathbf{Z})$

and the trivial bundle represents 0. So there must be an even number of non-trivial bundles.

Prove the converse by example, by pairing up the nontrivial bundles and making each pair the boundary of a nontrivial bundle over a cylinder.   □

Proposition 7.2.6 below shows part of Theorem 7.2.1 for $n = 2$. The rest of the proof of Theorem 7.2.1 will be delayed until Theorem 7.5.2.

Proposition 7.2.6 *If* $\mathfrak{T} = \{V_i, \mathcal{A}_i, p_i\}_{i=0}^3$ *is a compact 3 dimensional well indexed resolution tower with boundary and bias 0 and type UM, then* $\rho_2(\partial\mathfrak{T}) = 0$.

Proof: Since ∂V_3 bounds, it has even Euler characteristic, hence $\eta_2(\partial\mathfrak{T}) = 0$. Now let S be a component of V_2. Let $U = p_{23}^{-1}(S)$. Now the mod 2 degree is a bordism invariant, so ω is the same on each component of ∂S. So to show $\eta_i(\mathfrak{T}) = 0$ for $i = 0, 1$ it suffices to show that there are an even number of components of ∂U with nontrivial normal bundle in ∂V_3. This follows from Lemma 7.2.5.

So we must only show that $\eta_{3abcdef}(\mathfrak{T}) = 0$ for each $a, b, c, d, e, f \in \mathbf{Z}/2\mathbf{Z}$. In other words, there are an even number of points $x \in \partial V_1$ so that $\xi(x) = (a, b, c, d, e, f)$.

Let T be any component of V_1 which has boundary. Then T is an interval with two endpoints which we call t_0 and t_1. We claim that $\xi(t_0) = \xi(t_1)$ which would immediately imply that $\eta_{3abcdef}(\mathfrak{T}) = 0$.

Let us first show that $\xi_0(t_0) = \xi_0(t_1)$. Let $f\colon Q \to V_3$ be the immersion associated to V_{13}. Let $Q' = f^{-1}p_{13}^{-1}(T)$ be the union of the components of Q which map to T. Then by Lemma 7.2.5 there are an even number of components of $\partial Q'$ with nontrivial normal bundle. So $\xi_0(t_0) = \xi_0(t_1)$.

Now let us show that $\xi_1(t_0) = \xi_1(t_1)$. But this is true since the mod 2 degree is a bordism invariant, the bordism being $p_{13}|\colon V_{13}^2 \cap p_{13}^{-1}(T) \to T$.

So it only remains to show that $\xi_{2ab}(t_0) = \xi_{2ab}(t_1)$ for each $a, b \in \mathbf{Z}/2\mathbf{Z}$, i.e., that there are an even number of points q of $p_{12}^{-1}(\{t_0, t_1\})$ so that $\zeta'(q) = (a, b)$. Let Q be any component of $p_{12}^{-1}(T)$ which has boundary. Then Q is an interval with boundary points q_0 and q_1. It suffices to show that $\zeta'(q_0) = \zeta'(q_1)$. Let R be any component of $p_{23}^{-1}(Q)$ with boundary points. Then R is an interval with boundary points r_0 and r_1. Then it suffices to show that $\zeta(r_0) = \zeta(r_1)$. Let K be the sheet in $\mathcal{A}_{13}$ so that $K \cap V_{23} \supset R$ and let L be the sheet in $\mathcal{A}_2$ containing Q. Then $\zeta(r_i) =$ the exponent of K in L for the map p_{23}. But this exponent is constant on the connected set R, so $\zeta(r_0) = \zeta(r_1)$.   □

We will now prove Theorem 7.2.2 for $n = 2$ which will be a consequence of Lemma 7.2.7 below.

Lemma 7.2.7 *Let* $\mathfrak{T} = \{V_i, \mathcal{A}_i, p_i\}_{i=0}^2 \in \mathfrak{TR}_2$ *be a compact well indexed resolution tower of type UM with bias 0 and without boundary. Let* $X = |\mathfrak{T}|$ *and let* X_i *denote the* i-*skeleton of* X. *Then:*

    a) $\chi(X_0) \equiv \sum_{a,b,c,d,e,f} \eta_{3abcdef}(\mathfrak{T})$ *mod* 2.

    b) $\chi(X_1) \equiv \sum_{a,b,c,d,e,f} (1 + c + d + e + f)\eta_{3abcdef}(\mathfrak{T})$ *mod* 2.

    c) $\chi(X_2) \equiv \eta_2(\mathfrak{T}) + \sum_{a,b,c,d,e,f} (1 + b + c + f)\eta_{3abcdef}(\mathfrak{T})$ *mod* 2.

    d) *For any* $a, b, c, d = 0, 1$ *we have* $\chi(\mathcal{E}_{abcd}(X)) \equiv \sum_{ef} \eta_{3efgahb}(\mathfrak{T})$ *mod* 2 *where* $g = |c - a|$ *and* $h = |d - b|$.

Proof: Let $W_i = V_i - |\mathcal{A}_i|$, $W_{ji} = V_{ji} - \bigcup_{m<j} V_{mi}$, $\pi_{ji}\colon U_{ji} \to W_j$ and $\rho_{ji}\colon U_{ji} \to [0, \infty)$ satisfy the conclusions of Lemma 4.3.2. Pick ϵ small and let $U'_{ji} = \rho_{ji}^{-1}([0, \epsilon]) \subset U_{ji}$. Let U''_{ji} denote the image of U'_{ji} under the quotient map $V_0 \cup V_1 \cup V_2 \to X$. It is a neighborhood of the j strata in the i skeleton.

Pick any $x \in V_0$. Then $\pi_{01}^{-1}(x) \cap U'_{01}$ is a collection of intervals $Q_1, \dots, Q_k$ where $k = \sharp p_{01}^{-1}(x)$. Let the endpoints of Q_i be denoted q_{1i} and q_{-1i} and let the midpoint $Q_i \cap p_{01}^{-1}(x)$ be denoted y_i. So $p_{01}^{-1}(x) = \{y_1, \dots, y_k\}$ and $\pi_{01}^{-1}(x) \cap \rho_{01}^{-1}(\epsilon) = \{q_{11}, q_{-11}, q_{12}, \dots, q_{-1k}\}$. Note that the link of x has a resolution tower $\mathfrak{T}' = \{V'_i, \mathcal{A}'_i, p'_i\}_{i=0}^1$ where $V'_0 = \{q_{11}, q_{-11}, q_{12}, \dots, q_{-1k}\}$, $V'_1 = \pi_{02}^{-1}(x) \cap \rho_{02}^{-1}(\epsilon)$, $|\mathcal{A}'_1| = V'_{01} = V'_1 \cap V_{12}$ and $p'_{01} = p_{12}|$. So the points $q_{\pm 1i}$ correspond to vertices in the link of x.

Let $e_{di} = \sharp(p_{12}^{-1}(y_i) \cap \zeta^{-1}(d))$ for $d = 1, 2$. Note that $e_{di} \equiv \zeta'_d(y_i)$ mod 2.

Assertion 7.2.7.1 $\kappa(q_{1i}) + \kappa(q_{-1i}) = 4e_{1i} + 4e_{2i}$ *and* $\kappa(q_{si}) \equiv 2e_{1i}$ *mod* 4 *for all* i *and* $s = \pm 1$.

Proof: Pick any $i = 1, \dots, k$. Note that $p_{12}^{-1}(Q_i)$ is a collection of intervals $R_1, \dots, R_m$ with midpoints $\{r_1, \dots, r_m\} = p_{12}^{-1}(y_i)$. If $\zeta(r_j) = 1$ then p_{12} maps R_j homeomorphically to Q_i since it is locally a map $t \mapsto \pm t^d$ for d odd. If $\zeta(r_j) = 2$ then p_{12} folds R_j onto one half of Q_i. since it is locally a map $t \mapsto \pm t^d$ for d even. Thus $R_j \cap p_{12}^{-1}(q_{si})$ is one point if $\zeta(r_j) = 1$ and it is either zero or two points if $\zeta(r_j) = 2$. Since there are two points in the link of q_{si} for every point of $p_{12}^{-1}(q_{si})$, the result follows. $\square$

Now let us prove d). Pick any $x \in V_0$. Let y_i, q_{si} and e_{di} be as above. If $\zeta'_2(y_i) = 0$ then e_{2i} is even so by Assertion 7.2.7.1 we must have either $\kappa(q_{1i}) \equiv \kappa(q_{-1i}) \equiv 2e_{1i}$ or $\kappa(q_{1i}) \equiv \kappa(q_{-1i}) \equiv 2e_{1i} + 4$ mod 8. If $\zeta'_2(y_i) = 1$ then e_{2i} is odd so we must have either $\kappa(q_{1i}) \equiv \kappa(q_{-1i}) + 4 \equiv 2e_{1i}$ or $\kappa(q_{1i}) \equiv \kappa(q_{-1i}) + 4 \equiv 2e_{1i} + 4$ mod 8. Consequently we have:

$$\alpha_0(x) \equiv \xi_{201}(x) \text{ mod } 2$$
$$\alpha_6(x) \equiv \xi_{211}(x) \text{ mod } 2$$
$$\alpha_0(x) + \alpha_4(x) \equiv 2\xi_{200}(x) + 2\xi_{201}(x) \text{ mod } 4$$
$$\alpha_2(x) + \alpha_6(x) \equiv 2\xi_{210}(x) + 2\xi_{211}(x) \text{ mod } 4$$

Thus d) holds.

Now a) is immediate since $\chi(X_0) = \chi(V_0)$. To prove b), note that

$$
\begin{aligned}
\chi(X_1) &= \chi(X_1 - X_0) + \chi(U_{01}'') - \chi(U_{01}'' - X_0) \\
&= \chi(V_1 - V_{01}) + \chi(X_0) - \chi(U_{01}' - V_{01}) \\
&\equiv \chi(V_{01}) + \chi(X_0) \bmod 2
\end{aligned}
$$

since the Euler characteristic of a union of circles with k points deleted is k and since $U_{01}' - V_{01}$ is an even number of lines since it has two components for each point of V_{01}. Now for each point $x \in X_0$ there are $\xi_{200}(x) + \xi_{210}(x) + \xi_{201}(x) + \xi_{211}(x)$ points in $p_{01}^{-1}(x)$ mod 2. Consequently b) holds.

Now let us prove c). We have

$$
\begin{aligned}
\chi(X) &= \chi(X - X_1) + \chi(U_{02}''' \cup U_{12}''') - \chi(U_{02}''' \cup U_{12}''' - X_1) \\
&= \chi(V_2 - (V_{02} \cup V_{12})) + \chi(X_1) - \chi(U_{02}' \cup U_{12}' - (V_{02} \cup V_{12})).
\end{aligned}
$$

But

$$
\chi(V_2) - \chi(V_{02} \cup V_{12}) = \chi(V_2 - (V_{02} \cup V_{12})) - \chi(U_{02}' \cup U_{12}' - (V_{02} \cup V_{12}))
$$

since $U_{02}' \cup U_{12}'$ deformation retracts to $V_{02} \cup V_{12}$. So

$$
\begin{aligned}
\chi(X) &= \chi(V_2) - \chi(V_{02} \cup V_{12}) + \chi(X_1) \\
&\equiv \chi(V_2) + \chi(V_{02} \cap V_{12}) + \chi(V_{02}^2) + \chi(X_1) \bmod 2.
\end{aligned}
$$

Now

$$
\begin{aligned}
\chi(V_{02}^2) &\equiv \sum_{x \in V_0} \xi_1(x) \bmod 2 \\
\chi(V_{02} \cap V_{12}) &\equiv \sum_{y \in V_{01}} \zeta_1'(y) + \zeta_2'(y) \equiv \sum_{x \in V_0} \xi_{201}(x) + \xi_{210}(x) \bmod 2
\end{aligned}
$$

so c) follows   □

For $n = 0, 1, 2$ if we define $\mathfrak{N}_n^T$ to be the cobordism group of n-dimensional Euler Thom stratified sets, generated by the relation: $X \sim Y$ if X and Y are P.L. homeomorphic to $|\mathfrak{T}|$ and $|\mathfrak{T}'|$ respectively, and $\mathfrak{T} \sqcup \mathfrak{T}'$ is a weak boundary of a compact resolution tower $\mathfrak{T}''$. Then from Theorem 7.2.1 and Propositions 7.1.8 and 7.1.9 we conclude:

$$
\mathfrak{N}_n^T = \begin{cases} \mathbf{Z}/2\mathbf{Z} & \text{if } n = 0 \\ \mathbf{Z}/2\mathbf{Z} & \text{if } n = 1 \\ \mathbf{Z}/2\mathbf{Z}^5 & \text{if } n = 2 \end{cases}
$$

For $n = 0$ the generator is the one point space. For $n = 1$ a generator is the figure 8, since it has odd Euler characteristic. For $n = 2$ generators are the projective plane $\mathbf{RP}^2$ and the spaces Y_0, Y_1, Y_2 and Y_3 of section 1.

Similarly we can define a cobordism group $\mathfrak{N}_n^S$ of n-dimensional Euler Thom stratified sets with all strata having trivial normal bundles, and which are realizations of resolution towers in $\mathcal{TR}_n'$. It is generated by the relation: $X \sim Y$ if X and Y are isomorphic to $|\mathfrak{T}|$ and $|\mathfrak{T}'|$ respectively, and $\mathfrak{T} \sqcup \mathfrak{T}'$ is a weak boundary

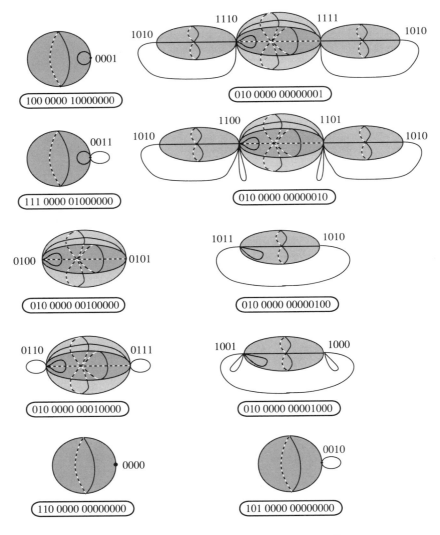

FIGURE VII.2.3. Some generators of $\mathfrak{N}_2^S$

of a compact resolution tower $\mathfrak{T}''$. Then:

$$\mathfrak{N}_n^S = \begin{cases} \mathbf{Z}/2\mathbf{Z} & \text{if } n = 0 \\ \mathbf{Z}/2\mathbf{Z}^2 & \text{if } n = 1 \\ \mathbf{Z}/2\mathbf{Z}^{15} & \text{if } n = 2 \end{cases}$$

The generators are those given for $\mathfrak{N}_n^T$ and the following: For $n = 1$, a circle stratified by two strata, a point and its compliment. For $n = 2$, the 10 spaces of Figure VII.2.3.

The boxed numbers give the value of ρ_2' for each figure.

Exercise: Prove that the spaces above are in fact generators of the various cobordism groups.   ◇

Note that transitivity of of the above equivalence relations are not obvious from definitions, they are implied by Theorem 7.2.1.

Let $\mathfrak{N}_n^R$ be the cobordism group generated by the relation: $\mathfrak{T} \sim \mathfrak{T}'$ in $\mathcal{TR}_n'$ if $\mathfrak{T} \sqcup \mathfrak{T}'$ is a weak boundary of a compact resolution tower. Then we get:

$$\mathfrak{N}_n^R = \begin{cases} \mathbf{Z}/2\mathbf{Z} & \text{if n=0} \\ \mathbf{Z}/2\mathbf{Z}^2 & \text{if n=1} \\ \mathbf{Z}/2\mathbf{Z}^{65} & \text{if n=2} \end{cases}$$

Exercise: Prove that $\mathfrak{N}_2^R = \mathbf{Z}/2\mathbf{Z}^{65}$. You may wish to use Theorem 7.2.1, Theorem 7.2.2, Proposition 7.2.4, Lemma 7.2.7 and Lemma 7.1.6a.   ◇

We can similarly define higher dimensional cobordism groups for resolution towers in general, but the computations of them becomes much more difficult. By generalizing a version of Theorem 7.2.1 one can define an obstruction theory for a stratified set in order to decide if it is a realization of a resolution tower. Roughly the obstructions lie in the cohomology group with coefficients in these cobordism groups [**AK9**].

3. Characterization in Dimension 3

In this section we prove Theorems 7.1.1 and 7.1.2. First we prove:

Lemma 7.3.1 *Let X be a three dimensional Zopen set. Then there is an algebraic stratification of X so that X and each $\mathcal{Z}_i(X)$ for $i = 0, \ldots, 11$ are Euler stratified sets.*

Proof: By Theorem 6.4.2 we know that there is an algebraic stratification of X which is isomorphic to the realization of a resolution tower of type UM. By Proposition 4.3.4 if U is any stratum and L_U is its link, there is a resolution tower with realization equal to cL_U. But then Theorems 7.2.1 and 7.2.2 imply that $\rho_i''(L_U) = 0$ if $\dim L_U = i$. But this is exactly what is needed to conclude that X and each $\mathcal{Z}_i(X)$ for $i = 0, \ldots, 11$ are Euler stratified sets.   □

Next we give a result which allows us to build up a resolution tower for an Euler Thom stratified set stratum by stratum. It has a number of technical conclusions which we will find useful, but which can be ignored on first reading. The basic idea is that you just glue in $T \times \mathfrak{T}''$ where $\mathfrak{T}''$ is a resolution tower of the cone on the link of T. If one assumes that X has a TCSS structure (see [**AK9**]), it is possible to prove a stronger statement which works for strata T of higher codimension if we assume the induced resolution tower on the link of T bounds a resolution tower whose realization is the cone on the link. In general, however, we expect that there are obstructions for a Thom stratified set with all normal bundles trivial to have a TCSS structure.

Proposition 7.3.2 *Let X be an Euler Thom stratified set with tubular data $(\pi_U, \rho_U)\colon X_U \to U \times \mathbf{R}$. Suppose for some closed union of strata $K \subset X$ we have a resolution tower $\mathfrak{T} = \{V_i, \mathcal{A}_i, p_i\}_{\mathfrak{I}}$ of type S and an isomorphism $h\colon |\mathfrak{T}| \to X - K$ so that the map $(\pi_U, \rho_U) \circ h\colon V_i \cap h^{-1}(X_U) \to U \times \mathbf{R}$ is smooth for each stratum U of X and each i. Also suppose that for each $j < i$ the map $\rho_U \circ h\colon V_i \cap h^{-1}(X_U) \to \mathbf{R}$ vanishes with order exactly 2 all along $V_{ji} \cap h^{-1}(X_U)$ where the stratum U is any component of $h(V_{ji} - \bigcup_{k<j} V_{ki})$ and U has codimension one.*

Let T be a stratum of K which is open in K so that the link L of T in X has dimension one or less and so that the normal bundle of T in X is trivial. Then there is a resolution tower $\mathfrak{T}' = \{V_i', \mathcal{A}_i', p_i'\}_{\mathfrak{I}'}$ of type S and an isomorphism $h'\colon |\mathfrak{T}'| \to (X - K) \cup T$ so that the map $(\pi_U, \rho_U) \circ h'\colon V_i' \to U \times \mathbf{R}$ is smooth for each stratum U of X and each i. Also, for each $j < i$ the map $\rho_U \circ h'\colon V_i' \cap h'^{-1}(X_U) \to \mathbf{R}$ vanishes with order exactly 2 all along $V_{ji}' \cap h^{-1}(X_U)$ where the stratum U is any component of $h(V_{ji} - \bigcup_{k<j} V_{ki})$ and U has codimension one. Furthermore, the restriction of $\mathfrak{T}'$ to $h'^{-1}(X - K)$ is just $\mathfrak{T}$ and the restriction of $\mathfrak{T}'$ to h'^{-1} of some neighborhood of T is a product $T \times \mathfrak{T}''$ for some resolution tower $\mathfrak{T}''$ with $|\mathfrak{T}''| = \mathfrak{c}L$ so that all exponents of $\mathfrak{T}''$ are ≤ 2. If $\dim L = 1$ then we may also ensure that $\xi_i(V_\ell'') = 0$ for $i = 0, 1$ where ℓ is the index so that $h'(V_\ell)$ is the vertex of $\mathfrak{c}L$.

Proof: Take a trivialization of the normal bundle of T in X. In particular, take a point $x_0 \in T$ and a small $\epsilon > 0$ and let $L = (\pi_T, \rho_T)^{-1}(x_0 \times \epsilon)$ be the canonical link of T. Then the normal bundle of T being trivial means there is a neighborhood Z_0 of $T \times$ vertex in $T \times \mathfrak{c}L$ and a Thom stratified set embedding $g\colon Z_0 \to X$ onto a neighborhood of T in X so that $(\pi_T, \rho_T) \circ g(x, (y, t)) = (x, t)$ for all $x \in T$ and $(y, t) \in \mathfrak{c}L$. Furthermore, $\pi_U \circ g(x, (y, t)) = g(x, (\pi_U(y), t))$ and $\rho_U \circ g(x, (y, t)) = \rho_U(y)$ for all strata U of X with $T \subset \mathrm{Cl}\, U$ and all $(x, (y, t))$ in a neighborhood of $Z_0 \cap T \times \mathfrak{c}(L \cap U)$.

For example, let X be the suspension of the figure 8 union a point and let $T = K = $ the two suspension points. We take one of the two possible resolution towers of $X - K$, as shown in Figure VII.3.1.

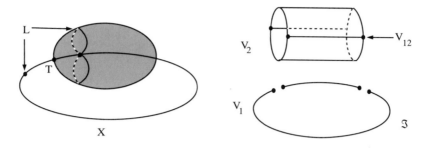

FIGURE VII.3.1. X and a resolution $\mathfrak{T}$ of $X - K$

The tower $\mathfrak{T}$ induces a resolution tower $\mathfrak{T}'''$ with $|\mathfrak{T}'''| = L$, namely $V_i''' = V_i \cap h^{-1}(L)$, $\mathcal{A}_i''' = \mathcal{A}_i \cap V_i'''$ and $p_i''' = p_i|$. See Figure VII.3.2. This works

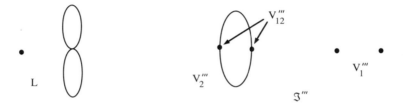

FIGURE VII.3.2. The link L and its induced resolution tower $\mathfrak{T}'''$

because $V_i''' = V_i \cap ((\pi_T, \rho_T) \circ h)^{-1}(x_0 \times \epsilon)$ and we hypothesized that $(\pi_T, \rho_T) \circ h$ is smooth, but we know it submerses each stratum of V_i because $\mathfrak{T}$ has type U and (π_T, ρ_T) submerses each stratum of X. So $(\pi_T, \rho_T) \circ h$ is a smooth submersion on each V_i.

Because $\mathfrak{T}'''$ has dimension one or less, the trivialization g induces a unique smooth trivialization on the part of $\mathfrak{T}$ in $h^{-1}(g(Z_0))$. In particular, we may suppose after restricting Z_0 that $Z_0 = \{ (x, (y, t)) \mid t < \delta(x) \}$ where $\delta : T \to (0, 1)$ is some smooth function. Then $g(Z_0) = \{ x \in X_T \mid \rho_T(x) < \delta \circ \pi_T(x) \}$. We may define a smooth $g' : T \times (0, 1) \times (V_i'' - |\mathcal{A}_i''|) \to V_i - |\mathcal{A}_i|$ by $g'(x, t, y) = h^{-1}g(x, (h(y), t\delta(x)))$. But g' extends continuously to an embedding $g' : T \times (0, 1) \times V_i'' \to V_i$ since $|\mathcal{A}_i''|$ is at worst a finite number of points in a one dimensional manifold. But g' is actually smooth because $\rho_U \circ h \circ g'(x, t, y) = \rho_U \circ h(y)$ and by our assumption on $\rho_U \circ h$, we know that $(\pi_T \circ h, \rho_T \circ h, \sqrt{\rho_U \circ h})$ is a local diffeomorphism for any codimension one stratum U.

We know that $|\mathfrak{T}'''| = L$ is an Euler stratified set, so by Theorem 7.2.2, $\rho_d(\mathfrak{T}''') = 0$ where d is the dimension of L. So by Theorem 7.2.1 and 7.5.1 below, there is a resolution tower $\mathfrak{T}''$ of type S so that $|\mathfrak{T}'''|$ is the weak boundary of $\mathfrak{T}''$, $|\mathfrak{T}''| = \mathfrak{c}|\mathfrak{T}'''|$, all exponents of $\mathfrak{T}''$ are ≤ 2, and there are diffeomorphisms $f_i : V_i''' \times (0, 1] \to V_i'' - V_{\ell i}''$ so that $f_i(V_{ji}''' \times (0, 1]) = V_{ji}'' - V_{\ell i}''$ and $p_{ji}'' f_i(y, t) =$

$f_j(p'''_{ji}(y), t)$ for all $(y, t) \in V'''_{ji} \times (0, 1]$, where ℓ is the index so that V_ℓ is the vertex of $c|\mathfrak{T}'''|$. See Figure VII.3.3.

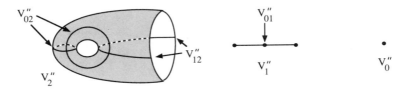

FIGURE VII.3.3. A resolution tower $\mathfrak{T}''$ for cL

Furthermore, if $d = 1$ we may assume $\xi_i(V''_\ell) = 0$ for $i = 0, 1$.

What is more, if $d = 0$ then each V''_i is a union of intervals $[-1, 1]$ with center $V_{\ell i}$ corresponding to $0 \in [-1, 1]$ and we may choose $f_i \colon \{-1, 1\} \times (0, 1] \to [-1, 1] - 0$ to be $f_i(s, t) = st^2$.

We now form $\mathfrak{T}'$ by setting $V'_i = V_i \cup T \times V''_i$ for $i = 0, 1$ where we identify $(x, f_i(y, t)) \in T \times V''_i$ with $g'(x, t, y) \in V_i$. We let $h'|_{V_i} = h|$ and $h'(x, y) = x$ for $(x, y) \in V''_i$. □

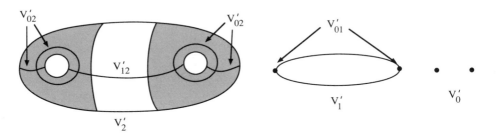

FIGURE VII.3.4. A resolution tower $\mathfrak{T}'$ for X

Corollary 7.3.3 *If X is a Thom Euler stratified set with trivial normal bundles and all strata have codimension ≤ 2 then X is the realization of a resolution tower of type S.*

We now prove Theorem 7.1.2.

Proof: (of Theorem 7.1.2) Note that b) follows from examples given in section 2 and c) is immediate from the definition. So let X be a three dimensional Euler Thom stratified set so that each $\mathcal{Z}_i(X)$ is an Euler space $i = 0, \ldots, 11$ and so that all normal bundles of components of strata are trivial.

Assertion 7.3.4.1 *The sets $\mathcal{E}_{abcd}(X)$ are all Euler spaces.*

Proof: Take any vertex v in X and let L be the link of v in X. For $a, b, c, d = 0, 1$ let n_{abcd} be the number of points in $\mathcal{E}_{abcd}(L)$. Because each $\mathcal{Z}_i(X)$ is an Euler space we know that each of the following numbers are even: $n_{1110} + n_{1111}$; $n_{ab00} + n_{ab01}$ for $ab = 01, 10, 11$; and n_{abc1} for any abc. In particular, $\sum_{c,d} n_{11cd}$ is even. But then by Lemma 7.1.6a we know that $\sum_{c,d} n_{abcd}$ is even for $a, b = 0, 1$ so $n_{ab10} + n_{ab11}$ is even for $ab = 10, 01$ or 11. So n_{abc0} is even for $ab = 10, 01$, or 11. So the only possible odd n_{abcd}'s are n_{0000} and n_{0010}. But by Lemma 7.1.6b, $\sum_{ab}(n_{ab00} + n_{ab11})$ is even so n_{0000} is even since all the other terms are even. But then n_{0010} must be even since the total number of points in L^0 is even. So all the n_{abcd}'s are even, but $\mathcal{E}_{abcd}(L)$ is the link of v in $\mathcal{E}_{abcd}(X)$ so all the $\mathcal{E}_{abcd}(X)$'s are Euler spaces.   □

We now set about finding a resolution tower $\mathfrak{T}$ whose realization is X. Let the i-skeleton of X be X_i. Let $\pi_U \colon X_U \to U$ and $\rho_U \colon X_U \to [0, \infty)$ be tubular data. For each stratum U, let L_U be the canonical link of U, let Z_U be a neighborhood of $U \times$ the vertex in $U \times_{\mathfrak{c}} L_U$, Y_U a neighborhood of U in X_U and let $\tau_U \colon Z_U \to Y_U$ be an isomorphism so that $\pi_U \tau_U(x, (y, t)) = x$ and $\rho_U \tau_U(x, (y, t)) = t$.

By repeated applications Proposition 7.3.2 (starting with $K = X$) we can find a well indexed resolution tower $\mathfrak{T} = \{V_i, \mathcal{A}_i, p_i\}_{i=1}^3$ of type S and an isomorphism $h \colon |\mathfrak{T}'| \to X - X_0$ so that the map $(\pi_U, \rho_U) \circ h \colon V_i' \to U \times \mathbf{R}$ is smooth for each stratum U of X and each i. Also, the restriction of $\mathfrak{T}'$ to h^{-1} of some neighborhood of U is a product $U \times \mathfrak{T}^U$ for some resolution tower $\mathfrak{T}^U$ with $|\mathfrak{T}^U| = \mathfrak{c}L_U$ so that all exponents of $\mathfrak{T}^U$ are ≤ 2. If $\dim L_U = 1$ then we may also ensure that $\xi_i(V_0^U) = 0$ for $i = 0, 1$.

We have an induced resolution tower $\mathfrak{T}^x$ on the canonical link L_x of any point x of X_0. Let us look at $\rho_2(\mathfrak{T}^x)$. By our construction, $\xi_i(y) = 0$ for all $i = 0, 1$ and $y \in V_1^x$. Hence if $ab = 01, 10$ or 11 we have $\eta_{3abcdef}(\mathfrak{T}^x) = 0$ for all $c, d, e, f \in \mathbf{Z}/2\mathbf{Z}$. But then by Lemma 7.2.7d, $\eta_{300cdef}(\mathfrak{T}^x)$ is the mod 2 number of points in $\mathcal{E}_{dfgh}(L_x)$ where $g \equiv c + d$ and $h \equiv e + f$. We showed in Assertion 7.3.4.1 that this is even, so $\eta_{3abcdef}(\mathfrak{T}^x) = 0$ for all $a, b, c, d, e, f \in \mathbf{Z}/2\mathbf{Z}$. But we must worry about η_i for $i = 0, 1, 2$. We will make them 0 by blowing up $\mathfrak{T}'$ along some curves of V_{13}' which lie over 1 strata of X.

Let d be 0 or 1 and let K_d be the closure of the union of the codimension one strata U of X so that $\kappa(U) > 0$ and $\kappa(U) \equiv 2d \bmod 4$. In other words, $K_1 = \mathrm{Cl}(\mathcal{K}_2(X) \cup \mathcal{K}_6(X))$ and $K_0 = \mathrm{Cl}(\mathcal{K}_0(X) \cup \mathcal{K}_4(X) - P)$ where P is the union of open codimension one strata of X, i.e., those two strata with empty link. Let A be any connected component of $K_d \cap X_1$ which contains points of X_0.

Assertion 7.3.4.2 $\sum_{x \in A \cap X_0} \eta_d(\mathfrak{T}^x) = 0$.

Proof: Let $K_d' = \mathrm{Cl}(\mathcal{K}_{2d}(X) \cup \mathcal{K}_{2d+4}(X))$, so $K_1' = K_1$ and $K_0' = K_0 \cup \mathrm{Cl}(P)$. Pick a compact neighborhood U of A in K_d' so that $U - X_1$ is a smooth codimension 0 manifold with boundary Y. Pick a small $\epsilon > 0$ and let $U_x = \rho_x^{-1}([0, \epsilon))$ for

$x \in A \cap X_0$. Let $U' = U - \bigcup_{x \in X_0 \cap A} U_x$. Then $C_2 = V_2' \cap h^{-1}(U')$ is a smooth compact codimension 0 submanifold of $V_2^{d\,\prime}$. Let $C_{23} = {p_{23}'}^{-1}(C_2)$. Then by Lemma 7.2.5 there are an even number of components of ∂C_{23} so that the restriction of the normal bundle of C_{23} is nontrivial. But by construction the normal bundle of $C_{23} - V_{13}'$ is trivial, since for each component T of $V_2' - V_{12}'$, a neighborhood of ${p_{23}'}^{-1}(T)$ in V_3' is obtained by taking $T \times$ a neighborhood of a tico in a one dimensional manifold, a resolution tower for the cone on the link of T. So since $\partial C_{23} = {p_{23}'}^{-1} h^{-1}(Y) \cup \bigcup_x {p_{23}'}^{-1} h^{-1} \rho_x^{-1}(\epsilon)$ and ${p_{23}'}^{-1} h^{-1}(Y) \subset C_{23} - V_{13}'$ we know that there are an even number of components of $\bigcup {p_{23}'}^{-1} h^{-1} \rho_x^{-1}(\epsilon)$ so that the restriction of the normal bundle of C_{23} is nontrivial. But this means precisely that $\sum_{x \in A \cap X_0} \eta_d(\mathfrak{T}^x) = 0$.   □

Assertion 7.3.4.3 *After blowing up $\mathfrak{T}'$ with centers lying over V_{13}', we may as well assume that $\eta_d(\mathfrak{T}^x) = 0$ for all $x \in A \cap X_0$.*

Proof: A is a one dimensional stratified set, hence a polyhedron. Pick a minimal tree $A' \subset A$ so that after replacing $\mathfrak{T}'$ with some blowups with centers lying over V_{13}', then $\eta_d(\mathfrak{T}^x) = 0$ for all $x \in (A - A') \cap X_0$. If $A' = \emptyset$ we are done so assume A' is not empty. By Assertion 7.3.4.2, $A' \cap X_0$ cannot have exactly one element, so it must have at least two elements. So we may pick x_1 and x_2 in $A' \cap X_0$ so that there is only one edge of A' coming into x_1 and that edge goes to x_2. The edge between x_1 and x_2 is a component T of a one stratum of X. We know that $\eta_d(\mathfrak{T}^{x_1}) = 1$ since otherwise A' would not be minimal.

Since $T \subset K_d$ we may pick a stratum T'' of $V_2^{d\,\prime} \cap V_{12}'$ so that $p_{12}'(T'') = T$. We may also pick a stratum T' of $V_{13}' \cap V_{23}'$ so that $p_{13}'(T') = T$ and $p_{23}'(T') = T''$. We now blow up V_3' with center T'. In the resolution tower $\mathfrak{T}^{x_1}$ this corresponds to blowing up a point in $V_{13}^{x_1} \cap V_{23}^{x_1}$. By Proposition 7.2.4c, the resulting resolution tower has $\eta_d = 0$. Unfortunately, ξ_1 has become 1. But by Proposition 7.2.4b, we can change ξ_1 back to 0 by blowing up a point in $V_{13}^{x_1}$. We can do this by blowing up a curve in V_{13}' lying over T. So the minimality of A' is violated.   □

So in light of Assertion 7.3.4.3 we may assume that for each $x \in X_0$ we have all components of $\rho_2(\mathfrak{T}^x)$ equal to 0 except possibly η_2. But by Lemma 7.2.7c we know that $\eta_2(\mathfrak{T}^x) = 0$, so $\rho_2(\mathfrak{T}^x) = 0$.

So by Lemma 7.5.2, each $\mathfrak{T}^x$ bounds a resolution tower $\mathfrak{T}^{x\,\prime}$ of type S. In fact we have diffeomorphisms $g_i^x \colon V_i^x \times (0,1] \to V_i^{x\,\prime} - V_{0i}^{x\,\prime}$ for $i = 1, 2, 3$ so that $g_i^x(V_{ji}^x \times (0,1]) = V_{ji}^{x\,\prime}$ and $p_{ji}^{x\,\prime} g_i^x(u,t) = g_j^x(p_{ji}^x(u), t)$ for all $u \in V_{ji}^x$ and $t \in (0,1]$. We then get a resolution tower $\mathfrak{T}$ of type S with $|\mathfrak{T}| = X$ by gluing the resolution towers $\mathfrak{T}^{x\,\prime}$ to $\mathfrak{T}'$. For example, suppose $L_x = \rho_x^{-1}(\epsilon_x)$. Then $\rho_x \circ h$ is a proper smooth submersion on each $V_i \cap h^{-1} \rho_x^{-1}((0, \epsilon_x])$, so there are diffeomorphisms $h_i^x \colon V_i^x \times (0, \epsilon_x] \to V_i \cap h^{-1} \rho_x^{-1}((0, \epsilon_x])$ compatible with the maps p_{ji}' so that $\rho_x h_i^x(x, t) = t$. So we could let $\mathfrak{T}$ be the resolution tower obtained from the disjoint union $\mathfrak{T}' \sqcup \bigsqcup_{x \in X_0} \mathfrak{T}^{x'}$ by identifying $g_i^x(x, t)$ with $h_i^x(t\epsilon_x)$ for $(x, t) \in V_i^x \times (0, 1)$. $\qquad\qquad\qquad\qquad\qquad\qquad\qquad\qquad\qquad\qquad\qquad$ □

We now prove Theorem 7.1.1.

Proof: (of Theorem 7.1.1) Conclusions b) and c) follow from Proposition 7.1.7.

So we only need prove a). If X is homeomorphic to a real algebraic set then Lemma 7.3.1 shows that X and $\mathcal{Z}_i(X)$ are Euler spaces for $i = 0, 1, 2, 3$. (We also know that X must be an Euler space by [**Su1**].)

Conversely suppose that X and $\mathcal{Z}_i(X)$ for $i = 0, 1, 2, 3$ are all Euler spaces. Triangulate X. Then Lemma 7.1.9 implies that the first barycentric subdivision is an Euler polyhedron and Proposition 7.1.8 and Theorem 7.1.2a imply that the second barycentric subdivision is isomorphic to a real algebraic set. $\qquad$ □

4. Algebraic Resolution of Real Algebraic Sets in Dimension Three

Our characterization of three dimensional algebraic sets has little hope of being extended to many more dimensions since it involves calculating cobordism groups which are already very complicated in low dimensions. Thus it would be nice to strengthen Theorems 6.4.2 and 5.0 so that they are converses of each other. To this end, we strengthen Theorem 6.4.2 sufficiently to characterize three dimensional real algebraic sets. T. C. Kuo has independently proven the essential part of this, although in a different context.

Theorem 7.4.1 *Any three dimensional real algebraic set X is the realization of a resolution tower $\mathfrak{T}$ of type RSF.*

Proof: At this point the easy proof is that by Theorem 7.1.1, X and each $\mathcal{Z}_i(X)$ for $i = 0, 1, 2, 3$ are Euler spaces. But then the proofs of Theorem 7.1.1 and 7.1.2 imply that X is the realization of a resolution tower of type S. Notice that type F is automatic because the dimensions are so low. However we will do a direct algebraic proof since it indicates how one might try to prove this in higher dimensions. The higher dimensional cases are a bit more complicated however since the local forms are not as simple.

Just as in Theorem 6.4.2 we may as well assume X is compact. The first step is to use Theorem 6.4.2 to get $X = |\mathfrak{T}|$ for some well indexed algebraic resolution tower $\mathfrak{T} = \{V_i, \mathcal{A}_i, p_i\}_{i=0}^3$ of type $FUNE$ with dim $V_i = i$. Actually, we only need type UM. A byproduct of the proof of Theorem 6.4.2 is that the complexification of $\mathfrak{T}$ is also a resolution tower of type U near the real points. For the most part $\mathfrak{T}$ has type S, the only possible points where some p_{ij} is not submersive are at $V_{23} \cap V_{03}$ where p_{23} may not be submersive. We call these points *bad points*. We will show that the image of the set of bad points is a finite set.

Our procedure is completely canonical. Let $Q \subset V_2$ be the image of the bad points. This Q is always finite. You first blow up any points necessary in V_{23} to make $p_{23}^*(\mathcal{J}(Q))$ locally principal. This is completely canonical since it is only locally not principal at isolated points, which you keep on blowing up until there aren't any. You now blow up V_2 at Q and lift the map p_{23} to the blowup. You then repeat the whole process. After a finite number of repetitions, it eventually terminates when there are no more bad points.

We know that V_{23} is an embedded dimension 2 submanifold of V_3 since if two intersecting sheets were in V_{23} their curve of intersection would have to submerse to the surface V_2 at almost all points. This is impossible for dimension reasons, c.f., Lemma 7.2.3.

The following gives the local canonical forms for the map p_{23}. If p is a polynomial, we let $ord(p)$ denote its order, i.e., the degree of the lowest degree term of p.

Assertion 7.4.1.1 *Pick any $q \in V_{23} \cap V_{03}$. Then after a coordinate change on V_{23} we may assume that locally $q = (0,0)$ and*

$$p_{23}(x,y) = (x^{ka}y^{kb}, \, p(x^a y^b) + x^c y^d)$$

where p is a polynomial, $ad - bc > 0$, and either

1) *a and b are both nonzero and relatively prime. Furthermore, p has degree $< d/b$. We call this the triple point case. In this case, p_{23} is submersive at q if and only if $p_{23}(q)$ is a double point of q and $p = 0$. Also if q is a bad point then the whole line $y = 0$ (and possibly $x = 0$ as well) consists of bad points.*

2) *$a = 1$, $b = 0$, $d = 1$. Furthermore, p has degree $\leq c$. We call this the double point case. In this case, p_{23} is submersive at q if and only if either $c = 0$ or else $p_{23}(q)$ is a double point of q and either $p = 0$ or $ord(p) = c$. Also, if q is a bad point the whole line $x = 0$ consists of bad points.*

So in any case there are no isolated bad points. Furthermore, p_{23} is locally constant on the bad points so the set of images of bad points is finite.

Proof: Pick analytic coordinates x, y for V_{23} near q and u, v for V_2 near $p_{23}(q)$ so that q is $(0,0)$, $p_{23}(q)$ is $(0,0)$, $V_{23} \cap V_{03}$ is locally $\{x = 0\}$ if q is a double point of $|\mathcal{A}_3|$ or $\{xy = 0\}$ if q is a triple point, V_{02} is $\{u = 0\}$ or $\{uv = 0\}$ and the map p_{23} is given by $p_{23}(x,y) = \left(x^{a'} y^{b'}, g(x,y)\right)$. Note $b' = 0$ if and only if q is a double point.

If $b' \neq 0$, let k be the greatest common divisor of a' and b'. If $b' = 0$, let $k = a'$. Let $a = a'/k$ and $b = b'/k$. So

$$p_{23}(x,y) = (x^{ka}y^{kb}, \, g(x,y)).$$

First assume we are in the case where q is a double point, i.e., $a = 1$, $b = 0$. Then the Jacobian determinant of p_{23} is $kx^{k-1}\partial g/\partial y$, and it only vanishes on

$x = 0$ even for complex x and y (since the complexification has type U). So $\partial g/\partial y = x^c u'(x,y)$ where $u'(0,0) \neq 0$ by Lemma 6.2.1. If we write $g(x,y) = \sum g_{ij} x^i y^j$ then

$$\partial g/\partial y = \sum j g_{ij} x^i y^{j-1}$$

so we see that $g_{ij} = 0$ if $j > 0$ and $i < c$ and also $g_{c1} \neq 0$. So

$$g(x,y) = p(x) + x^c y u(x,y)$$

where $p(x) = \sum_{i=1}^c g_{i0} x^i$ is a polynomial and u is a unit, i.e., $u(0,0) \neq 0$. After a coordinate change $y \mapsto yu$ we get

$$g(x,y) = p(x) + x^c y$$

as required.

Suppose q is not a bad point. If $p_{23}(q)$ is not a double point of V_{02} we must have p_{23} submerse $x = 0$ to $u = 0$, in particular $\partial g/\partial y(0,0) \neq 0$. So we would need $c = 0$. So now suppose $p_{23}(q)$ is a double point of V_{02}. Then to be submersive at q we must have $p(x) = 0$ or $p(x)$ a monomial of degree c.

Now suppose q is a triple point so a and b are nonzero. The Jacobian determinant of p_{23} is

$$(ay\partial g/\partial y - bx\partial g/\partial x) \cdot kx^{ka-1}y^{kb-1}.$$

By type U, p_{23} must be a submersion when $xy \neq 0$. So the Jacobian determinant must be nonzero when $xy \neq 0$ even for complex x and y (since the complexification has type U). So Lemma 6.2.1 implies that

$$ay\partial g/\partial y - bx\partial g/\partial x = x^c y^d u'(x,y)$$

where $u'(0,0) \neq 0$. If we write $g(x,y) = \sum g_{ij} x^i y^j$ then

$$ay\partial g/\partial y - bx\partial g/\partial x = \sum (aj - bi) g_{ij} x^i y^j$$

so we see that we must have $ad - bc \neq 0$ and $g_{cd} \neq 0$. So we may write

$$g(x,y) = p(x^a y^b) + x^c y^d u(x,y)$$

where p is a polynomial and u is a unit. The reason we may take p to be a polynomial is that any term of the form $(x^a y^b)^n$ with $n \geq \max(c/a, d/b)$ may be absorbed into the unit u. After a coordinate change $x \mapsto xu^{-b/(ad-bc)}$, $y \mapsto yu^{a/(ad-bc)}$ we may assume

$$g(x,y) = p(x^a y^b) + x^c y^d.$$

By switching x and y we may also assume $c/a < d/b$, i.e., $ad - bc > 0$.

Now suppose that q is not a bad point. Then the point stratum q must submerse to its target, so $p_{23}(q)$ is a double point of V_{02}. But then the only way to be submersive is when $p = 0$.

Suppose now that q is a bad point. Now look at a nearby double point $(w, 0)$ with $w \neq 0$. Then locally near $(w, 0)$, p_{23} looks like $(y^{kb}, p(y^b) + w^{c-ad/b}y^d + xy^d)$ after a change of coordinates $x \mapsto x^{c-ad/b} - w^{c-ad/b}$, $y \mapsto x^{a/b}y$. By part 2), if $(w, 0)$ is not bad then either $d = 0$ or $p_{23}(q)$ is a double point and $p = 0$. But $d > bc/a \geq 0$ so $d > 0$ and thus $(w, 0)$ is a bad point. Likewise $(0, w)$ will be a bad point unless $c = 0$.   □

In light of Assertion 7.4.1.1, we only need concentrate on eliminating the bad double points, the bad triple points will then be eliminated automatically. We will assign a complexity (m, n) to each double point as follows. Pick a local expression $p_{23}(x, y) = (x^k, p(x) + x^c y)$ as in Assertion 7.4.1.1. If either $p = 0$ or $ord\,(p) = c$ we take $m = 0$ and take $n = c/k$ if $p_{23}(q)$ is not a double point of V_{02} and $n = 0$ otherwise. If $p \neq 0$ and $ord\,(p) < c$ we take $m = c - ord\,(p)$ and $n = ord\,(p) + k$. This complexity might vary with the expression chosen but we say the complexity of p_{23} at q is the lexicographical minimum of the pair (m, n) over all the above local expressions for p_{23}. Actually, most expressions for p_{23} will give the minimal complexity anyway. The only exception is if $ord(p)$ is a multiple of k and $p_{23}(q)$ is not a double point of V_{02}, in which case a coordinate change will get rid of the initial term of p. Note that a double point is bad if and only if it has complexity $> (0, 0)$.

What we claim is that whenever we blow up the images of the bad points the maximum of this complexity decreases by at least an integral amount. Consequently by continuing to blow up the images of the bad points we eventually get rid of them all.

We will first illustrate this by an example. Suppose $V_{23} = \mathbf{R}^2$, $V_2 = \mathbf{R}^2$, $V_{23} \cap V_{03} = \mathbf{R}_1^2$, $V_{02} = \mathbf{R}_1^2$ and $p_{23}(x, y) = (x^4, x^2 + x^3 y)$. Then the line $x = 0$ consists of all bad points and their image is the single point $(0, 0)$. The complexity of $(0, 0)$ is $(1, 6)$. The pullback ideal is locally principal at the point $(0, 0)$, so we may blow up V_2 at $(0, 0)$ and p_{23} will lift. The lifted map will be $p'_{23}(x, y) = (x^2/(1+xy), x^2 + x^3 y)$ to one of the charts obtained by Lemma 2.5.1. In this new chart, $V_{02} = \mathbf{R}_1^2 \cup \mathbf{R}_2^2$. The complexity at $(0, 0)$ is now $(1, 4)$.

The line $x = 0$ still consists of bad points, so blow up V_2 again. The lifted map will be $p''_{23}(x, y) = (x^2/(1 + xy), (1 + xy)^2)$. After changing coordinates in V_2 so that the image of the bad point is $(0, 0)$, say by the change $(u, v) \mapsto (uv^{1/2}, v-1)$, and changing coordinates in V_{23} say by $(x, y) \mapsto (x, 2y + xy^2)$ our map has the form $p'''_{23}(x, y) = (x^2, xy)$ and the image $(0, 0)$ of the bad points is not a double point. The complexity is now $(0, 1/2)$.

The map p'''_{23} will not lift to the blowup of V_2 so we must first blow up V_{23} at $(0, 0)$. By Lemma 2.5.1 you get two charts A and B and your map $p'''_{23} \circ \pi$ is $(x^2, x^2 y)$ on chart A and $(x^2 y^2, xy^2)$ on chart B. These maps now lift to the blowup of V_2, one gets the map (x^2, y) on chart A which is submersive and on chart B one gets the map (x, xy^2) which is submersive since it maps from a triple point to a double point.

Now we return to the proof and see what happens to the complexity of a bad double point after we blow up its image in V_2. We assume that p_{23} lifts to the blowup. Of course it might not, we might have to first blow up points in V_{23} to obtain a lifting, but we will see in Assertion 7.4.1.3 that whenever one has to blow up the source, the maximal complexity does not decrease, or at least if it does it is still $\leq (0,1)$. Since complexities are positive and decrease by at least integral amounts, after a finite number of steps there will be no bad points left.

Assertion 7.4.1.2 *Suppose $q \in V_{23}$ is a double point and we blow up $p_{23}(q)$ in V_2 and the map p_{23} lifts to a map $p'_{23} \colon V_{23} \to \mathcal{B}(V_2, p_{23}(q))$. Suppose q is a bad point of p'_{23}. Then the complexity of p'_{23} at q is smaller than that of p_{23} at q. Moreover, if the complexity of p_{23} at q is $(0,\alpha)$ then the complexity of p'_{23} at q is $\leq (0, \alpha - 1)$.*

Proof: There will be two charts for this blowup, on chart C the blowup map is $(u,v) \mapsto (u, uv)$ and on chart D the blowup map is $(u,v) \mapsto (uv, v)$. On chart D, V_{02} is the intersecting curves $u = 0$ and $v = 0$. On chart C, V_{02} is either the curve $u = 0$ (if $p_{23}(q)$ is not a double point of V_{02}) or the two curves $u = 0$ and $v = 0$ (if $p_{23}(q)$ is a double point of V_{02}). So in what follows D is the only chart we can count on to have a double point.

Let $p_{23}(x,y) = (x^k,\, p(x) + x^c y)$ be an expression for p_{23} with minimal complexity. Let $e = ord(p)$ if $p \neq 0$.

First suppose $p = 0$ or $e = c$. In other words $p_{23}(x,y) = (x^k,\, x^c(y + y_0))$ for some constant y_0. Then the complexity of p_{23} at q is $(0, c/k)$ if $p_{23}(q)$ is not a double point and otherwise it is $(0,0)$.

If $k = c$ then $p'_{23}(x,y) = (x^k,\, y + y_0)$ mapping to chart C which is submersive.

If $k > c$ and $y_0 \neq 0$ then $p'_{23}(x,y) = (x^{k-c}/(y + y_0),\, x^c(y + y_0))$ mapping to chart D which is submersive since it maps to a double point.

If $k > c$ and $y_0 = 0$ then the lifting p'_{23} does not exist so we can ignore this case, but keep it in mind when we prove Assertion 7.4.1.3.

If $k < c$, then $p'_{23}(x,y) = (x^k,\, x^{c-k}(y + y_0))$ mapping to chart C and it has complexity $(0, c/k - 1)$ if $p_{23}(q)$ is not a double point and otherwise it is $(0,0)$.

Now we suppose that $p \neq 0$ and $c > e$. The complexity of p_{23} at q is $(c - e, e + k)$.

If $k < e$, then we have

$$p'_{23}(x,y) = (x^k,\, p(x) \cdot x^{-k} + x^{c-k} y)$$

mapping to chart C. This has complexity $(c - e, e)$ which is less than $(c - e, e + k)$.

If $k = e$ then

$$p'_{23}(x,y) = (x^k,\, p(x) \cdot x^{-k} + x^{c-k} y)$$

mapping to chart C. But the second coordinate is nonzero at q, so one must change variables in V_2, which has the effect of eliminating the constant term. So if e' is the next lowest degree of a monomial of p we see the complexity is at most $(c - k - (e' - k), e') = (c - e', e')$ if $e' < c$. If $e' = c$ then the new complexity would be $(0, c/k - 1)$ or $(0, 0)$ instead.

If $k > e$ then

$$p'_{23}(x, y) = \left(x^{k-e}/u(x, y), \, p(x) + x^c y\right)$$

mapping to chart C where u is the unit $u(x, y) = p(x) \cdot x^{-e} + x^{c-e}y$. After a coordinate change, this becomes $\left(x^{k-e}, p'(x) + x^c y\right)$ for some polynomial p' with the same order as p. Thus the complexity is $(c - e, k)$.   □

From the above proof we saw one kind of point where the map might not lift, when $p_{23}(x, y) = (x^k, x^c y)$ and $k > c$. But there are other points we did not look at, namely the triple points. So suppose we have a triple point q and we blow up its image. Suppose $p_{23}(x, y) = (x^{ka}y^{kb}, p(x^a y^b) + x^c y^d)$ is the local expression we get from Assertion 7.4.1.1. Then the only time p_{23} will not lift to the blowup is when either $p = 0$ and $c/a < k < d/b$ or else $p \neq 0$, $e > c/a$ and $k > c/a$.

We see below that if we blow up V_{23} at one of these three types of bad points, the maximal complexity does not increase unless it was already less than $(0, 1)$.

Assertion 7.4.1.3 Suppose $r \in V_{02}$. Suppose p_{23} does not lift to the blowup $\mathfrak{B}(V_2, r)$. Let $Q \subset p_{23}^{-1}(r)$ be the set of points at which $p_{23}^* \mathcal{I}(r)$ is not locally principal. If $\pi = \pi(V_3, Q)$ is the blowup map, then the maximal complexity of $p_{23} \circ \pi|$ at any double point is either $\leq (0, 1)$ or not greater than that of p_{23}. Moreover, after a finite number of such blowups, p_{23} will lift.

Proof: Note that Q is finite since we saw in the proof of Assertion 7.4.1.2 that the only double points which are in Q are isolated in Q, since p_{23} looks locally like $(x^k, x^c y)$. Then the fact that a lift will occur after a finite number of such blowups follows from Chapter 0, Section 5 of [H], c.f., Proposition 6.2.7. In this simple case, it is not too hard to prove it explicitly also.

Pick any $q \in Q$. First let us do the case where q is a double point which does not lift. Then locally, $p_{23}(x, y) = (x^k, x^c y)$ and $k > c$. Its complexity is either $(0, c/k)$ or $(0, 0)$. The strict transform of V_{23} in the blowup is just the blowup of V_{23} at q. By Lemma 2.5.1 we get two charts A and B for $\mathfrak{B}(V_{23}, Q)$. The map $p_{23} \circ \pi$ looks like $(x^k, x^{c+1}y)$ on chart A and $(x^k y^k, x^c y^{c+1})$ on chart B. Points not in $\pi^{-1}(q)$ will not have their complexity change since π is a diffeomorphism nearby. So we only need look at double points in $\pi^{-1}(q)$. The new double points are the points in chart A with $x = 0$ and $y \neq 0$. They all have complexity $(0, 0)$ or $(0, (c+1)/k) \leq (0, 1)$.

Now let us look at when q is a triple point. Pick a local expression

$$\left(x^{ka}y^{kb}, \, p(x^a y^b) + x^c y^d\right)$$

for p_{23} near q as in Assertion 7.4.1.1. Pick it so that $e = ord(p)$ is as large as possible if $p \neq 0$. Since p_{23} will not lift to the blowup we know that either $p = 0$ and $c/a < k < d/b$ or else $p \neq 0$, $e > c/a$ and $k > c/a$.

Then the complexity of p_{23} at points $(w, 0)$ with $w \neq 0$ is $(0, 0)$ if $p = 0$ and r is a double point of V_{02}, it is $(0, d/(kb))$ if $p = 0$ and r is not a double point of V_{02} and it is $(d - eb, eb + kb)$ if $p \neq 0$.

By Lemma 2.5.1 we get two charts A and B for $\mathfrak{B}(V_{23}, Q)$. On chart A, $p_{23} \circ \pi$ is

$$\left(x^{ka} y^{ka+kb}, \, p\left(x^a y^{a+b} \right) + x^c y^{d+c} \right)$$

and on chart B, $p_{23} \circ \pi$ is

$$\left(x^{ka+kb} y^{kb}, \, p\left(x^{a+b} y^b \right) + x^{c+d} y^d \right).$$

We only need check double points in $\pi^{-1}(q)$. These are points in chart B with $x = 0$ and $y \neq 0$. So after a change of coordinates near such a point, $p_{23} \circ \pi$ will be the map $(x^{ka+kb}, p(x^{a+b}) + x^{c+d} y)$.

If $p \neq 0$ and $e(a + b) \geq c + d$ then the complexity of $p_{23} \circ \pi$ at $(0, y)$ is $(0, 0)$ or $(0, (c+d)/(ka+kb))$ but this is less than the complexity of p_{23} at $(w, 0)$ so maximal complexity has not increased. If $p \neq 0$ and $e(a + b) < c + d$ then the complexity is

$$(d + c - eb - ea, \, (a+b)(e+k)) = (d - eb + a(c/a - e), \, (a+b)(e+k))$$

which is less than that of p_{23} at $(w, 0)$ since $e > c/a$. If $p = 0$ and r is not a double point of V_{02} then the complexity of $p_{23} \circ \pi$ at $(0, y)$ is $(0, (c+d)/(ka+kb))$ which is less than $(0, d/(kb))$ since $c/a < d/b$. If $p = 0$ and r is a double point of V_{02} then the complexity of $p_{23} \circ \pi$ at $(0, y)$ is $(0, 0)$. So in any case we see that the new double points have either smaller than maximal complexity or complexity $\leq (0, 1)$.

□

It is amusing to point out that what we end up doing here is resolving the singularity of the curve $t \mapsto (t^k, p(t))$ in V_{02}. So another way to look at what we are doing is to look at the collection of such curves, one for each bad point with $p \neq 0$. After blowing up enough to resolve each singularity, we see that p_{23} has a simpler local form, one where $p = 0$. At this point the maximum complexity is of the form $(0, \alpha)$. One could then finish off the proof fairly readily. □

5. Bounding Resolution Towers

In this section we complete the proof of Theorem 7.2.1 and give the more technical version of results which were useful in the proof of Theorem 7.1.1.

For convenience we use the following notation in this section. If n is a positive integer, we let $\mathbf{n}$ denote the set $\mathbf{n} = \{1, 2, \ldots, n\}$. We also let $\mathbf{J}$ denote the interval $[-1, 1]$.

First we show that ρ_1 determines when a one dimensional tower bounds.

Theorem 7.5.1 *Suppose $\rho_1(\mathfrak{T}) = 0$ for some $\mathfrak{T} \in \mathfrak{TR}_1'$. Then $\mathfrak{T}$ is a weak boundary of a resolution tower $\mathfrak{T}'$ of type S so that $|\mathfrak{T}'| = \mathfrak{c}|\mathfrak{T}|$ and all exponents of $\mathfrak{T}'$ are ≤ 2. Furthermore, if k is the index so that V_k' is the vertex of the cone, then $\xi_i(V_k) = 0$ for $i = 0, 1$ and there are diffeomorphisms $g_i \colon V_i \times (0, 1] \to V_i' - V_{ki}'$ so that $g_i(V_{ji} \times (0, 1]) = V_{ji}' - V_{ki}'$ and $p_{ji}' g_i(y, t) = g_j(p_{ji}(y), t)$ for all $(y, t) \in V_{ji} \times (0, 1]$.*

Proof: For convenience, assume that $\mathfrak{T} = \{V_i, \mathcal{A}_i, p_i\}_{i=1}^2$ has bias 1, then we wish to find $\mathfrak{T}' = \{V_i', \mathcal{A}_i', p_i'\}_{i=0}^2$ and the k in the statement above is 0. As we construct $\mathfrak{T}'$ we will also construct vector fields $\mathfrak{v}_i$ on V_i' which will integrate to give the g_i's and the cone structure on $|\mathfrak{T}'|$.

Since $\rho_1(\mathfrak{T}) = 0$, V_1 has an even number of points and also V_1 has an even number of points q for which $p_{12}^{-1}(q)$ is odd. So we may pair up the points of V_1 as $\{q_{11}, q_{-11}, \ldots, q_{1n}, q_{-1n}\}$. so that for some $n' \leq n$, the cardinality of $p_{12}^{-1}(q_{ji})$ is odd if and only if $i \leq n'$. Let $V_0' = $ a point, $V_1' = \mathbf{J} \times \mathbf{n}$, $V_{01}' = 0 \times \mathbf{n}$. Define the vector field $\mathfrak{v}_1$ on V_1' by $\mathfrak{v}_1(t, i) = -t$.

Note that each $p_{12}^{-1}(\{q_{1i}, q_{-1i}\})$ has an even number of points. So we may pair up the points of V_{12} as $\{r_{11}, r_{-11}, \ldots, r_{1m}, r_{-1m}\}$ so that there are functions $\alpha \colon \mathbf{m} \to \mathbf{n}$, $\sigma \colon \mathbf{m} \to \{1, 2\}$ and $\sigma' \colon \mathbf{m} \to \{-1, 1\}$ with $p_{12}(r_{ji}) = q_{\ell\alpha(i)}$ where $\ell = \sigma'(i) j^{\sigma(i)}$. Furthermore, we may as well also have $\sigma(i) = \sigma'(i) = 1$ and $\alpha(i) = i$ for $i \leq n'$ and have $\sigma(i) = 2$ for $i > n'$. In other words, $p_{12}(r_{ji}) = q_{ji}$ for $i \leq n'$ and $p_{12}(r_{1i}) = p_{12}(r_{-1i}) = q_{\sigma'(i)\alpha(i)}$ if $i > n'$.

For example, consider the tower $\mathfrak{T}$ given in Figure VII.5.1, then we could

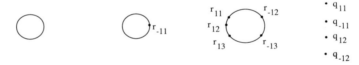

FIGURE VII.5.1. Pairing up points of V_1 and V_{12}

pair up the points as shown. V_{12} consists of the points r_{ji} and $p_{12}^{-1}(q_{11}) = \{r_{11}, r_{12}, r_{-12}\}$, $p_{12}^{-1}(q_{-11}) = \{r_{-11}\}$, $p_{12}^{-1}(q_{12}) = \{r_{13}, r_{-13}\}$ and $p_{12}^{-1}(q_{-12}) = \emptyset$. The realization $|\mathfrak{T}|$ is shown in Figure VII.5.2. In this example, $n' = 1$.

FIGURE VII.5.2. The realization of $\mathfrak{T}$

We will not yet construct all of V_2' but will construct more and more pieces of it by adding handles until we finally have all of it. For the sake of clarity, we will not bother to round off corners, specify a smooth structure or smooth out the vector field $\mathfrak{v}_2$ (or even make it continuous, except qualitatively) where the pieces fit together. However, the vector field $\mathfrak{v}_2$ we define below will certainly be integrable, so its flow will define a smooth structure on V_2', by declaring the flow to be smooth and declaring the interiors of the pieces we add to be smooth.

Pick disjoint embeddings $\mu_{ji} \colon (\mathbf{J}, 0) \to (V_2, r_{ji})$. Let $C_2^0 = V_2 \times [1/2, 1]$ and $C_2^1 = \mathbf{J} \times \mathbf{J} \times \mathbf{m}$ and glue them together by attaching $(j, v, i) \in \pm 1 \times \mathbf{J} \times \mathbf{m} \subset C_2^1$ with $(\mu_{ji}(v), 1/2) \in C_2^0$. Let $C_{12}^0 = V_{12} \times [1/2, 1] \subset C_2^0$ and $C_{12}^1 = \mathbf{J} \times 0 \times \mathbf{m} \subset C_2^1$ and $C_{02}^0 = \emptyset$ and $C_{02}^1 = 0 \times \mathbf{J} \times \mathbf{m} \subset C_2^1$.

Temporarily denote $C_2 = C_2^0 \cup C_2^1$, $C_{12} = C_{12}^0 \cup C_{12}^1$ and $C_{02} = C_{02}^1$. As we go on we will add more handles to C_2, still calling it C_2, until finally we have all of V_2'. We have actually already defined all of V_{12}', it will be C_{12}. We define $p_{12}' \colon C_{12}^0 \to V_1'$ by $p_{12}'(x, t) = (jt, i)$ if $p_{12}(x) = q_{ji}$ and define $p_{12}' \colon C_{12}^1 \to V_1'$ by $p_{12}'(u, 0, i) = (\sigma'(i) u^{\sigma(i)}/2, \alpha(i))$. Define $\mathfrak{v}_2$ on C_2^0 by $\mathfrak{v}_2(x, t) = (0, -t)$ and define $\mathfrak{v}_2$ on C_2^1 by $\mathfrak{v}_2(u, v, i) = (-u/\sigma(i), v)$. Note that $\mathfrak{v}_2$ is tangent to C_{12} and $dp_{12}'(\mathfrak{v}_2|_{C_{12}}) = \mathfrak{v}_1$. Also $\mathfrak{v}_2$ points outward on $\partial C_2 - V_2 \times 1$.

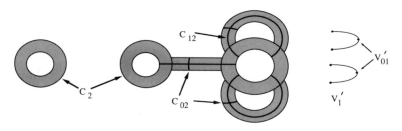

FIGURE VII.5.3. Starting to make $\mathfrak{T}$ a weak boundary

We now add handles to reduce to the case where each component T of $\partial C_2 - V_2 \times 1$ satisfies $T \cap C_{02} =$ one point. So take any component T of $\partial C_2 - V_2 \times 1$ and suppose $T \cap C_{02}$ has more than one point. Let x_1, x_{-1} be two of those points. Pick disjoint embeddings $\varphi_j \colon (\mathbf{J}, 0) \to (T, x_j)$ so that $\varphi_j^{-1}(C_{02}) = 0$ and so that with respect to some orientation of T, φ_1 is orientation preserving and φ_{-1} is orientation reversing. Let $C_2^2 = \mathbf{J} \times \mathbf{J}$ and attach it to C_2 by identifying $(j, v) \in \pm 1 \times \mathbf{J} \subset C_2^2$ with $\varphi_j(v) \in T$. Let $C_{02}^2 = 0 \times \mathbf{J} \cup \mathbf{J} \times 0$. Extend $\mathfrak{v}_2$ to C_2^2 by $\mathfrak{v}_2(u, v) = (-u, v)$.

Note that adding this handle split T into two circles, each with fewer intersections with C_{02}. Thus after doing this process a number of times we may assume that each component T of $\partial C_2 - V_2 \times 1$ intersects in zero or one point.

If the intersection is empty, we may pick disjoint embeddings $\varphi_j \colon \mathbf{J} \to T$ so that with respect to some orientation of T, φ_1 is orientation preserving and φ_{-1} is orientation reversing. Let $C_2^3 = \mathbf{J} \times \mathbf{J}$ and attach it to C_2 by identifying $(j, v) \in \pm 1 \times \mathbf{J} \subset C_2^3$ with $\varphi_j(v) \in T$. Let $C_{02}^2 = 0 \times \mathbf{J}$. Extend $\mathfrak{v}_2$ to C_2^2 by

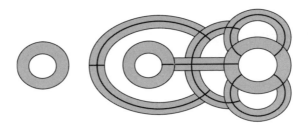

FIGURE VII.5.4. Handles added until each $T \cap C_{02}$ has at most one point

$\mathfrak{v}_2(u, v) = (-u, v)$. Note that adding this handle split T into two circles, each intersecting C_{02} in a single point $(0, \pm 1)$.

So we have reduced to the case where each component T of $\partial C_2 - V_2 \times 1$ satisfies $T \cap C_{02} =$ one point.

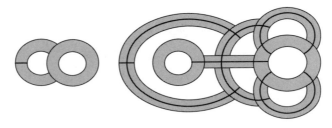

FIGURE VII.5.5. Handles added until each $T \cap C_{02}$ has one point

Since the number of points in $\partial C_2 \cap C_{02}$ is even, we may pair up the components of $\partial C_2 - V_2 \times 1$ as $\{T_{11}, T_{-11}, \ldots, T_{1k}, T_{-1k}\}$. Choose diffeomorphisms $\theta_{ji} \colon S^1 \to T_{ji}$ so that $\theta_{ji}(1) = T_{ji} \cap C_{02}$. We now add round handles to finish up the construction of $\mathfrak{T}'$. Let $C_2^4 = \mathbf{J} \times S^1 \times \mathbf{k}$ and attach it to C_2 by identifying $(j, x, i) \in \pm 1 \times \mathbf{J} \times \mathbf{k} \subset C_2^4$ with $\theta_{ji}(x) \in C_2$. Let $V_{02} = C_{02} \cup 0 \times S^1 \times \mathbf{k} \cup \mathbf{J} \times 1 \times \mathbf{k}$. Extend $\mathfrak{v}_2$ to C_2^4 by $\mathfrak{v}_2(u, x, i) = (-u, 0)$.

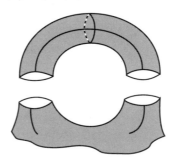

FIGURE VII.5.6. Adding a round handle

We will now show that $|\mathfrak{T}'| = \mathfrak{c}|\mathfrak{T}|$. Let $\psi_i\colon U_i \to V_i'$, $i = 0,1$ be the flow generated by $\mathfrak{v}_i$, where $U_i \subset V_i' \times \mathbf{R}$ is as large as possible. (For example, $\psi_1((t,i),s) = (te^{-s}, i)$.) Note that since $p_{12*}'(\mathfrak{v}_2) = \mathfrak{v}_1$ we have

$$p_{12}'\psi_2(x,s) = \psi_1(p_{12}'(x), s) \qquad\qquad *$$

Since $\mathfrak{v}_i$ points inward on $\partial V_i'$ is tangent to the ticos V_{02}' and V_{12}' and limits on V_{0i}', the ψ_i's restrict to diffeomorphisms $\psi_i|\colon \partial V_i \times [0,\infty) \to V_i' - V_{0i}'$. Because of $(*)$, these factor to an isomorphism $\psi\colon |\mathfrak{T}| \times [0,\infty) \to |\mathfrak{T}'| - V_0'$ which then gives an isomorphism $\psi'\colon \mathfrak{c}|\mathfrak{T}| \to |\mathfrak{T}'|$. Specifically, $\psi'(x,t) = \psi_2(x, 1/t - 1)$ for $x \in V_2$, $t \in (0,1]$, $\psi'(x,t) = \psi_1(x, 1/t - 1)$ for $x \in V_1$, $t \in (0,1]$ and $\psi'(x,0) =$ the point in V_0.

By Lemma 7.2.4, we may assume after blowing up $\mathfrak{T}'$ that $\xi_0(V_0) = \xi_1(V_0) = 0$.
$\square$

We now look at two dimensions. We proceed as in the proof of Theorem 7.5.1, the only difference is that things are a bit more complicated.

Theorem 7.5.2 *Suppose* $\rho_2(\mathfrak{T}) = 0$ *for some* $\mathfrak{T} \in \mathcal{TR}_2'$. *Then* $\mathfrak{T}$ *is the weak boundary of some resolution tower* $\mathfrak{T}'$ *of type S so that* $|\mathfrak{T}'| \approx \mathfrak{c}\,(|\mathfrak{T}|)$ *and all exponents of $\mathfrak{T}'$ are* ≤ 2. *Furthermore, if k is the index so that V_k' is the vertex of the cone, there are diffeomorphisms* $g_i\colon V_i \times (0,1] \to V_i' - V_{ki}'$ *so that* $g_i(V_{ji} \times (0,1]) = V_{ji}' - V_{ki}'$ *and* $p_{ji}'g_i(y,t) = g_j(p_{ji}(y),t)$ *for all* $(y,t) \in V_{ji} \times (0,1]$.

Proof: For convenience, assume that $\mathfrak{T} = \{V_i, \mathcal{A}_i, p_i\}_{i=1}^3$ has bias 1, then we wish to find $\mathfrak{T}' = \{V_i', \mathcal{A}_i', p_i'\}_{i=0}^3$ and the k in the statement above is 0. As we construct $\mathfrak{T}'$ we will also construct vector fields $\mathfrak{v}_i$ on V_i' which will integrate to give the g_i's and the cone structure on $|\mathfrak{T}'|$.

Since $\rho_2(\mathfrak{T}) = 0$ we may pair up the points of V_1 as

$$V_1 = \{\, q_{11}, q_{-11}, \ldots, q_{1n}, q_{-1n} \,\}$$

so that

$$\xi(q_{1i}) = \xi(q_{-1i}) \quad \text{for} \quad i = 1, \ldots, n.$$

Let V_0' be a point and let

$$V_1' = \mathbf{J} \times \mathbf{n}$$

where $\mathbf{n} = \{1, 2, \ldots, n\}$. Let $\mathcal{A}_1 = \mathcal{A}_{01} = \{\, (0,i) \mid i \in \mathbf{n} \,\}$. Since V_0' is a point there is only one possibility for each map p_{0i}', so we will refrain from defining these maps explicitly. Define $h_1\colon V_1 \to \partial V_1'$ by $h_1(q_{ji}) = (j,i)$. So h_1 identifies $-1 \times i \in \partial V_1'$ with q_{-1i} and $1 \times i \in \partial V_1'$ with q_{1i}. We define $\mathfrak{v}_1$ on V_1' to be $\mathfrak{v}_1(t,i) = -t$.

We will not yet construct all of V_2' and V_3' but will construct more and more pieces of them until we finally have all of them. For the sake of clarity, we will not bother to round off corners, specify a smooth structure or smooth out the

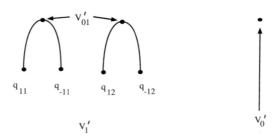

FIGURE VII.5.7. Resolving the cone on the 0-skeleton

vector fields $\mathfrak{v}_i$ (or even make them continuous, except qualitatively) where the pieces fit together. However, the vector fields we define below will certainly be integrable, so their flow will define a smooth structure on V_i', by declaring the flow to be smooth and declaring the interiors of the pieces we add to be smooth.

First, we will lay a bit of groundwork.

For each $a, b = 0, 1$ and $i \in \mathbf{n}$ we have $\xi_{2ab}(q_{1i}) = \xi_{2ab}(q_{-1i})$. In other words, there are an even number of points t of $p_{12}^{-1}(\{q_{1i}, q_{-1i}\})$ with $\zeta'(t) = (a, b)$. Thus we may pair off the points in V_{12} as

$$V_{12} = \{\, t_{11}, t_{-11}, \dots, t_{1m}, t_{-1m} \,\}$$

so that

$$\zeta'(t_{1i}) = \zeta'(t_{-1i}) \quad \text{for} \quad i = 1, \dots, m$$

and there are maps $\alpha\colon \mathbf{m} \to \mathbf{n}$ and $\delta\colon \mathbf{m} \to \{1, 2\}$ and $\delta'\colon \mathbf{m} \to \{-1, 1\}$ so that for all $i \in \mathbf{m}$ and $j = \pm 1$,

$$p_{12}(t_{ji}) = q_{\ell\alpha(i)} \text{ where } \ell = \delta'(i)j^{\delta(i)}$$

In other words, $\{t_{1i}, t_{-1i}\} \cap p_{12}^{-1}(\{\, q_{1k}, q_{-1k} \,\})$ is either empty (if $k \neq \alpha(i)$) or $\{t_{1i}, t_{-1i}\}$ (if $k = \alpha(i)$). If $\delta(i) = 1$ then p_{12} maps t_{1i} and t_{-1i} to different points and if $\delta(i) = 2$, p_{12} maps t_{1i} and t_{-1i} to the same point.

Since $\zeta'(t_{1i}) = \zeta'(t_{-1i})$, we know that for each $d = 1, 2$, there are an even number of points u in $p_{23}^{-1}(\{\, t_{1i}, t_{-1i} \,\})$ with $\zeta(u) = d$. Hence we may pair off the points of $V_{23} \cap V_{13}$ as

$$V_{23} \cap V_{13} = \{\, u_{11}, u_{-11}, \dots, u_{ik}, u_{-1k} \,\}$$

so that

$$\zeta(u_{1i}) = \zeta(u_{-1i}) \quad \text{for} \quad i \in \mathbf{k}$$

and there are maps $\beta\colon \mathbf{k} \to \mathbf{m}$, $\epsilon\colon \mathbf{k} \to \{1, 2\}$ and $\epsilon'\colon \mathbf{k} \to \{-1, 1\}$ so that

$$p_{23}(u_{ji}) = t_{\ell\beta(i)} \text{ where } \ell = \epsilon'(i)j^{\epsilon(i)}$$

Let $\zeta_i = \zeta(u_{1i}) = \zeta(u_{-1i})$, $i \in \mathbf{k}$.

Pick disjoint open embeddings $\mu_{ji} \colon \mathbf{R} \to V_2$ for $j = \pm 1$ and $i \in \mathbf{m}$ so that $\mu_{ji}(0) = t_{ji}$. Since p_{23} is a tico map, we may find disjoint embeddings $\eta_{ji} \colon \mathbf{J} \times \mathbf{J} \to V_3$ and maps $\sigma \colon \mathbf{k} \to \{0, 1\}$ and $\sigma' \colon \mathbf{k} \to \{-1, 1\}$ so that

$$
\begin{aligned}
\eta_{ji}(0) &= u_{ji} \\
\eta_{ji}^{-1}(V_{13}) &= 0 \times \mathbf{J} \\
\eta_{ji}^{-1}(V_{23}) &= \mathbf{J} \times 0 \\
p_{23}\eta_{ji}(\boldsymbol{v}, \mathbf{0}) &= \mu_{\ell\beta(i)}\left(\sigma'(i)j^{\sigma(i)}v^{\zeta_i}\right) \quad \text{where} \quad \ell = \epsilon'(i)j^{\epsilon(i)}
\end{aligned}
$$

(It may be necessary to rescale μ_{ji} to get the last condition.) Note that if $\zeta_i = 1$, we may compose η_{ji} with reflection about $0 \times \mathbf{J}$ if necessary to absorb the $\pm$ sign $\sigma'(i)j^{\sigma(i)}$, so if $\zeta_i = 1$ we may as well assume that $\sigma(i) = 0$ and $\sigma'(i) = 1$.

Our first two pieces of V_2' will be $C_2^0 = V_2 \times [1/2, 1]$ and the one handles $C_2^1 = \mathbf{J} \times \mathbf{J} \times \mathbf{m}$. We will glue them together by identifying

$$(j, u, i) \in C_2^1 \quad \text{with} \quad (\mu_{ji}(u), 1/2) \in C_2^0 \quad \text{for} \quad u \in \mathbf{J}, \ j = \pm 1.$$

Define $V_{j2}' \cap C_2^i = C_{j2}^i$ by

$$
\begin{aligned}
C_{12}^0 &= V_{12} \times [1/2, 1] \\
C_{12}^1 &= \mathbf{J} \times 0 \times \mathbf{m} \\
C_{02}^0 &= \emptyset \\
C_{02}^1 &= 0 \times \mathbf{J} \times \mathbf{m}
\end{aligned}
$$

Define $p_{12}'|_{C_{12}^0}$ by $p_{12}'(x, s) = (js, i)$ if $p_{12}(x) = q_{ji}$ and define $p_{12}'|_{C_{12}^1}$ by letting $p_{12}'(s, 0, i) = (\delta'(i)s^{\delta(i)}/2, \alpha(i))$. So for each $i \in \mathbf{m}$, $t_{1i} \times 1$ and $t_{-1i} \times 1$ are endpoints of an interval in $C_{12}^0 \cup C_{12}^1$ and p_{12}' maps this interval homeomorphically to $\mathbf{J} \times \alpha(i)$ if $\delta(i) = 1$ and folds it to half of $\mathbf{J} \times \alpha(i)$ if $\delta(i) = 2$.

Let the embedding $h_2 \colon V_2 \to \partial C_2^0$ be defined by $h_2(x) = (x, 1)$. Note that $h_2^{-1}(C_{12}^0) = V_{12}$ and $p_{12}'h_2 = h_1 p_{12}$. Define $\mathfrak{v}_2$ restricted to C_2^0 by $\mathfrak{v}_2(x, t) = (0, -t)$ and define $\mathfrak{v}_2$ restricted to C_2^1 by $\mathfrak{v}_2(s, u, i) = (-s/\delta(i), u)$. Note that $\mathfrak{v}_2$ points outward on $\partial(C_2^0 \cup C_2^1) - V_2 \times 1$, points inward on $V_2 \times 1$, is tangent to $C_{12}^0 \cup C_{12}^1$ and satisfies $dp_{12}'(\mathfrak{v}_2) = \mathfrak{v}_1$.

At this point we have constructed a neighborhood of $V_{12}' \cup \partial V_2'$ in V_2'. We now temporarily halt our construction of V_2' and start constructing V_3'. In particular we will construct a neighborhood of $\partial V_3' \cup (V_{23}' \cap V_{13}')$ in V_3'.

Let

$$
\begin{aligned}
C_3^0 &= V_3 \times [1/2, 1] \\
C_3^1 &= \mathbf{J} \times \mathbf{J} \times \mathbf{J} \times \mathbf{k} \\
C_3^2 &= \mathbf{J} \times \mathbf{J} \times \mathbf{J} \times \sigma^{-1}(1)
\end{aligned}
$$

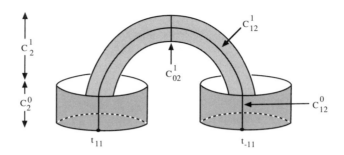

FIGURE VII.5.8. The first two pieces of V_2'

We will identify $(j, v, w, i) \in C_3^1$ with $(\eta_{ji}(v, w), 1/2) \in C_3^0$ for $j = \pm 1$, $i \in \mathbf{m}$ and $v, w \in \mathbf{J}$. We will identify $(j, v, w, i) \in C_3^2$ with $(v, j, w, i) \in C_3^1$ for $j = \pm 1$, $i \in \sigma^{-1}(1)$ and $v, w \in \mathbf{J}$. Thus we add a single one handle for each $i \in \sigma^{-1}(0)$ and add two one handles for each $i \in \sigma^{-1}(1)$.

Define $V_{j3}' \cap C_3^i = C_{j3}^i$ by

$$
\begin{aligned}
C_{j3}^0 &= V_{j3} \times [1/2, 1] \\
C_{23}^1 &= \mathbf{J} \times \mathbf{J} \times 0 \times \mathbf{k} \\
C_{13}^1 &= \mathbf{J} \times 0 \times \mathbf{J} \times \mathbf{k} \\
C_{03}^1 &= 0 \times \mathbf{J} \times \mathbf{J} \times \mathbf{k} \\
C_{23}^2 &= \mathbf{J} \times \mathbf{J} \times 0 \times \sigma^{-1}(1) \\
C_{13}^2 &= \emptyset \\
C_{03}^2 &= 0 \times \mathbf{J} \times \mathbf{J} \times \sigma^{-1}(1) \cup \mathbf{J} \times 0 \times \mathbf{J} \times \sigma^{-1}(1)
\end{aligned}
$$

Define $p_{j3}'|_{C_{j3}^0}$ by $p_{23}'(x, s) = (p_{23}(x), s) \in C_2^0$ and $p_{13}'(x, s) = (\ell s, i)$ if $p_{13}(x) = q_{\ell i}$. Define $p_{j3}'|_{C_{j3}^1}$ by

$$
\begin{aligned}
p_{23}'(u, v, 0, i) &= (\epsilon'(i)u^{\epsilon(i)}, \sigma'(i)u^{\sigma(i)}v^{\zeta_i}, \beta(i)) \in C_2^1 \\
p_{13}'(u, 0, w, i) &= p_{12}'p_{23}'(u, 0, 0, i) = (\delta'\beta(i)\epsilon'(i)^{\delta\beta(i)}u^{\delta_i}/2, \alpha\beta(i))
\end{aligned}
$$

where $\delta_i = \epsilon(i)\delta\beta(i)$. Define $p_{23}'|_{C_{23}^2}$ by

$$
p_{23}'(u, v, 0, i) = (\epsilon'(i)v^{\epsilon(i)}u^2, \sigma'(i)v, \beta(i)) \in C_2^1.
$$

Let $h_3: V_3 \to \partial C_3^0$ be defined by $h_3(x) = (x, 1)$. Define $\mathfrak{v}_3$ restricted to C_3^0 by $\mathfrak{v}_3(x, t) = (0, -t)$, define $\mathfrak{v}_3$ restricted to C_3^1 by

$$
\mathfrak{v}_3(u, v, w, i) = (-u/\delta_i, v(1 + \sigma(i)/\delta_i)/\zeta_i, w)
$$

and define $\mathfrak{v}_3$ restricted to C_3^2 by

$$
\mathfrak{v}_3(u, v, w, i) = (-u(1/\delta\beta(i) + \epsilon(i))/2, v, w).
$$

Note that $dp_{23}'(\mathfrak{v}_3) = \mathfrak{v}_2$ and $dp_{12}'(\mathfrak{v}_3) = \mathfrak{v}_1$.

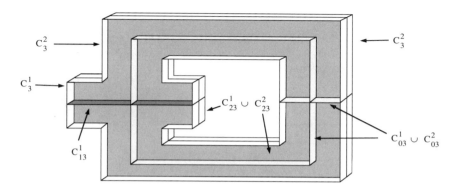

$$\text{FIGURE VII.5.9. The first pieces of } V_3'$$

We temporarily denote $C_3 = C_3^0 \cup C_3^1 \cup C_3^2$, (as we go along we will enlarge C_3). Likewise, let C_{j3} temporarily denote $C_{j3}^0 \cup C_{j3}^1 \cup C_{j3}^2$, let C_{j2} temporarily denote $C_{j2}^0 \cup C_{j2}^1$ and let C_2 temporarily denote $C_2^0 \cup C_2^1$. Let $\partial_- C_i$ denote $\partial C_i - V_i \times 1$ and let $\partial_- C_{ji}$ denote $\partial C_{ji} - V_{ji} \times 1$.

Notice now that $p_{23}' : C_{23} \to C_2$ is proper, i.e., $p_{23}'^{-1}(\partial C_2) = \partial C_{23}$. Furthermore, $p_{23}'^{-1}(\partial_- C_2) = \partial_- C_{23}$. Also the map $p_{23}'| : \partial_- C_{23} \to \partial_- C_2$ is a tico map between one dimensional manifolds and a perusal through our construction shows that its exponents are 1 and 2, nothing larger. Note that $\mathfrak{v}_3$ points outward on $\partial_- C_3$.

We have finished constructing V_2' in a neighborhood of $V_{12}' \cup \partial V_2'$ and V_3' in a neighborhood of $(V_{23}' \cap V_{13}') \cup \partial V_3'$. We will now complete the construction of V_2' and of V_3' in a neighborhood of V_{23}' by adding more and more handles to C_2, adding corresponding handles to C_3 in such a way that C_{23} and p_{23}' extend and p_{23}' is proper.

In order to see when we can add such handles, we need a definition. For each component L of $\partial_- C_2 - C_{02}$) let $\lambda'(L)$ denote the number of points in $p_{23}'^{-1}(x)$ for any $x \in L$. Since p_{23}' is a submersion over L this is well defined. Pick any $x \in \partial C_2 \cap C_{02}$ and let L_0 and L_1 be the two components of $\partial_- C_2 - C_{02}'$ on either side of x. Note that $\lambda'(L_1) \equiv \lambda'(L_0)$ mod 2 by degree theory. Define $\lambda(x) \in \mathbf{Z}/2\mathbf{Z}^2$ by

$$\lambda(x) = (\lambda'(L_0), (\lambda'(L_0) + \lambda'(L_1))/2).$$

Assertion 7.5.2.1 *Suppose $\lambda(x_1) = \lambda(x_{-1})$ for distinct $x_i \in \partial C_2 \cap C_{02}$. Suppose we attach a one handle $\mathbf{J} \times \mathbf{J}$ to C_2 at $x_{\pm 1}$, i.e., we have disjoint embeddings $\mu_i \colon \mathbf{J} \to \partial_- C_2$, $i = \pm 1$ so that $\mu_i(0) = x_i$ and $\mu_i^{-1}(C_{02}) = 0$ and we identify $(i, v) \in \mathbf{J} \times \mathbf{J}$ with $\mu_i(v) \in C_2$, $i = \pm 1$. Extend the vector field $\mathfrak{v}_2$ to $\mathbf{J} \times \mathbf{J}$ by setting $\mathfrak{v}_2(u, v) = (-u, v)$. Then we may attach one handles to C_3 and extend p_{23}' and $\mathfrak{v}_3$ so that p_{23}' is still proper, $dp_{23}'(\mathfrak{v}_3) = \mathfrak{v}_2$, p_{23}' is a tico map and $\mathfrak{v}_3$ points outward on the boundary of $C_3 \cup$ one handles $-V_3 \times 1$.*

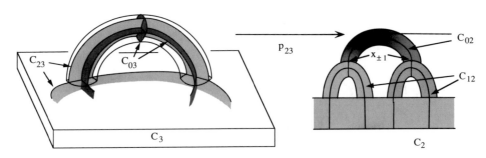

FIGURE VII.5.10. Covering a new handle in C_2 **by one** in C_3

Proof: Let C_2^3 denote the handle $\mathbf{J} \times \mathbf{J}$. We let $C_{02}^3 = 0 \times \mathbf{J} \cup \mathbf{J} \times 0$ be the extension of the tico C_{02}.

Let $p_{23}'^{-1}(x_i) = \{r_{i1}, \ldots, r_{ia_i}\}$ and pick disjoint embeddings $\theta_{ij} \colon \mathbf{R}^2 \to \partial C_3$ so that $\theta_{ij}^{-1}(C_{23}) = \mathbf{R}_2^2$ and $\theta_{ij}^{-1}(C_{03}) = \mathbf{R}_1^2$ and $\theta_{ij}(0) = r_{ij}$ and $p_{23}'\theta_{ij}(u, 0) = \mu_i(\kappa_{ij}' u^{\kappa_{ij}})$ for $|u| \leq 1$, $\kappa_{ij}' = \pm 1$, $\kappa_{ij} = 1$ or 2, $i = \pm 1$, $j = 1, \ldots, a_i$. We may as well take $\kappa_{ij}' = 1$ if $\kappa_{ij} = 1$ and after reordering assume that $\kappa_{ij}' = \kappa_{ij} = 1$ for all $j \leq b_i$, $\kappa_{ij}' = 1$ and $\kappa_{ij} = 2$ for all $b_i < j \leq c_i$ and $\kappa_{ij}' = -1$ and $\kappa_{ij} = 2$ for all $c_i < j \leq a_i$.

Note that the number of points in $p_{23}'^{-1}\mu_i(1)$ is $b_i + 2(c_i - b_i)$ and the number of points in $p_{23}'^{-1}\mu_i(-1)$ is $b_i + 2(a_i - c_i)$. So $\lambda(x_i) = (b_i, a_i)$.

If some $b_i \geq 2$, we can add a handle $C_3^3 = \mathbf{J} \times \mathbf{J} \times \mathbf{J}$ to C_3, attaching $(1, v, w)$ to $\theta_{i1}(v, w)$, attaching $(-1, v, w)$ to $\theta_{i2}(v, w)$, letting $C_{23}^3 = \mathbf{J} \times \mathbf{J} \times 0$ and $C_{03}^3 = 0 \times \mathbf{J} \times \mathbf{J} \cup \mathbf{J} \times 0 \times \mathbf{J}$, extending p_{23}' by setting $p_{23}'(u, v, 0) = (iu^2, v) \in C_2^3$ and extending $\mathfrak{v}_3$ to C_3^3 by setting $\mathfrak{v}_3(u, v, w) = (-u/2, v, w)$. So we may as well assume each $b_i < 2$. But since $b_1 \equiv b_{-1} \bmod 2$ we must then have $b_1 = b_{-1}$.

If $b_1 > 0$ or $b_{-1} > 0$ then $b_{-1} = b_1 = 1$ and we can add a handle $C_3^4 = \mathbf{J} \times \mathbf{J} \times \mathbf{J}$ to C_3, attaching $(\pm 1, v, w)$ to $\theta_{\pm 1,1}(v, w)$, letting $C_{23}^4 = \mathbf{J} \times \mathbf{J} \times 0$ and $C_{03}^4 = 0 \times \mathbf{J} \times \mathbf{J} \cup \mathbf{J} \times 0 \times \mathbf{J}$, extending p_{23}' by setting $p_{23}'(u, v, 0) = (u, v) \in C_2^3$ and extending $\mathfrak{v}_3$ to C_3^4 by setting $\mathfrak{v}_3(u, v, w) = (-u, v, w)$. So we may as well assume the b_i are both zero.

If $c_i > 0$ and $a_i - c_i > 0$ we can add a handle $C_3^5 = \mathbf{J} \times \mathbf{J} \times \mathbf{J}$ to C_3, attaching $(1, v, w)$ to $\theta_{i1}(v, w)$, attaching $(-1, v, w)$ to $\theta_{i,c_i+1}(v, w)$, letting $C_{23}^5 = \mathbf{J} \times \mathbf{J} \times 0$ and $C_{03}^5 = 0 \times \mathbf{J} \times \mathbf{J} \cup \mathbf{J} \times 0 \times \mathbf{J}$, extending p_{23}' by setting $p_{23}'(u, v, 0) = (iu^2, uv^2) \in C_2^3$ and extending $\mathfrak{v}_3$ to C_3^5 by setting $\mathfrak{v}_3(u, v, w) = (-u/2, 3v/4, w)$. Now p_{23}' is not proper so we add another handle $C_3^{5'} = \mathbf{J} \times \mathbf{J} \times \mathbf{J}$ to $C_3 \cup C_3^5$, attaching $(j, v, w) \in C_3^{5'}$ to $(v, j, w) \in C_3^5$, $j = \pm 1$, letting $C_{23}^{5'} = \mathbf{J} \times \mathbf{J} \times 0$ and $C_{03}^{5'} = 0 \times \mathbf{J} \times \mathbf{J} \cup \mathbf{J} \times 0 \times \mathbf{J}$, extending p_{23}' by setting $p_{23}'(u, v, 0) = (iu^2v^2, v) \in C_2^3$ and extending $\mathfrak{v}_3$ to $C_3^{5'}$ by setting $\mathfrak{v}_3(u, v, w) = (-3u/2, v, w)$. Now p_{23}' is proper. So we may as well assume that for each $i = \pm 1$, either $c_i = 0$ or $a_i = c_i$.

If $c_i \geq 2$ we may add a handle $C_3^6 = \mathbf{J} \times \mathbf{J} \times \mathbf{J}$ to C_3, attaching $(1, v, w)$ to $\theta_{i1}(v, w)$, attaching $(-1, v, w)$ to $\theta_{i2}(v, w)$, letting $C_{23}^6 = \mathbf{J} \times \mathbf{J} \times 0$ and $C_{03}^6 =$

$0 \times \mathbf{J} \times \mathbf{J} \cup \mathbf{J} \times 0 \times \mathbf{J}$, extending p'_{23} by setting $p'_{23}(u,v,0) = (iu^2, v^2) \in C_2^3$ and extending $\mathfrak{v}_3$ to C_3^6 by setting $\mathfrak{v}_3(u,v,w) = (-u/2, v/2, w)$. So we may as well assume $c_i = 0$ or 1. Likewise we may assume $a_i - c_i = 0$ or 1.

If $c_1 = c_{-1} = 1$ we can add a handle $C_3^7 = \mathbf{J} \times \mathbf{J} \times \mathbf{J}$ to C_3, attaching $(\pm 1, v, w)$ to $\theta_{\pm 1,1}(v,w)$, letting $C_{23}^7 = \mathbf{J} \times \mathbf{J} \times 0$ and $C_{03}^7 = 0 \times \mathbf{J} \times \mathbf{J} \cup \mathbf{J} \times 0 \times \mathbf{J}$, extending p'_{23} by setting $p'_{23}(u,v,0) = (u, v^2) \in C_2^3$ and extending $\mathfrak{v}_3$ to C_3^7 by setting $\mathfrak{v}_3(u,v,w) = (-u, v/2, w)$. We can do likewise if $a_1 - c_1 = a_{-1} - c_{-1} = 1$.

So the only possibilities left consistent with the assumption $\lambda(x_1) = \lambda(x_{-1})$ (i.e., $a_1 \equiv a_{-1} \bmod 2$) are $a_i = c_i = 0$ (in which case we are done) or $a_1 = a_{-1} = c_{-1} = 1$, $c_1 = 0$ or $a_1 = a_{-1} = c_1 = 1$, $c_{-1} = 0$.

The last two cases are symmetric, so assume the third possibility. We add a handle $C_3^8 = \mathbf{J} \times \mathbf{J} \times \mathbf{J}$ to C_3, attaching $(\pm 1, v, w)$ to $\theta_{\pm 1,1}(v,w)$, letting $C_{23}^8 = \mathbf{J} \times \mathbf{J} \times 0$ and $C_{03}^8 = 0 \times \mathbf{J} \times \mathbf{J} \cup \mathbf{J} \times 0 \times \mathbf{J}$, extending p'_{23} by setting $p'_{23}(u,v,0) = (u, uv^2) \in C_2^3$ and extending $\mathfrak{v}_3$ to C_3^8 by setting $\mathfrak{v}_3(u,v,w) = (-u,v,w)$. The resulting map is not proper, so we add another handle $C_3^{8'} = \mathbf{J} \times \mathbf{J} \times \mathbf{J}$ to C_3, attaching $(\pm 1, v, w)$ to $(v, \pm 1, w) \in C_3^8$, letting $C_{23}^9 = \mathbf{J} \times \mathbf{J} \times 0$ and $C_{03}^9 = 0 \times \mathbf{J} \times \mathbf{J} \cup \mathbf{J} \times 0 \times \mathbf{J}$, extending p'_{23} by setting $p'_{23}(u,v,0) = (u^2 v, v) \in C_2^3$ and extending $\mathfrak{v}_3$ to $C_3^{8'}$ by setting $\mathfrak{v}_3(u,v,w) = (-u,v,w)$. □

Assertion 7.5.2.2 *Suppose L_i, $i = \pm 1$ are two components of $\partial C_2 - C_{02}$ and $\lambda'(L_0) = \lambda'(L_1)$. Suppose we attach a one handle $\mathbf{J} \times \mathbf{J}$ to C_2 at $L_{\pm 1}$, i.e., we have disjoint embeddings $\mu_i \colon \mathbf{J} \to \partial C_2$, $i = \pm 1$ so that $\mu_i(0) \in L_i$ and $\mu_i^{-1}(C_{02}) = \emptyset$ and we identify $(i,v) \in \mathbf{J} \times \mathbf{J}$ with $\mu_i(v) \in C_2$, $i = \pm 1$. Extend the vector field $\mathfrak{v}_2$ to $\mathbf{J} \times \mathbf{J}$ by setting $\mathfrak{v}_2(u,v) = (-u,v)$. Then we may attach one handles to C_3 and extend p'_{23} and $\mathfrak{v}_3$ so that p'_{23} is still proper, $dp'_{23}(\mathfrak{v}_3) = \mathfrak{v}_2$, p'_{23} is a tico map and $\mathfrak{v}_3$ points outward on the boundary of $C_3 \cup$ one handles $-V_3 \times 1$.*

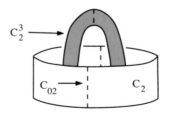

FIGURE VII.5.11. Another type of handle we may add to C_2

Proof: Let C_2^3 denote the handle $\mathbf{J} \times \mathbf{J}$. We let $C_{02}^3 = 0 \times \mathbf{J}$ be the extension of the tico C_{02}.

Let $p'^{-1}_{23}(\mu_i(0)) = \{r_{i1}, \ldots, r_{ia_i}\}$ and pick disjoint embeddings $\theta_{ij} \colon \mathbf{R}^2 \to \partial C_3$ so that $\theta_{ij}^{-1}(C_{23}) = \mathbf{R}_2^2$, $\theta_{ij}^{-1}(C_{03}) = \emptyset$, $\theta_{ij}(0) = r_{ij}$ and $p'_{23}\theta_{ij}(u,0) = \mu_i(u)$, for $i = \pm 1$ and $j = 1, \ldots, a_i$. Since $\lambda'(L_0) = \lambda'(L_1)$ we know that $a_1 \equiv a_{-1} \bmod 2$.

Thus if a_1 is odd then a_{-1} is odd and we can add a handle $C_3^3 = \mathbf{J} \times \mathbf{J} \times \mathbf{J}$ to C_3, attaching $(\pm 1, v, w)$ to $\theta_{\pm 1,1}(v, w)$, letting $C_{23}^3 = \mathbf{J} \times \mathbf{J} \times 0$ and $C_{03}^3 = 0 \times \mathbf{J} \times \mathbf{J}$, extending p'_{23} by setting $p'_{23}(u, v, 0) = (u, v) \in C_2^3$ and extending $\mathfrak{v}_3$ to C_3^3 by setting $\mathfrak{v}_3(u, v, w) = (-u, v, w)$. So we may as well assume the a_i are both even.

If some $a_i > 0$ then $a_i \geq 2$ and we can add a handle $C_3^4 = \mathbf{J} \times \mathbf{J} \times \mathbf{J}$ to C_3, attaching $(1, v, w)$ to $\theta_{i1}(v, w)$, attaching $(-1, v, w)$ to $\theta_{i2}(v, w)$, letting $C_{23}^4 = \mathbf{J} \times \mathbf{J} \times 0$ and $C_{03}^4 = 0 \times \mathbf{J} \times \mathbf{J}$, extending p'_{23} by setting $p'_{23}(u, v, 0) = (iu^2, v) \in C_2^3$ and extending $\mathfrak{v}_3$ to C_3^4 by setting $\mathfrak{v}_3(u, v, w) = (-u/2, v, w)$. $\square$

We wish to reduce to the case where for each component T of $\partial_- C_2$, $T \cap C_{02}$ is exactly one point. Suppose this is not true for some component T. Note that the first coordinate of $\lambda(x)$ is the same for all $x \in T \cap C_{02}$. So if $T \cap C_{02}$ has more than two points, we may find two distinct points x_1 and x_{-1} in $T \cap C_{02}$ so that $\lambda(x_1) = \lambda(x_{-1})$. Likewise, if $T \cap C_{02}$ is just two points $x_{\pm 1}$ then $\lambda(x_1) = \lambda(x_{-1})$. In either case, by Assertion 7.5.2.1, we may as well add a handle at $x_{\pm 1}$ and this breaks up T into two components, each with fewer elements in $T \cap C_{02}$. So after doing this repeatedly we may as well assume that the number of points in $T \cap C_{02}$ is either zero or one. If it is zero we may attach a one handle to T by Assertion 7.5.2.2 and thus break it up into two components, each with one point of intersection with C_{02}.

So we have reduced to the case where for each component T of $\partial_- C_2$, $T \cap C_{02}$ is exactly one point.

We have an involution $\tau \colon \partial C_2 \cap C_{02} \to \partial C_2 \cap C_{02}$ defined as follows.

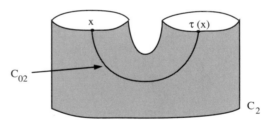

FIGURE VII.5.12. The involution τ

Pick any $x \in \partial C_2 \cap C_{02}$. Let J be the component of the sheet of C_{02} which contains x. Then J is an interval and $\partial J = \{x, y\}$ for some $y \in \partial C_2 \cap C_{02}$. We let $\tau(x) = y$.

Assertion 7.5.2.3 $\lambda(x) = \lambda(\tau(x))$.

Proof: Let S be the component of C_2 containing x. Then $\lambda(x) = (d, d)$ where d is the mod 2 degree of the map $p'_{23}| \colon p'_{23}{}^{-1}(S) \to S$. But $\tau(x) \in S$ also. $\square$

Assertion 7.5.2.4 *After adding 1-handles to C_2 and C_3 we may as well assume that $p'_{23}| \colon \partial_- C_{23} \to \partial_- C_2$ is a submersion, i.e., a covering projection.*

Proof: Pick any $x_1 \in \partial C_2 \cap C_{02}$. Let $x_{-1} = \tau(x_1)$. Attach a 1-handle between x_1 and x_{-1}. Let us see what happens when we add 1-handles to C_3 to cover it as we do in the proof of Assertion 7.5.2.1. We let notation be as in the proof of Assertion 7.5.2.1. Also, let T_i be the component of $\partial_- C_2$ containing x_i.

Now p'_{23} properly submerses $p'^{-1}_{23}(T_i - x_i)$ so the number of inverse images of any point of $T_i - x_i$ is the same, so

$$b_i + 2(a_i - c_i) = b_i + 2(c_i - b_i).$$

Then looking at the proof of Assertion 7.5.2.1, we are finished after adding the handles of type C^3_3, C^4_3, C^5_3 and $C^{5'}_3$.

We now attach another one handle $C^4_2 = \mathbf{J} \times \mathbf{J}$ to C_2 and attach it by identifying $(\pm 1, v) \in C^4_2$ with $(v, \pm 1) \in C^3_2$. Extend $\mathfrak{v}_2$ to C^4_2 by $\mathfrak{v}_2(u, v) = (-u, v)$ and extend C_{02} by $C^4_{02} = 0 \times \mathbf{J} \cup \mathbf{J} \times 0$.

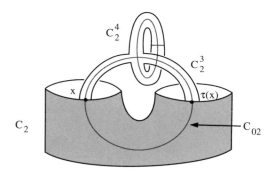

FIGURE VII.5.13. A handle pair added to C_2

We must now cover this handle by adding handles to C_3. We will attach a handle to C_3 for each handle we attached to cover C^3_2.

To a one handle of type C^3_3, we add a one handle $C^6_3 = \mathbf{J} \times \mathbf{J} \times \mathbf{J}$ to C_3, attaching $(\pm 1, v, w)$ to $(v, \pm 1, w) \in C^3_3$, letting $C^6_{23} = \mathbf{J} \times \mathbf{J} \times 0$ and $C^6_{03} = 0 \times \mathbf{J} \times \mathbf{J} \cup \mathbf{J} \times 0 \times \mathbf{J}$, extending p'_{23} by setting $p'_{23}(u, v, 0) = (u, iv^2) \in C^4_2$ and extending $\mathfrak{v}_3$ to C^6_3 by setting $\mathfrak{v}_3(u, v, w) = (-u, v/2, w)$.

To a one handle of type C^4_3, we add a one handle $C^7_3 = \mathbf{J} \times \mathbf{J} \times \mathbf{J}$ to C_3, attaching $(\pm 1, v, w)$ to $(v, \pm 1, w) \in C^4_3$, letting $C^7_{23} = \mathbf{J} \times \mathbf{J} \times 0$ and $C^7_{03} = 0 \times \mathbf{J} \times \mathbf{J} \cup \mathbf{J} \times 0 \times \mathbf{J}$, extending p'_{23} by setting $p'_{23}(u, v, 0) = (u, v) \in C^4_2$ and extending $\mathfrak{v}_3$ to C^7_3 by setting $\mathfrak{v}_3(u, v, w) = (-u, v, w)$.

To a handle pair of type C^5_3 and $C^{5'}_3$, we add a one handle $C^8_3 = \mathbf{J} \times \mathbf{J} \times \mathbf{J}$ to C_3, attaching $(\pm 1, v, w)$ to $(v, \pm 1, w) \in C^{5'}_3$, letting $C^8_{23} = \mathbf{J} \times \mathbf{J} \times 0$ and $C^8_{03} = 0 \times \mathbf{J} \times \mathbf{J} \cup \mathbf{J} \times 0 \times \mathbf{J}$, extending p'_{23} by setting $p'_{23}(u, v, 0) = (u, iv^2) \in C^4_2$ and extending $\mathfrak{v}_3$ to C^8_3 by setting $\mathfrak{v}_3(u, v, w) = (-u, v/2, w)$.

Now we look at the map p'_{23} on the new boundary and we see that it is a submersion on the new part of the boundary $\mathbf{J} \times \pm 1 \times 0$ in C_3^6, C_3^7 and C_3^8. So if we do this process for each pair $x, \tau(x)$ in $\partial C_2 \cap C_{02}$ we are done. $\qquad\square$

Assertion 7.5.2.5 *After adding handles, we may assume that each component of $\partial_- C_{23}$ is mapped homeomorphically by p'_{23} to a component of $\partial_- C_2$.*

Proof: For each component T of $\partial_- C_2$, let $d(T)$ be the maximum degree of the covering of T by a connected component of $p_{23}^{-1}(T)$ and let $e(T)$ be the number of connected components of $p_{23}^{-1}(T)$ which cover T with degree $d(T)$. Let (d, e) be the lexicographical maximum of $(d(T), e(T))$ over all components T of $\partial_- C_2$. Let f be the number of components T of $\partial_- C_2$ so that $d(T) = d$ and $e(T) = e$. Our proof will be by induction on (d, e, f) with lexicographical ordering. If $d = 1$ we are done, so assume $d > 1$. Take a component T of $\partial_- C_2$ with $d(T) = d$ and $e(T) = e$. Choose a component S of $p_{23}^{-1}(T)$ which is a d-fold cover of T. Pick $y_1, y_2 \in S \cap C_{03}$.

We now redo the process described in the proof of Assertion 7.5.2.4, with a slight variation and show that after doing so, (d, e, f) will have improved. Let the notation be as in Assertion 7.5.2.4, (which implies some notation from Assertion 7.5.2.1).

First we choose x_1 so $x_1 \in T$, i.e., $\{x_1\} = T \cap C_{02}$. Note that $p'_{23}(y_i) = x_1$, $i = 1, 2$. We now choose the points r_{1i} in a special way. We choose $r_{11} = y_1$ and $r_{12} = y_2$. Note that $a_i = b_i = c_i$. By the process described in Assertion 7.5.2.4, we will attach a handle of type C_3^3 between y_1 and y_2. When it comes time to add a handle C_3^6 to this C_3^3 we attach $(j, v, w) \in C_3^6$ to $(jv, j, w) \in C_3^3$ instead of to (v, j, w) as in Assertion 7.5.2.4. Other than this, everything is done the same as in Assertion 7.5.2.4. In the end you get a new ∂C_{23} which is almost isomorphic to the old ∂C_{23}, the only difference is that S has been split into two components, each of which maps to T with degree smaller than d (the two degrees are positive and add up to d). So the new (d, e, f) is smaller than the

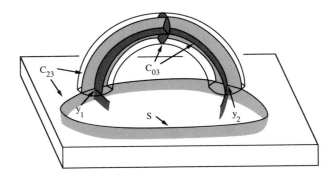

FIGURE VII.5.14. Adding a handle to C_3 which reduces (d, e, f)

old one and we are through by induction.   □

Let D^i, $i = 0, 1$ be the union of the connected components S of C_2 so that $p_{23}'^{-1}(S) \to S$ has mod 2 degree i. Let $\partial_- D_i = \partial D_i - V_2 \times 1$. Since $x \in D^i \cap \partial C_2 \cap C_{02}$ implies $\tau(x) \in D^i$ and each connected component of $\partial_- C_2$ contains exactly one point of C_{02}, we know that the number of connected components of each $\partial_- D^i$ is even. Consequently, the number of connected components of each $p_{23}'^{-1}(\partial_- D^i)$ is even by Assertion 7.5.2.4.

Since $\eta_i(\mathfrak{T}) = 0$, $i = 0, 1$ we know that the number of components of each $p_{23}'^{-1}(\partial D^i) \cap V_{23} \times 1$ which have nontrivial normal bundle in ∂C_3 is even. Hence by Lemma 7.2.5, the number of components of each $p_{23}'^{-1}(\partial_- D^i)$ which have nontrivial normal bundle in ∂C_3 is even. Hence the number of components of each $p_{23}'^{-1}(\partial_- D^i)$ which have trivial normal bundle in ∂C_3 is even also.

So we may pair up the points of $\partial_- C_2 \cap C_{02}$ as $\{s_{11}, s_{-11}, \dots, s_{1b}, s_{-1b}\}$ where for some $b' \leq b$, $s_{ji} \in D^0$ if and only if $i \leq b'$ and also so that if T_{ji} is the connected component of ∂C_2 containing s_{ji} then the mod 2 number of components of $p_{23}'^{-1}(T_{1i})$ with nontrivial normal bundle is the same as that for $p_{23}'^{-1}(T_{-1i})$. Consequently, we may pair up the points of $\partial C_{23} \cap C_{03}$ as $\{v_{11}, v_{-11}, \dots, v_{1d}, v_{-1d}\}$ so there are functions $\kappa \colon \mathbf{d} \to \mathbf{b}$, $\gamma' \colon \mathbf{d} \to \{1, -1\}$ and $\gamma \colon \mathbf{d} \to \{1, 2\}$ and a $d' \leq d$ so that if S_{ji} is the component of ∂C_{23} containing v_{ji} then S_{ji} has trivial normal bundle if and only if $i \leq d'$ and so that $p_{23}'(v_{ji}) = s_{\ell \kappa(i)}$ where $\ell = \gamma'(i) j^{\gamma(i)}$.

Let $S^1 = \{z \in \mathbf{C} \mid |z| = 1\}$ be the unit circle and pick diffeomorphisms $\omega_{ji} \colon S^1 \to T_{ji}$ onto the components of $\partial_- C_2$ so that $\omega_{ji}(1) = s_{ji}$.

We now define $C_2^9 = \mathbf{J} \times S^1 \times \mathbf{b}$, set $V_2' = C_2 \cup C_2^9$, identifying $(j, x, i) \in C_2^9$ with $\omega_{ji}(x) \in C_2$. Let $V_{12}' = C_{12}$ and $V_{02}' = C_{02} \cup \mathbf{J} \times 1 \times \mathbf{b} \cup 0 \times S^1 \times \mathbf{b}$. Extend $\mathfrak{v}_2$ to C_2^9 by setting $\mathfrak{v}_2(u, x, i) = (-u, 0)$.

We must now add round handles to C_3 to cover this.

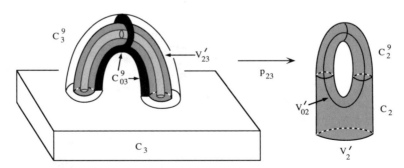

FIGURE VII.5.15. Covering with a round handle

Let F denote the Möbius band, thought of as the nontrivial closed interval bundle over S^1. Let $\pi \colon F \to S^1$ be the bundle projection, $G = \pi^{-1}(1)$ and let $E \subset F$ be the zero section. Let $F_i = F$, $G_i = G$, $E_i = E$ and $\pi_i = \pi$ if $i > d'$. Let $F_i = S^1 \times \mathbf{J}$, $G_i = 1 \times \mathbf{J}$, $E_i = S^1 \times 0$ and define $\pi_i \colon F_i \to S^1$ by $\pi_i(x, t) = x$ if $i \le d'$. We may choose disjoint embeddings $\varphi_{ji} \colon F_i \to \partial_- C_3$ so that $\varphi_{ji}(E_i) = S_{ji}$, $\varphi_{ji}^{-1}(C_{03}) = G_i$ and $\varphi_{ji}^{-1}(C_{13}) = \emptyset$. Let $C_3^9 = \bigcup_{i \le d} \mathbf{J} \times F_i \times i$. Glue C_3^9 to C_3 by identifying $(j, x, i) \in C_3^9$ with $\varphi_{ji}(x) \in C_3$. Let $C_{23}^9 = \bigcup_{i \le d} \mathbf{J} \times E_i \times i$ and $C_{03}^9 = \bigcup_{i \le d} (\mathbf{J} \times G_i \times i \cup 0 \times F_i \times i)$. Extend p'_{23} to C_{23}^9 by setting $p'_{23}(y, x, i) = (\gamma'(i)y^{\gamma(i)}, \pi_i(x), \kappa(i)) \in C_2^9$. Extend $\mathfrak{v}_3$ to C_3^9 by setting $\mathfrak{v}_3(y, x, i) = (-y/\gamma(i), \nu_i(x))$ where ν_i is a vector field on F_i which is 0 on E_i, points outward on ∂F_i and satisfies $d\pi_i(\nu_i) = 0$.

So we have now constructed all of V_2'. We have also constructed all of V_{23}', set $V_{23}' = C_{23} \cup C_{23}^9$. We will now complete the construction of V_{13}'.

Let V_{13}^2 be the double point set of V_{13}. Since $\xi_1(q_{1i}) = \xi_1(q_{-1i})$ we may pair up the points of V_{13}^2 as

$$\{w_{11}, w_{-11}, \dots, w_{1e}, w_{-1e}\}$$

so that there are functions $\varphi \colon \mathbf{e} \to \mathbf{n}$, $\pi \colon \mathbf{e} \to \{1, 2\}$ and $\pi' \colon \mathbf{e} \to \{-1, 1\}$ so that $p_{13}(w_{ji}) = q_{\ell\varphi(i)}$ where $\ell = \pi'(i)j^{\pi(i)}$. Choose disjoint embeddings $\theta_{ji} \colon \mathbf{R}^2 \to \partial C_3 \cap C_3^0 = \partial C_3 \cap V_3 \times 1/2$ so that $\theta_{ji}^{-1}(C_{13}) = 0 \times \mathbf{R} \cup \mathbf{R} \times 0$. We now attach $C_3^{10} = \mathbf{J} \times \mathbf{J} \times \mathbf{J} \times \mathbf{e}$ to C_3 by identifying $(j, v, w, i) \in C_3^{10}$ with $\theta_{ji}(v, w) \in C_3$, $j = \pm 1$. Let $C_{13}^{10} = \mathbf{J} \times 0 \times \mathbf{J} \times \mathbf{e} \cup \mathbf{J} \times \mathbf{J} \times 0 \times \mathbf{e}$ and $C_{03}^{10} = 0 \times \mathbf{J} \times \mathbf{J} \times \mathbf{e}$. Extend p'_{13} by setting $p'_{13}(u, v, w) = (\pi'(i)u^{\pi(i)}/2, \varphi(i))$. Extend $\mathfrak{v}_3$ by setting $\mathfrak{v}_3(u, v, w) = (-u/\pi(i), v, w)$.

At this point, $\partial_- C_3$ contains no points of C_{23} and no double points of C_{13}. So $\partial_- C_{13}$ is a union of circles embedded in ∂C_3. Let us look carefully at the map p'_{13} on one of these circles T. Pick $i \le m$ so that $p'_{13}(T) \subset \mathbf{J} \times i$. Let $T \cap C_{03}$ have c points. If $c = 0$ then p'_{13} is constant on T, it is $(\pm 1/2, i)$. If $c > 0$ then T is divided up into $2c$ intervals which alternately map to a constant $(\pm 1/2, i)$ or homeomorph or fold to $[-1/2, 1/2] \times i$ or the half intervals $[-1/2, 0] \times i$ or $[0, 1/2] \times i$. We say a point $x \in T \cap C_{03}$ has type h if p'_{13} homeomorphs a neighborhood of x in T and we say x has type f if p'_{13} folds a neighborhood of x in T.

Assertion 7.5.2.6 *After adding handles to C_3 we may as well assume that for each connected component T of $\partial_- C_{13}$:*

   a) *$T \cap C_{03}$ is two points.*

   b) *T has trivial normal bundle in ∂C_3.*

Proof: The first step will actually be to add handles until for each connected component T of ∂C_{13}, $T \cap C_{03}$ is one point, and this point has type f.

Pick any component T of ∂C_{13}. Pick $i \le m$ so that $p'_{13}(T) \subset \mathbf{J} \times i$. If $T \cap C_{03}$ has a point of type h then it must have another point of type h since p'_{13} is continuous, so we may pick two points $x_{\pm 1} \in T \cap C_{03}$ which both have type h.

In this case, let $\epsilon = 1$ and $\sigma = 1$. Suppose $T \cap C_{03}$ has two or more points, but none are of type h. Then by continuity they must all fold to the same half interval. Let $\sigma = 1$ if they fold to $[0, 1/2] \times i$ and let $\sigma = -1$ if they fold to $[-1/2, 0] \times i$. Also let $\epsilon = 2$ and pick any two points $x_{\pm 1} \in T \cap C_{03}$.

We may choose disjoint embeddings $\varphi_j \colon \mathbf{J} \times \mathbf{J} \to \partial_- C_3$ for $j = \pm 1$ so that

$$
\begin{aligned}
\varphi_j(0,0) &= x_j \\
\varphi_j^{-1}(C_{03}) &= 0 \times \mathbf{J} \\
\varphi_j^{-1}(C_{13}) &= \mathbf{J} \times 0 = \varphi_j^{-1}(T) \\
p_{13}'\varphi_j(u,0) &= (\sigma u^\epsilon / 2, i)
\end{aligned}
$$

In case $\epsilon = 2$ we may also assume that for some orientation of T, $\varphi_1|_{\mathbf{J} \times 0}$ is orientation preserving and $\varphi_{-1}|_{\mathbf{J} \times 0}$ is orientation reversing. We now attach a handle $C_3^{11} = \mathbf{J} \times \mathbf{J} \times \mathbf{J}$ to C_3 by identifying $(j, v, w) \in C_3^{11}$ with $\varphi_j(v, w) \in C_3$ for $j = \pm 1$, setting $C_{03}^{11} = 0 \times \mathbf{J} \times \mathbf{J} \cup \mathbf{J} \times 0 \times \mathbf{J}$ and $C_{13}^{11} = \mathbf{J} \times \mathbf{J} \times 0$ extending p_{13}' to C_{13}^{11} by $p_{13}'(u, v, 0) = (\sigma v^\epsilon u^2 / 2, i)$ and extending $\mathfrak{v}_3$ by $\mathfrak{v}_3(u, v, w) = (-u, v/\epsilon, w)$.

If $\epsilon = 2$ then T is broken up into two circles and if $\epsilon = 1$ then the two points of type h are replaced by points of type f (and T might be broken up as well). So after repeating this process as often as we can we reduce to the case where for each component T of ∂C_{13}, either $T \cap C_{03}$ is empty or it consists of one point of type f.

If $T \cap C_{03}$ is empty, pick disjoint embeddings $\varphi_j \colon \mathbf{J} \times \mathbf{J} \to \partial C_3$ so that $\varphi_j^{-1}(C_{03}) = \emptyset$, $\varphi_j^{-1}(C_{13}) = \mathbf{J} \times 0 = \varphi_j^{-1}(T)$ and for some orientation of T, $\varphi_1|_{\mathbf{J} \times 0}$ is orientation preserving and $\varphi_{-1}|_{\mathbf{J} \times 0}$ is orientation reversing. Pick $\sigma = \pm 1$ and $i \le n$ so that $p_{13}'(T) = (\sigma/2, i)$. Attach a handle $C_3^{12} = \mathbf{J} \times \mathbf{J} \times \mathbf{J}$ to C_3 by identifying $(j, v, w) \in C_3^{12}$ with $\varphi_j(v, w) \in C_3$, setting $C_{03}^{12} = 0 \times \mathbf{J} \times \mathbf{J}$ and $C_{13}^{12} = \mathbf{J} \times \mathbf{J} \times 0$ extending p_{13}' to C_{13}^{12} by $p_{13}'(u, v, 0) = (\sigma u^2 / 2, i)$ and extending $\mathfrak{v}_3$ by $\mathfrak{v}_3(u, v, w) = (-u/2, v, w)$. Then T has been broken into two circles, each with one point of intersection with C_{03}.

So we have completed our first step. Next we claim that for each $i \in \mathbf{n}$ there are an even number of components T of ∂C_{13} with $p_{13}'(T) \subset \mathbf{J} \times i$. This is because each such component corresponds to a point of $\partial C_3 \cap C_{03} \cap p_{13}'^{-1}(\mathbf{J} \times i)$ and there are an even number of such points because they are the boundary of $C_{03} \cap p_{13}'^{-1}(\mathbf{J} \times i)$ which is a tico in $p_{13}'^{-1}(\mathbf{J} \times i)$. We also know by Lemma 7.2.5 and the fact that $\xi_0(q_{1i}) = \xi_0(q_{-1i})$ that for each $i \in \mathbf{n}$ there are an even number of components T of ∂C_{13} with $p_{13}'(T) \subset \mathbf{J} \times i$ and the normal bundle of T in ∂C_3 nontrivial.

Now we will add handles to get conclusions a) and b). Pick any component T_1 of $\partial C_3 \cap C_{13} - V_3 \times 1$. Pick $i \le n$ so that $p_{13}'(T_1) \subset \mathbf{J} \times i$. Then by our claims above we know there is another component T_{-1} of $\partial C_3 \cap C_{13}$ so that $p_{13}'(T_{-1}) \subset \mathbf{J} \times i$ and so that the normal bundle of T_{-1} is trivial if the normal bundle of T_1 is trivial and the normal bundle of T_{-1} is nontrivial if the normal bundle of T_1 is nontrivial. Let $x_j = T_j \cap C_{03}$ for $j = \pm 1$. We may choose disjoint embeddings $\varphi_j \colon \mathbf{J} \times \mathbf{J} \to \partial C_3$, $j = \pm 1$ so that $\varphi_j(0,0) = x_j$, $\varphi_j^{-1}(C_{03}) = 0 \times \mathbf{J}$, $\varphi_j^{-1}(C_{13}) = \mathbf{J} \times 0 = \varphi_j^{-1}(T_j)$ and so that $p'_{13}\varphi_j(u,0) = (\sigma' j^\sigma u^2/2, i)$ for some $\sigma' = \pm 1$ and $\sigma = 1$ or 2. We now attach a handle $C_3^{13} = \mathbf{J} \times \mathbf{J} \times \mathbf{J}$ to C_3 by identifying $(j, v, w) \in C_3^{13}$ with $\varphi_j(v, w) \in C_3$, setting $C_{03}^{13} = 0 \times \mathbf{J} \times \mathbf{J} \cup \mathbf{J} \times 0 \times \mathbf{J}$ and $C_{13}^{13} = \mathbf{J} \times \mathbf{J} \times 0$ extending p'_{13} to C_{13}^{13} by $p'_{13}(u, v, 0) = (\sigma' v^2 u^\sigma/2, i)$ and extending $\mathfrak{v}_3$ by $\mathfrak{v}_3(u, v, w) = (-3u/\sigma, v, w)$. This handle combines T_1 and T_{-1} into a single circle $(T_1 - \varphi_1(\mathbf{J} \times 0)) \cup (T_{-1} - \varphi_{-1}(\mathbf{J} \times 0)) \cup \mathbf{J} \times \pm 1 \times 0$ with trivial normal bundle and two points of intersection with C_{03}. Now do this process for every component of $\partial C_3 \cap C_{13}$.   □

At this point it is easy to complete the construction of V'_{13}. Take any component T of $\partial_- C_{13}$. Pick $i \leq n$ so that $p'_{13}(T) \subset \mathbf{J} \times i$. Let D be the unit square $D = \mathbf{J} \times \mathbf{J}$. We have an embedding $\varphi \colon \partial D \times \mathbf{J} \to \partial_- C_3$ so that

$$\begin{aligned}
\varphi^{-1}(C_{13}) &= \varphi^{-1}(T) = \partial D \times 0 \\
\varphi^{-1}(C_{03}) &= 0 \times \pm 1 \times \mathbf{J} \\
p'_{13}\varphi(u, v, 0) &= (\sigma u^\epsilon/2, i)
\end{aligned}$$

for some $\sigma = \pm 1$, $\epsilon = 1$ or 2 and all $(u, v) \in \partial D$. We now just add the two handle $C_3^{14} = D \times \mathbf{J}$, identifying $(u, v, w) \in \partial D \times \mathbf{J}$ with $\varphi(u, v, w) \in \partial C_3$. Let $C_{13}^{14} = D \times 0$, $C_{03}^{14} = 0 \times \mathbf{J} \times \mathbf{J}$ extend p'_{13} by $p'_{13}(u, v, 0) = (\sigma u^\epsilon/2, i)$ and extend $\mathfrak{v}_3$ by $\mathfrak{v}_3(u, v, w) = (-u/\epsilon, -v, w)$.

After doing this for all components of $\partial C_3 \cap C_{13} - V_3 \times 1$, we have completed the construction of a neighborhood of $\partial V'_3 \cup V'_{23} \cup V'_{13}$.

It remains to fill in the last bit. Since $\rho_2(\mathfrak{T}) = 0$ we know by Lemma 7.5.3 below that $(V_3, \mathcal{A}_3)$ is the boundary of a compact tico. Hence $(\partial_- C_3, \partial C_{03})$ is the boundary of a compact tico. But then Lemma 7.5.3b implies that there is a tico $\mathcal{B}$ in a compact manifold T and a diffeomorphism $h \colon (\partial_- C_3 - C_{03}) \times (0, 1] \to T - |\mathcal{B}|$ so that $(\partial T, \partial \mathcal{B}) = (\partial_- C_3, \partial C_{03})$ and $h(x, 1) = x$ for all $x \in \partial_- C_3 - C_{03}$. We now let $V'_3 = C_3 \cup T$ and $V'_{03} = C_{03} \cup |\mathcal{B}|$. Extend $\mathfrak{v}_3$ by setting $\mathfrak{v}_3 h(x, t) = dh(0, -t)$. We have not actually defined $\mathfrak{v}_3$ on $|\mathcal{B}|$, but that is all right since we only need $\mathfrak{v}_3$ defined on $V'_3 - V'_{03}$.

We now construct g_i as we did in Theorem 7.5.1.   □

Lemma 7.5.3 *Let $\mathcal{A}$ be a tico in a closed smooth surface S. Suppose $\mathcal{A}$ has just one sheet. Then the following are equivalent:*

   a) *There is a tico $\mathcal{B}$ in a compact manifold T so that $(\partial T, \partial \mathcal{B}) = (S, \mathcal{A})$.*

   b) *There is a tico $\mathcal{B}$ in a compact manifold T and a diffeomorphism $h \colon (S - |\mathcal{A}|) \times (0, 1] \to T - |\mathcal{B}|$ so that $(\partial T, \partial \mathcal{B}) = (S, \mathcal{A})$ and $h(x, 1) = x$ for all $x \in S - |\mathcal{A}|$.*

c) S has even Euler characteristic, $\mathcal{A}$ has an even number of double points and if $f: Q \to S$ is the immersion associated to $|\mathcal{A}|$, then there are an even number of components of Q for which the normal bundle of the immersion is nontrivial.

Proof: That b) implies a) is trivial. To see that a) implies c), note that the double points of $|\mathcal{A}|$ are the boundary of the double curves of $|\mathcal{B}|$, hence there are an even number. The normal bundle condition follows from Lemma 7.2.5 and the observation that f is the boundary of the immersion associated to $|\mathcal{B}|$. That S has even Euler characteristic follows from the fact that S bounds.

We now show that c) implies a). Let $T_0 = S \times [1/2, 1]$. Pair up the double point set of $|\mathcal{A}|$ as $\{q_{11}, q_{-11}, \ldots, q_{1n}, q_{-1n}\}$. Pick disjoint embeddings $\mu_{ji}: \mathbf{J} \times \mathbf{J} \to S$ so that $\mu_{ji}^{-1}(|\mathcal{A}|) = \mathbf{J} \times 0 \cup 0 \times \mathbf{J}$ and $\mu_{ji}(0,0) = q_{ji}$. Let $T_1 = \mathbf{J} \times \mathbf{J} \times \mathbf{J} \times \mathbf{n}$ where $\mathbf{n} = \{1, \ldots, n\}$ and attach T_1 to T_0 by identifying $(j, v, w, i) \in T_1$ with $\mu_{ji}(v, w) \times 1/2$. We have a tico $\mathcal{B}_1$ on $T_0 \cup T_1$ with one sheet

$$|\mathcal{B}_1| = |\mathcal{A}| \times [1/2, 1] \cup \mathbf{J} \times \mathbf{J} \times 0 \times \mathbf{n} \cup \mathbf{J} \times 0 \times \mathbf{J} \times \mathbf{n}.$$

Then $\partial(T_0 \cup T_1) = S \times 1 \cup S_1$ where $S_1 \cap \mathcal{B}_1$ has no double points, hence is a regular tico. We also know by Lemma 7.2.5 that the number of components of $S_1 \cap |\mathcal{B}_1|$ with nontrivial normal bundle is even. Hence by Lemma 7.2.5 there is a manifold Q_2 and a line bundle $\pi_2: E_2 \to Q_2$ and an embedding of $\pi_2^{-1}(\partial Q_2)$ onto a neighborhood of $S_1 \cap |\mathcal{B}_1|$ which sends ∂Q_2 to $S_1 \cap |\mathcal{B}_1|$. Let $T_2 = T_0 \cup T_1 \cup E_2$ glued by this embedding and let $\mathcal{B}_2$ be the tico with one sheet $|\mathcal{B}_1| \cup Q_2$. Then $\partial T_2 = S \times 1 \cap S_2$ and $S_2 \cap |\mathcal{B}_2|$ is empty. Since S has even Euler characteristic, so does S_2 and hence $S_2 = \partial T_3$ for some compact 3-manifold T_3. We now let $T = T_2 \cup T_3$ and let $\mathcal{B}$ be the the tico with one sheet $|\mathcal{B}_2|$ and a) is proven. To go all the way to proving b) it is necessary to add some handles to T as was done in Proposition 2.8.13. The reader could do this as an exercise or refer to the proof in [**AK9**]. $\qquad \square$

Bibliography

[AK1]   S. Akbulut and H. King, The topology of real algebraic sets with isolated singularities, Annals of Math. 113 (1981) 425-446.

[AK2]   S. Akbulut and H. King, Real algebraic structures on topological spaces, Publ. I.H.E.S. 53 (1981) 79-162.

[AK3]   S. Akbulut and H. King, A relative Nash theorem, Transactions of the A.M.S. 267 (1981) 465-481.

[AK4]   S. Akbulut and H. King, Lectures on topology of real algebraic varieties, I.A.S. Lecture Notes (1980-1981), preprint.

[AK5]   S. Akbulut and H. King, The topology of real algebraic sets, A.M.S. Proc. Symp. in Pure Math. 40 (1983) 641-654.

[AK6]   S. Akbulut and H. King, The topology of real algebraic sets, L'Enseignement Math., 29 (1983) 221-261.

[AK7]   S. Akbulut and H. King, Submanifolds and homology of nonsingular algebraic varieties, American Journal of Math. (1985) 45-83.

[AK8]   S. Akbulut and H. King, A resolution theorem for homology cycles of real algebraic varieties, Invent. Math. 79 (1985) 589-601.

[AK9]   S. Akbulut and H. King, The topology of resolution towers, Transactions of the A.M.S., Vol. 302, no. 2 (1987) 497-521.

[AK10] S. Akbulut and H. King, Polynomial equations of immersed surfaces, Pac. Jour. of Math. 131 (no. 2) (1988) 209-217.

[AK11] S. Akbulut and H. King, All compact manifolds are homeomorphic to totally algebraic real algebraic sets, Comment. Math. Helv., 66 (1991) 139-149.

[AK12] S. Akbulut and H. King, On approximating submanifolds by algebraic sets and a solution to the Nash conjecture, (to appear in Invent. Math.).

[AK13] S. Akbulut and H. King, Algebraicity of immersions, (to appear in Topology).

[AK14] S. Akbulut and H. King, Resolution Towers, M.S.R.I. preprint 13812-85 (1985).

[AK15] S. Akbulut and H. King, Algebraic Structures on Resolution Towers, M.S.R.I. preprint 13012-85 (1985).

[AK16] S. Akbulut and H. King, Resolution tower structures on algebraic sets, M.S.R.I. preprint 13212-85 (1985).

[AK17] S. Akbulut and H. King, The topological classification of 3-dimensional algebraic sets, M.S.R.I. preprint 12912-85 (1985).

[AK18] S. Akbulut and H. King, All knots are algebraic, Comment. Math. Helv., 56 (1981) 339-351.

[AT]   S. Akbulut and L. Taylor, A topological resolution theorem, Publ. I.H.E.S. 53 (1981) 163-195.

[B]   G. Brumfiel, Partially Ordered Rings and Semi-Algebraic Geometry, Cambridge University Press (1979).

[BCR]   J. Bochnak, M. Coste, M. F. Roy, Géometrie algébrique réele, Ergeb. der Math. (12) Springer (1987).

[BD1]   R. Benedetti and M. Dedo, Counterexamples to representing homology classes by real algebraic subvarieties up to homeomorphism, Comp. Math. 53 (1984) 143-151.

[BD2]   R. Benedetti and M. Dedo, The topology of two-dimensional real algebraic varieties, Annali di Math. pura ed applicata 127 (1981) 141-171.

[BKS]   J. Bochnak, W. Kucharz and M. Shiota, On Algebraicity of global real analytic sets and functions (I), Invent. Math. 70 (1) (1982) 115-156.

[BM]   E. Bierstone and P. Milman, A simple constructive proof of canonical resolution of singularities, preprint.

[BT]   R. Benedetti and R. Tognoli, On real algebraic vector bundles, Bull. des Sci. Math. 2 serie' 104 (1980) 89-112.

[CF]   P. Conner and E. Floyd, Differentiable periodic maps, Ergeb. der Math. und ihrer Grenz. (33) Springer-Verlag, Berlin (1964).

[CK]   M. Coste and K. Kurdyka, On the link of a stratum in a real algebraic set, Topology (1991).

[G]   M. Goresky, Triangulation of stratified objects, PAMS 72 (1978) 193-200.

[GWPL] C. G. Gibson, K. Wirthmuller, A. A. du Plessis, E. J. N. Looijenga, Topological Stability of Smooth Mappings, Lecture Notes in Math (552) Springer-Verlag (1970).

[H]   H. Hironaka, Resolution of singularities of an algebraic variety over a field of characteristic zero, Annals of Math. 79 (1964) 109-326.

[I]   N. V. Ivanov, An improvement of the Nash-Tognoli theorem, Issled. po topologii, Steklov Math. Inst. (1982) 66-71.

[J]   F. E. A. Johnson, On the triangulation of stratified sets and singular varieties, TAMS 275 (1983) 333-343.

[K1]   H. King, Topological invariance of intersection homology without sheaves, Topology and its Applications 20 (1985) 149-160.

[K2]   H. King, Survey on the topology of real algebraic sets, Rocky Mount. Jour. of Math. Vol. 14, no. 4 (1984) 821-830.

[Ku]   W. Kucharz, Topology of real algebraic threefolds, Duke Math. Jour. (53) (No.4) (1986) 1073-1079 .

[L1]   S. Lojasiewicz, Sur le problème de la division, Razprawy Math. 22 (1961).

[L2]   S. Lojasiewicz, Triangulation of semianalytic sets, Ann. Sc. Norm. Sup. Pisa 18 (1964) 449-474.

[Ma]   B. Malgrange, Ideals of Differentiable Functions, Oxford University Press.

[Mi]   J. Milnor, On the Stiefel-Whitney numbers of complex manifolds and spin manifolds, Topology (3) (1965) 223-230.

[M]   D. Mumford, Algebraic Geometry I Complex Projective Varieties, Springer-Verlag (1976).

[N]   J. Nash, Real algebraic manifolds, Annals of Math. 56 (1952) 405-421.

[P]   R. Palais, Real Algebraic Differential Topology, Publish or Perish (1981).

[Sh]   I. Shaferevich, Basic Algebraic Geometry, Springer-Verlag (1977).

[Si]   L. Siebenmann, Deformation of Homeomorphisms on Stratified Sets, Comm. Math. Helv. 47 (1972) 123-163.

[Sp]   E. Spanier, Algebraic Topology, McGraw-Hill (1966).

[Su1]   D. Sullivan, Invariants of analytic spaces, Proc. Liverpool Singularities I, Lecture Notes 192, Springer-Verlag (1971) 165-168.

[Su2]   D. Sullivan, Singularities in spaces, Proc. Liverpool Singularities II, Lecture Notes 209, Springer-Verlag (1971) 196-206.

[Th]   R. Thom, Quelques propriétés globales de variétés différentiables, Comm. Math. Helv. 28 (1954) 17-86.

[To1]   A. Tognoli, Su una congettur di Nash, Ann. Sc. Norm. Sup. Pisa 27 (1973) 167-185.

[To2]   A. Tognoli, Any compact differentiable submanifold of $\mathbf{R}^n$ has algebraic approximation in $\mathbf{R}^n$, Topology 27 (1988) 205-210.

[To3]   A. Tognoli, Algebraic Geometry and Nash Functions, Instituto Nazionale di Alta Matematica, Academic Press (1978).

[V]   A. Verona, Stratified Mappings - Structure and Triangulability, Lecture Notes 1102, Springer-Verlag (1984).

[Wa]   C.T.C. Wall, Regular stratifications, Dynamical Systems – Warwick 1974, Lecture Notes 468, Springer-Verlag (1975) 332-344.

[W]   H. Whitney, Elementary structure of real algebraic varieties, Annals of Math. 66 (1957) 545-556.

Index